Fiber Optic Communications Design Handbook

Robert J. Hoss

PRENTICE HALL, Englewood Cliffs, New Jersey 07632

Library of Congress Cataloging-in-Publication Data

Hoss, Robert J.
    Fiber optic communications design handbook / Robert J. Hoss.
      p.   cm.
    ISBN 0-13-308586-4
    1. Optical communications.   2. Fiber optics.   3. Telecommunication
systems—Design and construction.   I. Title.
TK5103.59.H67   1980
621.382′8—dc20   89-36566
                                                                      CIP

Editorial/production supervision
  and interior design: *Laura A. Huber*
Cover design: *Wanda Lubelska*
Manufacturing buyer: *Kelly Behr*

> *This book is dedicated to my wife Lynne and my children Kristen and Melanie, all of whom displayed much patience during the long process.*

© 1990 by Prentice-Hall, Inc.
A Division of Simon & Schuster
Englewood Cliffs, New Jersey 07632

The publisher offers discounts on this book when ordered
in bulk quantities. For more information, write:

> Special Sales/College Marketing
> College Technical and Reference Division
> Prentice Hall
> Englewood Cliffs, New Jersey 07632

Printed in the United States of America

10 9 8 7 6 5 4 3 2 1

ISBN 0-13-308586-4

PRENTICE-HALL INTERNATIONAL (UK) LIMITED, *London*
PRENTICE-HALL OF AUSTRALIA PTY. LIMITED, *Sydney*
PRENTICE-HALL CANADA INC., *Toronto*
PRENTICE-HALL HISPANOAMERICANA, S.A., *Mexico*
PRENTICE-HALL OF INDIA PRIVATE LIMITED, *New Delhi*
PRENTICE-HALL OF JAPAN, INC., *Tokyo*
SIMON & SCHUSTER ASIA PTE. LTD., *Singapore*
EDITORA PRENTICE-HALL DO BRASIL, LTDA., *Rio de Janeiro*

Contents

2 DEFINING REQUIREMENTS AND STANDARDS   35

Preface

Optical-fiber technology has been in use for more than 20 years. However, it has only been since the early 1980s that fiber optics has truly transformed the nature of operating transmission networks, changing forever the standards of quality and the economies of bandwidth. Fiber optics has become a standard telecommunications transmission approach and the fiber itself a commodity. Although fiber optics continues to evolve, the basic systems-design practices as presented here remain pertinent.

This book covers the technology through the next stage of evolution, synchronous transmission, which does not significantly alter the basic engineering practices, but provides more opportunities and products to interface directly in the optical domain by defining standard synchronous optical- and electrical-signal interfaces.

The generation following synchronous transmission will be the productization of coherent transmission using advanced single-mode fiber designs and integrated optics terminals and repeaters. Although this technology has been well demonstrated, wide-scale productization will not begin before well into the 1990s. From an engineering standpoint, it represents a very different set of practices, since the optical carrier itself is treated as a discrete carrier frequency and signaling is modulated onto that carrier much as with amplitude-, frequency-, and phase-modulated radio transmission. The technology and its application is discussed here, but the specifics of coherent modulation is the subject for a different text.

The book is intended to present practical or standard design practices that can be used to develop telecommunications networks today and in the future. Although it is based on the near-term state of the technology, most of the practices are generic to systems design and therefore will not be affected by the evolution of the technology. The scope is the presentation of basic performance relationships associated with fiber technology and the methodology for transmission-systems design regardless of the state of the art of the component design.

The design methodology presented is unique to this book, although it is compatible in principle to ANSI and IEEE approaches. The methodology was developed with the communications-systems engineer in mind, as a product of a series of professional courses that have been presented over the last 10 years at George Washington University and at professional engineering organizations. Every attempt has been made to condense the rather detailed and lengthy process into a "cookbook" methodology.

Chapter 1 introduces the systems designer to a general overview of the technology and the basic methodology used to design fiber into the network. Chapter 2 then discusses the application of the technology and the standards employed. Chapter 3 familiarizes the reader with the fiber-optics components and their key performance parameters. An in-depth treatment of signal-to-noise theory and relationships pertinent to the fiber-optics system is then presented in Chapter 4. This is followed in Chapter 5 with an in-depth treatment of signal modulation, encoding, and multiplexing. Chapter 6 then integrates the tools presented in Chapters 1 through 5 into a total signal-performance analysis using a cookbook style design methodology. Chapter 7 follows with a discussion of systems architecture and evolving systems standards. Finally, Chapter 8 rounds out the formal design process with a discussion of reliability and cost analysis.

Robert J. Hoss

ACKNOWLEDGMENTS

The author wishes to acknowledge the following individuals for their direct contribution to this book: Tran Muoi for material on detectors and receiver design; J. Richard Jones for his contribution on local-area network design; Dr. C. J. Hwang for materials on optical source and transmitter design; Dennis Knecht for materials on connector and splice insertion loss; and Dave Charlton for information on fiber performance and the history of the fiber industry. The author also wishes to thank the following for providing their time and information in support of this effort: Art Nelson for his support in the area of optical couplers; Dan Berg for information on cable design; and Gordon Day for fiber-measurement information.

1

Introduction to Fiber-Optic Transmission-Systems Design

1.1 THE EVOLUTION OF FIBER OPTICS*

In the early years, from Alexander Graham Bell's experiment with his Photophone in 1880 up until 1950, optical communications remained in the concept stage. In the early 1950s, B. O'Brian, Sr., at American Optical developed the first optical-fiber bundles for image transmission. Due to high loss, these were not practical for communications. Attenuation remained in the 1000 dB/km range even into the mid-1960s.

It was not until 1960, when T. H. Maiman at Hughes Research announced operation of the first laser, that optical communications began to be more practical. The first communications systems were atmospheric, but the atmosphere made a poor communications medium due to scatter and unpredictable weather interference. Interest in fiber as the medium began in 1966 when C. Kao and G. A. Hockham at Standard Telecommunications Laboratory predicted that by removing the impurities in glass, 20 dB/km attenuations would be achievable. At this level, fiber becomes a practical communications medium. In 1970, Maurer, D. B. Keck, and P. C. Schultz at Corning broke the 20 dB/km barrier. The fiber-optics revolution had begun.

The first fiber-optics systems in the 1970s were short and used principally for military applications, where harsh electromagnetic environments could benefit from the dielectric properties. The first prototype telephone systems were implemented by

* The author wishes to acknowledge David Charlton of Corning Glass Works, Corning, New York for helping to compile this historical perspective on fiber-optic technology.

1

the Bell System and General Telephone in 1977. In 1977, Corning and NTT reported achieving 0.5 dB/km at 1200 nm. Systems development based on graded index fiber at 3 to 5 dB/km began in 1978 for both military and telephone applications. The Army and Marine Corps developed rugged fibers and systems for tactical field application, and the Air Force developed fiber-optic databus systems and other flight systems. The Navy developed components for both airborne and shipboard databus systems. Standards activity began in the military, the SAE, NATO, and the EIA. Although installation of 45 and 90 Mb/s systems for telephone entrance links and interoffice trunks commenced between 1980 and 1982, fiber was not yet cost effective for full-scale trunking, so the market remained small.

The capacity and lower attenuation of single-mode fiber were needed in order to make long-haul trunking applications practical. In 1983, a number of factors came together to revolutionize the fiber industry. Deregulation and technology merged and the first volume purchases of single-mode fiber came from MCI at the same time as the first major production of single-mode fiber was occurring at AT&T. The price of single-mode fiber was driven almost overnight from about $1.50 to less than $0.50 per meter. The wide-scale commercial fiber revolution had truly begun.

With AT&T, MCI, and other carriers, such as US Sprint and the resellers entering the field, the fibering of the United States began. Although there were still military programs, the emphasis on military product faded as the suppliers geared up for the exploding commercial telephone market. Cabled fiber prices fell into the $0.30 per meter range, at first held higher by demand, but then falling below that figure as the rush began to subside. By 1987, most of the major construction had subsided or leveled off. Fiber began to become a commodity and many companies merged or scaled down operations to stay alive.

By 1985, long-wavelength operation at 1550 nm had become a practical option, although most systems were still designed for 1300 nm. The early systems of 1983 operated at 90 and 135 Mb/s, with some 405 Mb/s coming from Japan. By 1986, product in the 560-Mb/s range was being delivered, and by 1988, 1.2 to 2.4 Gb/s became a possibility. By 1985, fiber had been installed by Warner Amex CATV in Dallas and New York City for multiple-channel supertrunking of video (8 channels per fiber). By 1987, companies were advertising up to 16 video channels per fiber. Many fiber-to-the-home experimental systems were implemented in the early to mid-1980s, however, none met the cost per subscriber goals to make it a competitor with coax for the U.S. CATV market. The later part of the 1980s emphasized product standardization, network management systems for fiber networks, and higher-speed multiplexing. Although all systems until 1989 had operated based on asynchronous multiplexing based principally on DS-1 and DS-3 electrical interface standards, no standard rates and formats existed at the optical interface.

In late 1988, the first-draft synchronous optical network (SONET) standards emerged. This standard began the era of synchronous optical communications. It recommended standard synchronous optical and electrical interfaces that drop and insert channels at multiples of the synchronous frame rate. This standard brings about equipment compatibility at the optical interface as well as opens the door for optical multiplexing and switching. In the same year, the FDDI 100 Mb/s LAN standard began to emerge, creating a high-speed backbone LAN that could extend throughout a building or into a metropolitan area. With these standards, open-systems architecture at the optical as well as the electrical interface is possible in the 1990s.

The next major evolutionary phase of fiber, coherent transmission, will come in the mid- to late 1990s. Here the source spectral width is in the megahertz range instead of the terahertz range, permitting the carrier to be modulated as one would a radio-wave carrier. Sensitivity improvements of 10 to 20 dB (25 to 50 km further distance) and data rates in the 10-Gb/s range become possible. This technology will potentially bring about a practical means for fiber to the residence and revolutionize undersea and long-haul trunking.

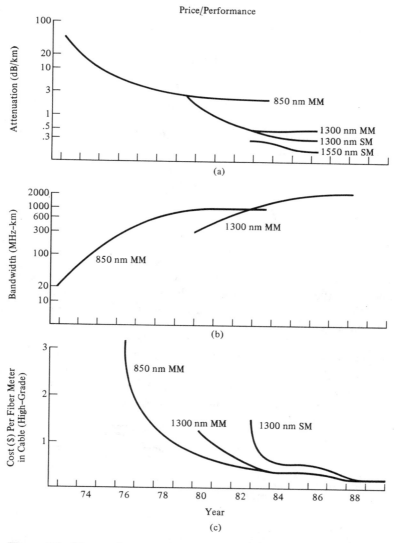

Figure 1.1 History of commercially available production fiber. (a) Attenuation history, (b) multimode bandwidth history, and (c) price history.

Figure 1.1 illustrates the cost and performance history of typical fiber product from the early 1970s to the present.

1.2 INTRODUCTION TO FIBER TECHNOLOGY

Optical fiber is a circular dielectric waveguide that propagates electromagnetic energy principally in the visible and infrared portions of the electromagnetic spectrum. When this optical energy is modulated, fiber optics can be used to transmit information over the length of the fiber.

1.2.1 Optical-Fiber Basic Operation

An optical fiber is produced by forming concentric layers of cladding glass around a core region. The core region maintains the low-optical-loss properties necessary for the propagation of the optical energy. The higher the refractive index, the slower optical energy propagates. When a high-refractive-index glass core is surrounded by the lower-index cladding material, the light energy is contained within the higher-index core due to the reflection at the interface of the two materials. The refractive index of the core and cladding is described by [6]:

$$n_r = n_1 \left[1 - 2 \Delta \left(\frac{r}{a} \right)^g \right]^{1/2} \qquad \text{for } r < a$$

$$n_r = n_2 = n_1 (1 - 2 \Delta)^{1/2} \qquad \text{for } r > a \qquad (1\text{-}1)$$

$$\Delta = \frac{(n_1^2 - n_2^2)}{2n_1^2}$$

where n_1 is the maximum core refractive index (center), n_2 is the cladding refractive index, a is the core radius, and g is the profile shape parameter.

Figure 1.2 illustrates the mechanism in the simplest class of fiber design, the step-index multimode, where $g \geq 10$. As a light ray strikes the surface of the fiber, it is refracted slightly toward the center of the core by an angle that is a function of the glass-to-air interface refractive-index difference. Once inside the core, it propagates and eventually strikes the core/cladding interface. If the angle is less than the critical angle defined by the refractive-index difference, it reflects back into the core and continues to propagate in this fashion. If the light ray strikes the core/clad inter-

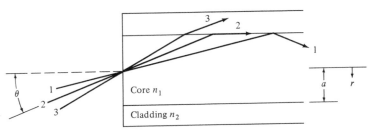

Figure 1.2   Numerical aperture defined for a step-index fiber.

face at an angle greater than the critical angle, it passes into the cladding. Since the cladding is coated with some lower-refractive-index plastic buffering material, the light propagating in the cladding is absorbed and lost.

The critical angle within the fiber translates to an acceptance angle at the fiber surface. The sine of this angle defines the numerical aperture (NA) of the fiber. Numerical aperture is therefore a parameter that defines the cone of optical-energy acceptance for the fiber. It is a parameter critical to the coupling efficiency and propagation properties of the fiber. The relationship for numerical aperture in a step-index fiber is as follows [1]:

$$NA = \sin \theta = k(n_1^2 - n_2^2)^{1/2} = kn_1(2 \Delta)^{1/2} \qquad (1\text{-}2)$$

where k equals 0.98 for the EIA-455-29 measurement method and 0.94 for EIA-455-44. The large acceptance angle of step-index multimode fiber makes it ideal for coupling to large-area Lambertian-emission-pattern light-emitting diodes (LEDs). The major limitation, however, is that the propagating rays follow many different path lengths. Rays from a light pulse entering the fiber at one instant in time are spread at the other end since some rays are delayed longer than others. This has a significant limiting effect on fiber bandwidth.

Because the large-angle rays interact with the core/cladding interface closer to the critical angle, and because they propagate over a longer path distance in the core, they are more likely to lose energy due to absorption in the fiber to be refracted into the cladding by perturbations at the interface. There therefore exists a trade-off between numerical aperture (coupling efficiency), bend sensitivity, and bandwidth. For practical designs, bandwidths are limited to a few megahertz-kilometers.

Although the diagram in Figure 1.2 illustrates ray propagation for simplicity, fiber is generally classified as to the modal propagation properties of the waveguide. There exists a rough relationship between the order of the mode propagated in the fiber and the angle of the light rays associated with the mode. Low-order modes correspond to the rays closer to zero angle from the axis of the core. High-order modes correspond roughly to the wider-angle rays. The number of modes propagating in the fiber is roughly determined by [2]:

$$M = 2\frac{g}{g + 2}\left(\frac{\pi a}{\lambda}\right)^2 (NA)^2 \qquad (1\text{-}3)$$

In order to improve the bandwidth of the fiber while maintaining a reasonable numerical aperture and bend sensitivity, a class of multimode fiber called graded index was developed. The propagation characteristics and core refractive-index characteristics are contrasted with the step index in Figure 1.3. Here the core-index profile is $3 > g > 1$ (roughly 2), or parabolic. Like a parabolic lens, it tends to curve the rays toward the axis of the core. The parabolic core index slows the rays propagating down the axis more than those of a higher angle, therefore equalizing the time it takes any ray to propagate the length of the fiber. As Chapter 3 indicates, this optimization is wavelength-dependent. The result is that at the design wavelength, bandwidth can theoretically be infinite. This is, of course, limited by practical limitations on profile design and production control.

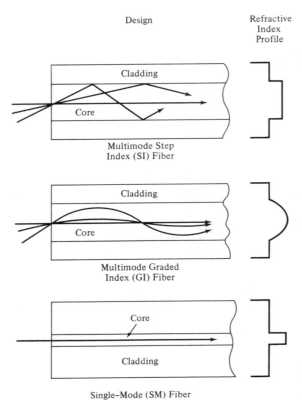

Design

Refractive
Index
Profile

Cladding

Core

Multimode Step
Index (SI) Fiber

Cladding

Core

Multimode Graded
Index (GI) Fiber

Core

Cladding

Single–Mode (SM) Fiber

Figure 1.3   Classification of optical
fibers by virtue of waveguide structure
and propagation modes.

As it turns out, practical design limits, tolerances, and production yields limit graded index to about 1 to 2 GHz-km for production fiber. This is not adequate for long-haul applications, where transmission rates of 560 Mb/s to 2.4 Gb/s are routinely required over distances of 30 km or more. The answer to this requirement is single-mode fiber. If the propagating energy in the fiber can be restricted to one mode, then there can be no differential path distances or time-delay differences between modes. The bandwidth, from the standpoint of waveguide design, would be infinite.

Equation (1-3) for propagating modes illustrates that by reducing core size a and making Δ or numerical aperture small, then there is a point at which the number of modes M is less than or equal to 1. This occurs when [2]:

$$\frac{2\pi a}{\lambda} n_1 \sqrt{2\,\Delta} \leq 2.4 \tag{1-4}$$

1.2.2 Fiber-Optic Systems Operation

The operation of a basic fiber-optic transmission link is illustrated in Figure 1.4. The process is that of near-linear conversion of drive current to optical power and of lin-

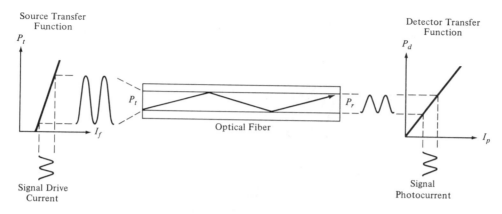

Source	Fiber	Detector	Preamp
LED Laser diode	Step index Graded index Single mode	PIN APD	Low impedance Integrating Transimpedance
GaAs 0.9 μm GaAlAs 0.8–0.85 μm	Plastic Plastic clad	Si 0.8–1.0 μm	Bipolar
InGaAsP 1.3–1.6 μm	Fused silica	InGaAsP 1–1.6 μm	FET

Figure 1.4   Basic link operation, the electrooptical conversion process.

ear conversion of detected optical power back to photocurrent. At the input is an optical source driver that converts the input signal to a drive current that intensity modulates the source. The optical source then generates the optical energy that is coupled into the fiber. The energy propagates down the fiber and is attenuated to a degree and distorted by the modal dispersive properties discussed in the previous sections. The energy exits the fiber at the other end and a majority is coupled into a photodetector. The light energy that is absorbed in the photodetector is converted to a photocurrent. This photocurrent is then amplified in the receiver electronics and converted to the proper signal format at the output.

Depending on the application and transmission requirements, the fiber link can be constructed of various elements. A few of the major ones are listed in Figure 1.4. The trade-offs involved in the proper selection of the component mix are discussed in detail in Chapter 3 as well as throughout most of the book.

The wavelength of the optical energy that is propagated in the fiber is dependent primarily on the region of the electromagnetic spectrum where the fiber core materials exhibit the lowest loss. With plastic, this is in the visible region. With silica glass, the most predominant material, the lowest loss is in the infrared region between wavelengths of 800 and 1600 nm. The relative location in the electromagnetic spectrum is illustrated in Figure 1.5. The typical loss spectrum of fused silica fiber is illustrated.

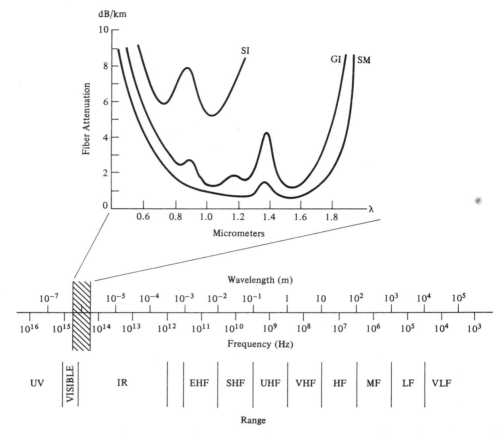

Figure 1.5 Fiber-optic operating spectrum in relation to the electromagnetic spectrum.

1.3 ADVANTAGES AND DISADVANTAGES OF FIBER TECHNOLOGY

Fiber is used in a particular application because it exhibits some advantage over an alternative material. Whenever the characteristics of fiber permit it to be the low-cost alternative, it is generally employed. Fiber is also used when it provides a unique feature or service not available with other materials. If fiber is used in applications where its inherent qualities (such as wide bandwidth and low attenuation) are not required, it is then a more expensive solution.

1.3.1 Advantages

Some of the more positive characteristics and applications follows.

Large-length bandwidth.   Because of its wide bandwidth (multigigahertz-kilometers) and low attenuation (less than 0.3 dB/km), fiber often is the lowest-cost

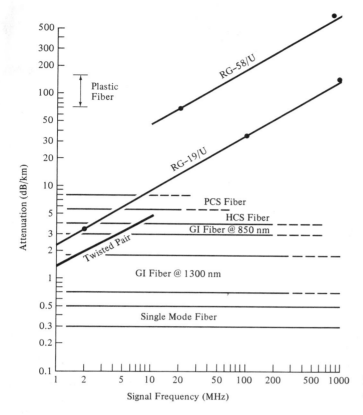

Figure 1.6 Comparison of fiber performance with coax and twisted pair.

transmission medium per channel kilometer. In Chapter 3, Figure 3.1 illustrates the bandwidth length capabilities typical of present fiber technology. Figure 1.6* compares the transmission properties of fiber with those of other conventional transmission media [3, 4]. Note that the attenuation of copper wire has a strong and direct dependency on signal frequency. The dependency in fiber is less a function of signal frequency except where it approaches the dispersion limits or at longer transmission distances, following Conradi et al. [5]:

Twisted Pair:

$$\text{Attenuation} = k\sqrt{\frac{wRC}{2}}D \qquad \text{at low frequency} \qquad (1\text{-}5)$$

$$\text{Attenuation} = k\frac{R}{2}\sqrt{\frac{C}{L}}D \qquad \text{at high frequency} \qquad (1\text{-}6)$$

Coax:

$$\text{Attenuation} = (x + y\sqrt{f} + zf)D \qquad (1\text{-}7)$$

*The author wishes to acknowledge and thank J. Richard Jones of Broadband Technologies for the materials for this section.

Fiber:

$$\text{Attenuation} = \alpha D + 10 \log \left[\left(\frac{f\Delta\lambda}{f_m}\right)^2 D^3 \right] + 10 \log \left[b\left(\frac{f}{f_w}\right)^2 D^c \right] \quad (1\text{-}8)$$

where k, x, y, and z are constants peculiar to the cable design

D is transmission distance

R is resistance, L inductance, and C capacitance

α is the fiber attenuation coefficient in dB/km

f is the signal frequency

f_m is the fiber-bandwidth limitation due to material dispersion (dispersion is given in picoseconds/nanometer/kilometer)

f_w is the fiber-bandwidth limitation due to waveguide dispersion (dispersion generally given in nanoseconds/kilometer)

b and c are constants relating to the degree of mode mixing present in multimode fiber: $b = 1$ and $c = 3$ for no mode mixing; $b = 3/2$ and $c = 2$ for mode mixing

Low installation and operating costs.   The wide bandwidth and low loss increase repeater spacing, thereby reducing cost in the outside plant. The elimination or reduction of repeaters reduces the maintenance, power, and operating expenses. The simplicity, low-power requirements, and very high reliability of the terminal equipment also reduce maintenance costs. Generally, fiber can be installed and maintained within the capabilities of an existing telecommunications organization, with little training and only a few added pieces of optical test equipment such as a power meter and possibly an OTDR (Optical Time Domain Reflectometer) and a splicer. Installation is often simpler than with copper since fiber requires no special balancing, conditioning, or environmental precautions (pressurization).

Small physical size.   The large capacity possible within the small dimension of the fiber provides advantages where space is a premium, such as in a congested duct system, within computer rooms, or within an aircraft or ship.

Light weight.   Because of the large capacity of an individual fiber, the weight advantage over a copper conductor is obvious. Also, because the fiber requires no shielding nor insulation (only a protective buffer), the entire cable structure is lightweight. Even high-tensile-strength cable can use Kevlar®, which is an extremely lightweight material for the tensile strength it provides. The weight advantage is particularly useful in such applications as aircraft, where weight is key to fuel consumption and cargo capacity.

Interference immunity.   Fiber is a dielectric waveguide. It therefore is not affected by, nor does it radiate, electromagnetic energy. It is immune to crosstalk from adjacent (coated) fibers and does not conduct ground currents that can effect the signal. It not only provides noise-free transmission quality, but it eliminates many of the problems experienced during installation and checkout of a system due to nonoptimal grounding.

High-quality transmission.   As a result of the noise immunity of the fiber transmission path, fiber routinely provides communications quality that is orders of magnitudes better than copper or microwave. The general standard for a fiber transmission link is a 10^{-9} BER minimum, with 10^{-11} or better as the norm. This is in comparison with 10^{-5} to 10^{-7} for copper and microwave systems.

Nonconductor.   Fiber does not conduct electricity. It therefore can be installed in applications where shock and explosive hazards exist. It does not conduct lightning and therefore is a means of protecting against outages in lightning-active regions.

Environmental stability.   Fiber retains its transmission characteristics virtually unaffected by environmental extremes encountered in normal installations. Only extreme cold (-20 to -40 °C) causes an increase in attenuation to various degrees, depending on cable design. With coax, temperature can have a continuous effect on performance. Water has no effect (except on lifetime if the fiber is under stress), whereas with copper cable, it causes rapid performance degradation.

Secure transmission and tempest.   Because fiber does not radiate a field of electromagnetic energy, it is inherently secure unless physically contacted or distorted. This is important in strategic or tactical installations where it is desirable that an RF signature is not produced by cable conductors. If the waveguide is contacted or distorted such that a portion of the optical energy field propagating in the cladding or core is diverted, then the security can be compromised. In this event, however, the fiber and terminating equipment can be designed to detect the loss of energy or resultant perturbations in the modal patterns and initiate an alarm.

Electromagnetic-pulse (EMP) immunity.   Since fiber is not a conductor, it provides immunity to EMP. EMP is the result of a high-altitude nuclear burst. The magnetic fields created by the high-speed atomic particles cause free electrons in the upper atmosphere to spin, in turn creating a short-duration but high-energy electromagnetic wave. When this wave hits a wire conductor, it can produce a short-duration current pulse of 1000 A or more. The result is the destruction of highly sensitive receiving electronics attached to the wire. To prevent damage, heavy filtering, arc arrestors, and surge protectors are used. Replacing the copper system with fiber not only provides EMP protection, but reduces weight and preserves signal quality. This is particularly important to military aircraft applications and tactical field telecommunications systems.

Rugged construction.   Fiber-optic cable has been designed in many cases to be more rugged than its copper cable counterpart. This is particularly true when comparing it with coax, where the transmission properties can be altered by extreme tension or bending. Fiber cable for outside plant installation generally has a tensile strength of 600 lb. Some tactical field-cable designs can withstand multiple tank and vehicle crossings, a property that copper cable cannot provide.

1.3.2. Disadvantages

Some limitations of fiber follow.

Radiation hardness.   Fiber darkens to some degree when bombarded by high-energy nuclear particles. If the event is a nuclear blast, the fiber first glows during the initial event, then darkens very rapidly to the point where it it loses useful transmission properties. It then gradually recovers to an attenuation some level greater than prior to the event. The recovery level (permanent damage) depends on the level and type of fiber dopant. Lightly doped $GeO_2 - SiO_2$ fused silica fibers such as single-mode or nondoped pure fused silica are affected less. The effect is also much less at longer wavelengths (1300 nm), where loss is very low. The total transient and permanent absorption is a function of dose and dose rate. For dosage in the 1000-rad range, induced loss of from 20 to 100 dB/km can be expected, and transient recovery can last for less than 1 second to 100 seconds, depending on operating wavelength and fiber doping. Temperature has a strong effect on recovery. Recovery can be 100 times faster for a 50 °C increase in temperature and 100 times slower for a 50 °C decrease.

Nonconductor.   Fiber cannot transmit electrical power. This limits its application where the receiving terminal (such as a telephone set) must be powered from the line. The cable must provide separate conductors for power.

Bend strength.   Most fiber is proof tested to between 50,000 and 100,000 psi. The pull tension on fiber, therefore, is adequate for rough handling when properly cabled and protected. It is difficult, by hand, to pull apart a single coated fiber. If a fiber is tightly bent, however, the stress on the outer surface can exceed the tensile strength and breakage can result. Generally, cable is designed to prevent such a small bend radius, and such breakage only occurs with single-fiber "pigtails" or by accident during splice operations.

Hydrogen absorption.   Molecular hydrogen can diffuse into silica fibers and produce attenuation change. Hydrogen can come from certain cable and fiber buffering materials and, since it is confined within the cable, can build up to significant concentrations. According to David Charlton of Corning Glass Works, the losses can come from two mechanisms: (1) unreacted H_2 that diffuses into the silica, creating reversible losses due to absorption, and (2) H_2 molecules reacting with sites in the silica matrix to form permanent hydroxyl ions to a degree dependent on fiber composition, temperature, and H_2 partial pressure. The loss spectrum from the diffusion mechanism has a dominant peak around 1.24 micrometers with a loss proportional to the hydrogen partial pressure, and reversible if the partial pressure decreases. The most vulnerable to the second mechanism is multimode fiber doped with a high degree of phosphorus. In single mode, any small effect is generally reversible. Under common conditions, the effect is generally negligible even in multimode. Charlton quotes some work performed by Rush of BTRL that indicates predicted increases of only 0.05 dB/km at 1300 nm under worst-case cabling conditions of 1000 ppm H_2 at 20 °C for 25 years.*

High cost in low-bandwidth applications.   Fiber is generally only cost effective when its capabilities for bandwidth and attenuation are required. If high-quality fiber is used in low-bandwidth short-distance applications, it may be a more costly solution than copper. Moderate- to high-quality fused-silica fiber can cost anywhere from 20 to 50 cents per cabled fiber meter, depending on quality, quantity, and fiber count. Lower-quality fiber can cost just as much or more due to lower production volumes. Copper wire, on the other hand, is in the 3 to 30 cents per conductor meter range (although high-quality coax can exceed this). If the application does not require the dielectric properties of fiber, and runs adequately on copper, then fiber can be a poor choice.

1.4 INTRODUCTION TO SYSTEMS-DESIGN METHODOLOGY

The basic approach to the design of a fiber transmission system is not much different than the approach to the design of any information-transmission system, whether radio, microwave, cable, or satellite. It is only the design of the fiber subsystem, and the different trade-offs that the fiber technology offers, that differs from other media.

The design of a fiber system is a lot like the design of a radio or microwave system in that receiver power levels are low, noise performance is focused on receiver design, and signal-modulation formats are similar. The difference is that until the next generation of fiber optics (coherent optics), the optical carrier is noncoherent, so that heterodyning the carrier is not possible. The optical carrier is either intensity modulated with the signal or some intermediate-frequency carrier transports the signal and in turn intensity modulates the optical carrier. Another key difference is that within the optical receiver, we deal with signal- and noise-current levels and ratios rather than voltages.

The design of a system involves more than the transmission medium. It involves signal-processing electronics, modulation approaches, multiplexing, encoding for transmission and error performance, reliability and availability considerations, and above all cost-versus-performance trade-offs. The book is structured to lead the designer through the various trade-offs and considerations surrounding the overall systems design from the perspective of the peculiar set of considerations that fiber-optics technology imposes.

Sections 1.5 through 1.7 define the fiber-optic transmission system and provide an introduction to the design methodology that is to be developed in detail in the remaining chapters.

---

*The author wishes to acknowledge David Charlton of CGW for his assistance in preparing this section.

1.5 FIBER-OPTIC SYSTEMS-DESIGN MODEL

A fiber-optic system design must consider not only the optical and electrooptical elements, but the signal-processing elements as well (modulation, multiplexing, encoding, etc.). The methodology presented in this book, therefore, approaches fiber-optic systems design from the standpoint of end-to-end baseband channel performance. The impact of fiber on the choice of signal modulation, multiplexing, and encoding is treated as a integral part of that overall systems design.

The design model presented here is intended to be somewhat generic to all systems architectures. The partitioning of the fiber-optic transmission systems model is defined roughly as in the 1988 proposed ANSI draft standard T1.105-1988 (or ECSA document T1X1/87-129R1), "Digital Heirarchy Optical Interface Rates and Formats (SONET)." Although this standard is intended for a specific synchronous digital transmission approach, the functional partitioning of the layers presented in this standard (path, line, section, and photonics) is clever and somewhat common to all designs. We therefore borrow but redefine these layers to more broadly describe the partitioning of asynchronous and analog transmission systems as well. Every attempt is made not to conflict with the definitions where synchronous digital systems are involved. The relationships with the layers as defined in this book and the structures of other associated ANSI and IEEE standards are illustrated in Figure 1.7. The ANSI SONET standard is described in Chapter 7.

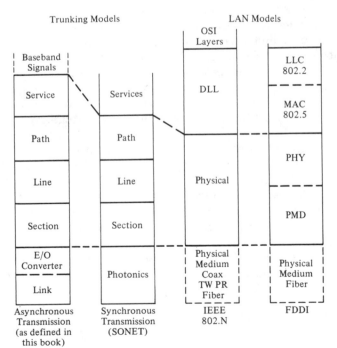

Figure 1.7 Relationships between functional-layer definitions.

1.5.1 Systems-Element Definitions

In simple terminology, a system is one or more fiber-optic links interconnected, along with appropriate modulation, multiplexing, and encoding electronics, (terminal equipment) in order to perform a service channel transmission function. A service in the context of this book is considered a baseband signal such as a voice, video, or digital channel. Processing of this signal may be necessary either prior to or within the path layer (or both) in order to present a standard channel format to the line-layer termination equipment. The line layer then combines (multiplexes) the channels into a single optical transmission section that contains the photonics equipment for fiber-optic transmission. A system can have multiple path-termination points (as in an add/drop or local-area network), or a single pair of path-termination equipments (a point-to-point network). Figure 1.8 illustrates the functional or logical relationship between the systems layers in a repeater section.

The following definitions are used in this book to partition the layers:

Service layer.   This layer is responsible for the presentation of channels to the path layer in a set of formats that may not be common, but that fall within a standard set that the path layer is designed to serve. Some processing may be required at this layer if the "baseband channels" at the service layer are not within the bounds of the standard format set of the path-termination equipment. For example, the SONET termination equipment, as well as most digital office termination equipment, handles a range of low-speed and high-speed DS-X standard signals, but does not handle di-

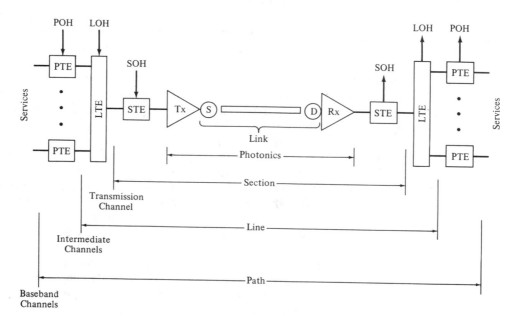

Figure 1.8   Definition of functional layers in a fiber transmission system.

rect analog voice. A PCM channel bank must be used to convert analog to this set of standard asynchronous interface signals. We consider that the PCM channel bank resides in the service layer.

Path layer.   The function of the path layer is the processing and mapping of channels from the service layer to achieve a channel format as required by the line layer. In an analog transmission system, this might consist of signal modulation, such as frequency modulation (FM). In a digital transmission system, this might involve the synchronization and framing of a standard channel unit. The path-termination equipment (PTE) may also perform a multiplexing function in order to map multiple service channels into the line-layer channel format. For example, in the ANSI T1.105 standard, this is where services such as DS-N, video, and others are encoded and multiplexed to form the standard frame format, synchronous transport-signal level 1 (STS-1). This layer also inserts and extracts the path overhead (POH) signals. Such functions as voice orderwire and multiplexer signaling are examples of POH. We define the channel at the path to line termination as an "intermediate channel."

Line layer.   The function of the line layer is to map the intermediate-channel format into a format consistent with the section layer and to multiplex (insert/extract) multiple intermediate channels along with any line overhead (LOH). This layer formats a "transmission-channel" signal that is given to the optical section layer. Examples of LOH signals are voice and maintenance orderwire and channel-protection signaling. Line-termination equipment (LTE) may perform such functions as frequency-division multiplexing (FDM) in analog transmission systems and time-division multiplexing (TDM) in digital transmission systems. In ANSI T1.105, this is where the STS-1 signals and LOH are mapped into the STS-N signal.

Section layer.   The section layer transmits to its peer section-termination equipments (STEs) the transmission channel multiplexed with the section overhead (SOH), if any, and performs the encoding necessary to form this composite signal into properly formatted pulses for optical transmission over the photonics layer. Signal encoding, such as NRZ to Manchester, for example, may take place within this layer to format the signal properly to comply with electrooptical circuitry limitations. In ANSI T1.105, this is the layer where the STS-N signal, error encoding, and the SOH are mapped into a frame, and that frame is scrambled into a pulse format compatible with the photonics layer. In an analog transmission system, this is where SOH signals would be frequency multiplexed, for example, with the intermediate-channel signals and signal-gain balancing or preemphasis would be controlled.

Photonics layer.   The photonics layer converts the electrical signal from the section layer to an optical signal for transmission over the fiber-optics physical medium. Although it is not a definition in the ANSI standard, we also define this layer as the point at which optical multiplexing can take place, i.e., multiple section-layer signals are converted to optical signals and wavelength multiplexed onto the same physical fiber-optics link. Physically, the photonics layer contains the fiber-optic source/driver, detector/preamp, and the passive fiber-optics link. In the ANSI standard, this is where the OC-N signals are created.

1.5.2 Physical Elements

The functional elements in Figure 1.8 are simply designated as path-, line-, and section-termination equipment (PTE, LTE, STE, respectively). In a practical system, these termination equipments perform such functions as encoding, carrier modulation, multiplexing, and other systems management and support functions. Some of the more common functions and physical realizations are as follows.

Modulator/demodulator. This function is typical of path-termination equipment in an analog transmission system. The baseband signal modulates a carrier in a fashion so as to contain the information content in a frequency and amplitude envelope better suited to the characteristics of the chosen multiplexing approach and fiber-optics medium. For fiber systems, bandwidth is generally traded for transmitted signal level in order to better suit the limited power conditions of a fiber link. This is known as a modulation improvement factor. Common modulation approaches are pulse-code modulation (PCM), frequency modulation (FM), amplitude modulation (AM), and phase modulation (PM).

Mutliplexer/demultiplexer. This function combines multiple baseband or intermediate channels into a composite signal for transmission and then extracts them individually at the receiving end. The most common approach for analog modulated signals is frequency-division multiplexing (FDM). The common approach for binary modulated signals is time-division multiplexing (TDM). The multiplexing function can exist at all layers. In the services layer, it can take the form of a PCM channel bank that PCM encodes 24 voice channels into a DS-1 signal compatible with path-termination equipment. At the path layer, a multiplexer can be used to combine multiple standard low-data-rate channels (DS-0, DS-1, etc.) to a standard channel format (DS-3, STS-1, etc.) for the line-termination equipment. A multiplexer at the line layer combines standard high-speed channels (DS-3, STS-1, etc.) to a high-speed channel for transmission over the fiber-optics section. The multiplexing function can also exist at the photonics layer in the form of wavelength-division multiplexing (WDM). Here multiple optical signals at differing wavelengths are combined optically onto a single fiber for transmission.

Encoder/decoder. Encoding is a term that describes the process of formatting the transmission channel for compatibility with the photonics layer. It can also be used to describe the process of analog-to-digital conversion. We call this process PCM encoding so as not to confuse the two. Fiber-optic transmitter and receiver electronics requires coding that is compatible with unipolar signaling, with timing-recovery requirements, and with capacitively coupled receiver electronics. This function as defined is generally a part of the transmitter, receiver, and regenerator electronics, and is generally found within the section layer, although it is sometimes a function of the multiplexer within the line-termination equipment.

Optical transmitter.   This transmitter contains the drive electronics and optical source. Its function is to directly convert the transmitted composite signal voltage to modulated optical power. Its effect on systems performance is to establish the coupled optical power transmitted signal level and to introduce source modal noise. This transmitter is found within the photonics layer.

Optical receiver.   This receiver contains a photodetector, amplification stages, and signal-recovery circuitry. Its function is to convert the received modulated optical power signal to a composite signal voltage. This receiver lies within the photonics layer.

Repeater or regenerator.   This contains an optical transmitter, receiver, and processing equipment necessary to restore the optical signal to a level necessary for retransmission down another fiber transmission link. At present, the repeater is a complete back-to-back receiver/transmitter generally containing functions of both the section and photonics layer. For analog transmission, the repeater simply receives, reamplifies, and retransmits the signal. In simple analog repeaters, where orderwire is not required, only the functions of the photonics layer can be employed. For binary transmission, the repeater has the additional function of regeneration and retiming of the binary waveform and the functions of drop/add of orderwire and other SOH signals.

Drop/insert repeater.   This is the same repeating function as before, however, in this case, the repeater contains the line layer and permits channels to be demultiplexed (dropped) and multiplexed (inserted) into the transmission section.

Optical link.   This contains the passive optical components that carry the optical signal from transmitter to receiver. Included are couplers, connectors, splices, and optical fiber.

1.6 SYSTEMS-PERFORMANCE MODEL

The last section described the general model of the fiber-optic transmission system. In order to analyze the performance of components and processing techniques selected to fit that model, some performance parameters are defined at various functional interfaces within the model. Although the definitions were developed specifically for this book, most are common to industry standards. In some cases, it is appropriate to define completely new terms where no adequate definitions existed. For example, minimum required received power (MRP) is a new term developed because the closest industry standard (received power) is not specific enough.

Performance is typically defined in terms of signal performance with reference to noise level, or by the signal-to-noise ratio (SNR), and spectral occupancy requirements, or bandwidth (BW). These terms (SNR and BW) are generally used in conjunction with analog transmission. When the transmission is in digital format, the relationship of signal to noise is converted to error probability of the detected signal, or bit-error rate (BER). With digital transmission, bit rate (BR) is used to relate to spectral occupancy instead of bandwidth.

There are many other terms used to define signal performance as well, including jitter (timing noise in digital transmission), harmonic distortion, intersymbol interference, and such. These are all addressed; however, the focus is on the principal criteria of SNR, BER, BW, and BR.

A single-section (unrepeatered) system with simple signal processing is used to illustrate the systems interfaces where signal-performance parameters are most commonly defined. The model is illustrated in Figure 1.9.

1.6.1 Electrical Versus Optical Parameters

The relationships between signal and noise and the measurement of bandwidth can differ in definition depending on whether they are measured on the electrical side (path, line, section, or photonics interface) or the optical side (link interface). The optical interface is at the optical-link physical layer, i.e., the output of the optical source or the input to the optical detector.

Electrical $(SNR)_e$ and $(CNR)_e$.   These are the signal-to-noise and carrier-to-noise ratios, respectively, as measured and defined on the electrical side of the electrooptical system. They are measured in terms of voltage, current, or electrical power ratios. The digital transmission parameter, the bit-error Rate (BER), is always measured after detection as an electrical parameter with a corresponding SNR parameter Q.

Electrical bandwidth, $(BW)_e$.   This is defined as a range of frequencies whereby signal response falls within specified limits based on measurement of electrical gain, voltage, or current ratios. The most common definition is the 3-dB bandwidth, where signal response falls to 0.707 of its value at zero frequency.

Optical $(SNR)_o$ and $(CNR)_o$.   These are the signal-to-noise and carrier-to-noise ratios, respectively, as measured and defined within the optical portion of the system. They are measured as the ratio of average received optical signal power, P_r, to optical noise equivalent power, $(NEP)_o$.

Optical bandwidth, $(BW)_o$.   This is the range of frequencies whereby optical-power levels fall within specified limits. The 3-dB optical bandwidth is defined as the modulation frequency, where optical power falls to 50% of its value at zero frequency.

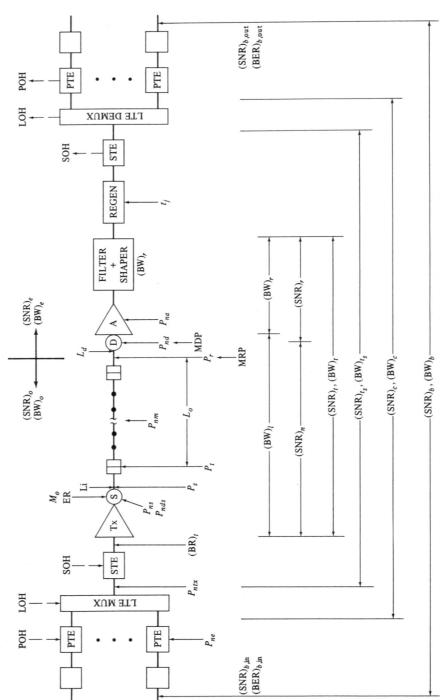

Figure 1.9 Transmission performance model.

1.6.2 Noise in the Optical Transmission System

Noise is introduced into the transmitted signal from the following sources:

Transmitted-noise power, P_{ntx}.   Transmitted noise originates as part of the transmitted signal. This is reflected in the signal-to-noise ratio of the baseband signal input $(SNR)_{b,in}$ and translates to the $(SNR)_{ts}$ at the section layer.

Encoding-noise power, P_{ne}.   Noise can be created in the process of digital encoding and signal formatting. This is generally treated as an SNR limitation imposed by the signal processing and is reflected in the SNR transfer function of that processing stage.

Crosstalk-noise power, P_{nct}, and reflection-noise power, P_{nr}.   These are the noises or unwanted signal elements introduced by crosstalk in couplers or reflections, respectively, in optical components.

Distortion-noise power, P_{nds}.   This is the distortion of the signal or creation of unwanted harmonic products due to nonlinearities at the transmitter or receiver.

Source-noise power, P_{ns}.   This is the optical source noise caused by quantum noise, modal noise, reflections, and other effects within the optical source. This is generally negligible for most single-mode systems.

Modal-noise power P_{nm}.   This is the modal noise caused by interference interactions of optical modes within the fiber medium or at joints and discontinuities. This is a significant source of noise in laser-driven multimode systems and can become significant in very long or very high-speed single-mode systems as well. It is negligible in LED-driven systems.

Detector-noise power P_{nd}.   This is the noise introduced by the photodetector, which includes quantum noise, leakage-current noise, and excess avalanche noise.

Amplifier-noise power, P_{na}.   This is the noise introduced by the preamplifier, which includes thermal noise in resistive elements and noise sources in active elements. This together with detector noise generally determines the basic noise-performance limits of fiber-optic systems.

1.6.3 Analog Transmission Parameters

When the transmission approach uses linear transmission elements or sinusoidal-carrier modulation, the transmission is defined as analog. The performance parameters within the various systems layers are as follows:

Baseband signal-to-noise ratio, $(SNR)_b$. This is generally related in terms of signal level or power to the root-mean-square (rms) noise level or power within the baseband signal bandwidth at the service layer.

Baseband bandwidth, $(BW)_b$. This is the band of frequencies occupied by the signal before processing at the path layer.

Intermediate-channel bandwidth, $(BW)_c$. This is the bandwidth requirements of any one of N channels after baseband signal processing but before line-layer processing. It generally refers to the channel bandwidth between the path and line layer. $(BW)_c$ is the per-channel bandwidth required of the line layer to pass the channel without distortion. $(BW)_c$ can be used to describe other intermediate channels between, say, the service layer processing and the path layer. In this case, modifiers such as $(BW)_{cs}$ are used.

$(SNR)_c$ and $(CNR)_c$. These are the signal-to-noise or carrier-to-noise ratios, respectively, of the received demultiplexed per-channel composite signal at the interface to the line-termination equipment measured in the intermediate-channel bandwidth, $(BW)_c$.

Transmission bandwidth, $(BW)_t$. This is defined as the required bandwidth within the photonics layer to pass the modulated and multiplexed composite transmission signal without distortion. If processing within the section layer is performed, such as adding overhead signals or altering modulation format, then a different transmission bandwidth, $(BW)_{ts}$, is defined end to end for the section.

$(SNR)_t$ and $(CNR)_t$. These are the signal-to-noise and the carrier-to-noise ratios, respectively, of the composite electrical signal measured at both the input and output of the photonics layer within the defined transmission bandwidth, $(BW)_t$. They are used to define the performance of the section layer.

Photonics SNR penalty, $(SNR)_n$. This is the noise source within the photonics layer that degrades SNR performance. The SNR requirement on the receiver, $(SNR)_r$, must be increased to compensate for the SNR limitations of the remaining photonics, $(SNR)_n$, so that the overall photonics-layer requirement, $(SNR)_t$, is met. Some of the noise sources are treated separately from the $(SNR)_n$ parameter as power penalties, P_p.

$(SNR)_r$. This is the signal-to-noise ratio measured within the receiver bandwidth at the photonics-layer receiver analog interface prior to regeneration.

Source optical power modulation index, M_o. This is the fraction of available power used by the modulating signal as referenced to the laser bias point (or average power point). M_o is generally selected (less than 1) based on distortion considerations.

1.6.4 Digital Transmission Parameters

When the transmission method is digital, binary encoding is generally used, whereby only two states exist at each interface, "on" and "off." Analog baseband signals presented at the services layer are binary encoded to a two-state signal, formatted, sometimes multiplexed, and mapped into an intermediate-channel format by the path-termination equipment.

Performance within the digital transmission system is quantified in terms of error performance (BER), bit rate (BR), and timing noise (jitter).

(BER)$_b$.   This is the bit-error rate of a digital baseband service.

(BR)$_b$.   This is the bit rate of a digital baseband service.

Intermediate-channel bit rate, (BR)$_c$.   This is the intermediate per-channel bit rate generally defined at the interface to the line layer. It results from a mapping of baseband channel bit rates and path overhead (POH) into the line-frame format. If the services layer employs pulse-code modulation (PCM) to digitally encode an analog baseband signal, (BR)$_c$ is determined by the baseband signal bandwidth and the PCM processing.

Intermediate-channel bit-error rate, (BER)$_c$.   The bit-error rate per channel is measured before multiplexing and after demultiplexing at the interface to the line layer. For ideal time-division multiplexing, per-channel (BER)$_c$ is essentially equivalent to (BER)$_t$ if errors are considered random.

Transmission bit rates, (BR)$_t$ and (BR)$_{ts}$.   (BR)$_t$ is defined here as the composite transmitted bit rate presented to the photonics layer after multiplexing and encoding. (BR)$_t$ is related to the number of multiplexed intermediate channels; the rate per channel, (BR)$_c$; and any additional bits added to the frame to accommodate line overhead (LOH). If the section layer performs an encoding function or adds overhead (SOH) that effectively changes the transmission bit rate, (BR)$_{ts}$ defines the bit rate at the section layer.

Transmission bit-error rates, (BER)$_t$ and (BER)$_{ts}$.   (BER)$_t$ defines the transmission bit-error rate at the photonics layer. The (BER)$_t$ requirement as defined here is based on the (BER)$_t$ measured at the output of the receiver. If the section layer performs an encoding function that effectively alters BER performance, then BER requirements for the section layer are redefined as (BER)$_{ts}$.

SNR penalty, Q_n.   This is the limitation on link SNR, related as parameter Q in a digital system and created by noise generated in the link, that is separate from that within the receiver.

Receiver SNR, Q_r.   In digital transmission systems, the required SNR of the receiver, $(SNR)_r$, is generally defined in terms of peak or average optical signal current to rms noise current, and is related to the required received $(BER)_t$ by the error-function relationship Q_r.

Extinction ratio, ER.   The extinction ratio is defined as the ratio of the average optical energy transmitted for a logical zero (0) level divided by the average optical energy transmitted for the opposite logical one (1) level. Like the modulation index (M_o), the extinction ratio is a measure of the efficiency of the transmitter in its use of total available optical power.

1.6.5 Photonics and Optical-Link Parameters

Transmitted optical power, P_t.   This is the optical power from the source that is coupled into the fiber medium. Where the optical source is factory coupled to a fiber pigtail, this is generally measured at the line side (output) of the transmitter unit connector. When the cable is connector interfaced directly to the fiber, this power is defined within the aperture of that coupling fiber. Transmitted power is generally expressed as mean or average coupled power. When total source power, P_s, is of interest, then the source-to-fiber coupling loss, or injection loss, L_i, is considered, where $P_t = P_s - L_i$.

Minimum required optical power, MRP. P_r is the received optical power coupled into the detector aperture (or pigtail) on the line side of the receiver equipment connector. MRP is the P_r required to achieve specified performance. They are generally expressed in terms of average or rms power. Link performance is generally measured by the difference in minimum optical power required by the receiver (MRP) versus the actual received power (P_r). Detected power (P_d), and minimum required detected power (MDP), relates to the power incident on the detector. MRP is related to the MDP by subtracting the detector coupling loss, L_d, between the fiber end and the detector as MDP = MRP − L_d.

Optical loss, L_o.   Optical power losses in the link are due to insertion losses of the couplers, connectors, and splices and the attenuation within the fiber.

Link bandwidth, BW_l.   This is the passband (bandwidth) of the electrooptical and passive optics portion of the photonics. It has three basic components: (1) the bandwidth of the optical fiber, $(BW)_f$, (2) the bandwidth of the source and driver, $(BW)_s$, and (3) the bandwidth of the detector, $(BW)_d$. The optical elements have the effect of limiting signal power at higher frequencies or causing pulse distortion in binary signals. This effect translates to a power penalty or reduction in the SNR that must be compensated by adding more power or increasing receiver filter bandwidth, $(BW)_r$.

1.7 SYSTEMS-DESIGN METHODOLOGY

The logical flow of the systems-design process is illustrated in Figure 1.10. As evidenced by the direction of the arrows, the process is iterative. Candidate approaches are analyzed and either modified or discarded until the design requirements are met.

The designer first starts with a set of systems requirements or specifications. Various possible approaches (candidate architectures) are then considered to meet the design requirements. Candidate components and subsystem elements are then chosen to satisfy the most promising architecture. The performance of the most promising systems realization is then analyzed against the signal-performance requirements. If the chosen design does not meet the criteria, then the designer must determine what to change. If the performance is significantly deficient, an alternate architecture may be tried. If it appears that the performance can be achieved with a different mix of components, then alternatives will be chosen and the performance analysis repeated.

Once performance analysis indicates an acceptable design candidate, analysis is performed against other systems parameters such as reliability and cost.

The analysis of systems reliability and availability involves not only the choice of components for acceptable reliability, but the choice of systems architecture to achieve the necessary redundancy. Changes in components and architecture, to satisfy reliability requirements, require another signal-performance analysis to be completed.

Once the signal and reliability performance are satisfied, the design is then tested to determine whether cost goals are achieved. This is a key test since the cost-versus-performance trade-off generally outweighs any other trade-off.

1.7.1 Systems-Requirements Definition

The first step in any systems design is to completely define the performance and physical requirements as well as the cost constraints. A general set of requirements is illustrated in Table 1.1.

1.7.2 Architecture Selection

Once the requirements are defined, the first step in the systems design is to select candidate architectures to meet the design requirements. Architecture, as used here, essentially means the functional block diagram of the system. The designer defines the elements represented in Table 1.2 and converts them to a functional block diagram of the system.

The designer generally makes these choices based on experience, engineering design aids, supplier information, and judgment. The objective of the follow-on design stages is to validate the architecture and define the functional and physical characteristics of each element. It is at this point, therefore, that the iterative design process begins.

Chapter 7 discusses typical architectural designs and systems standards for fiber-optic transmission systems using present-day technology. It can be used as an aid in this step of the design process.

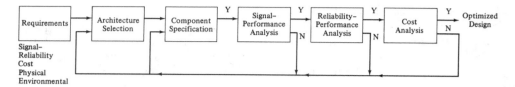

Figure 1.10 Design methodology.

1.7.3. Component (Functional-Element) Specification

Once the conceptual design is complete, then the performance of each functional block and each functional interface is defined. The process of component, or functional-element, specification involves determining the performance of that element in the overall system. The process is therefore iterative and can involve a number of cycles through the entire systems-design process in order to fully define each element. The first time through the process, however, only the basic characteristics of each element or component need be defined. Refinement comes as the cycles of optimization proceed.

The objective is that each functional block be characterized in terms of:

(a) input and output interface,
(b) transfer function or performance characteristics, and
(c) other applicable specifications.

If the functional elements are specified in enough detail (by allocation of performance, reliability, and cost), the signal performance, availability, and cost analysis can be completed without selecting specific components. Components can then be selected or developed after the fact to satisfy the requirements of each individual functional element. This is an approach often used in major systems-development programs. More commonly, available components or classes of components are selected from supplier information and analysis proceeds on this basis. Chapter 3 discusses the process of component specification and selection for the fiber-optics subsystem.

1.7.4 Signal-Performance Analysis

Once the initial development of the architecture is completed and the functional elements basically defined, the design is then analyzed with regard to signal performance. This process results in a validation of the design and a further refinement of the specification of each functional element. Chapter 6 is devoted to this process.

TABLE 1.1   GENERAL SYSTEMS-REQUIREMENTS DEFINITION

Requirements category	Requirement	Detailed specification
Functional	Systems configuration	Number of terminals Number drop/add points Topology
	Transmission distances	Between term and drop/add nodes Between repeaters End-to-end
	Circuit specifications	Analog or digital interface Voice, video, data Interface standards
	Design capacity	Installed requirement Installed capacity Ultimate design capacity
	Physical	Building/node locations Cable route
	Network management	Centralized/local Alarm/alert reporting Control capability Security required Billing system Reports
Signal Performance	Analog signal	Voice or video? Performance standard Signal-to-noise performance Bandwidth performance Intermodulation or distortion Other parameters
	Digital signal	Data, encoded analog? Protocol Performance standard Encoding: NRZ, RZ, other Bit-error rate Bit rate Timing jitter Waveform distortion Other
Environmental	Temperature	Operating long term Operating short term Temperature cycling and shock Storage temperature
	Humidity	Relative humidity Temperature/humidity cycle Condensation? Water blocking?
	Vibration and shock	
	Exposure	Salt atmosphere/spray Sand and dust

TABLE 1.1 GENERAL SYSTEMS-REQUIREMENTS DEFINITION (*cont.*)

Requirements category	Requirement	Detailed specification
	Nuclear	Blast damage Radiation: Permanent damage Radiation: Transient effect
	Cable/connector mechanical	Tension, crush, twist, bend Water and gas blocking Abrasion, rodent protection Fire resistance Filler drip test Other
	EMI and RFI	Emission and susceptibility Tempest
Availability Reliability	Outage definition	Catastrophic Bit-error rate maximum Hits or errored seconds Mean time between outage
	Availability of system	% availability per year or month Mean time to restore
	Reliability of equipment	Mean time between maintenance event Mean time between failure Mean time to repair
Physical	Terminals	Dimensions and footprint Configuration Front/rear access
	Cable plant	Fiber count Sheath size Configuration (armor, jacket strength member, filler) Fiber buffer type, size
	Terminations/connectors	Connectors Patch panels Enclosures Cross connects
	Splices	Enclosures Organizers
Network Management	Alarm alert control	Maintenance philosophy Command center systems Network level alarms/controls Equipment local level Reports: Real time; history Trouble tickets
	Billing	Configuration management and tracking Billing system
	Access	Security: Data and building Access levels

TABLE 1.1 GENERAL SYSTEMS-REQUIREMENTS DEFINITION (*cont.*)

Requirements category	Requirement	Detailed specification
Facilities	Description	Terminal facilities Repeater sites Network control
	Specifications	Floor space layout Environ controls Power (primary, battery, backup systems) Ventilation Cable runs (raised floor, cable trays, conduit, riser) Access and security Lighting Fire control Personnel space, storage
Installation	Terminal equipment	Mounting Power and signal wiring Grounding Optical termination
	Cable plant	Construction (aerial, burial, duct, undersea) Tensile loading and bend radius during and after installation Construction practices
	Splicing	Technology (fusion, bonded, or mechanical) Splicing practices Splice Equipment and Van Insertion-Loss specification Measurement method Enclosure and organizer Placement and protection
Acceptance	Terminals, equipment, repeaters, net management	Quality control data/tests Factory and receiving Activation Checkout and setup Acceptance-test procedures and parameters Burn-in period Reports
	Cable plant	Factory and receiving tests Span performance data End-to-end performance tests: attenuation, dispersion/BW
	Documentation	Test reports As-built drawings Software Manuals

TABLE 1.1 GENERAL SYSTEMS-REQUIREMENTS DEFINITION (*cont.*)

Requirements category	Requirement	Detailed specification
Maintenance	Maintenance support	Equipment Software Cable plant
	Sparing philosophy	On site Depot 24-h replacement
	Help desk Staffing and dispatch	Vendor, owner
	Procedures	Preventative Demand maintenance
	Building Access Modularity	Repair levels Module access
	Diagnostics	Central, local, remote
Implement	Installation milestones	Design Procurement Production (if applicable) Facilities, right of way Installation Acceptance Traffic cutover
	Technology timing	Immediate, near term, long term
	Staffing	Design and installation Operation
Cost	First costs	Research and development Nonrecurring engineering Procurement Production startup cost Equipment and software costs Facilities and right of way Installation Activation and startup costs
	Recurring costs	Recurring production Operations and quality control Maintenance cost Right of way Lease, utilities, etc.
	Payback	Return on investment Payback period

TABLE 1.2 SYSTEMS-ARCHITECTURE DEFINITION

Requirements category	Requirement	Detailed specification	
Physical	Physical topology	Node interconnection Path, line, and span redundancy Physical route redundancy	
Signal interfaces	Baseband signals	Path-termination interface point Interface parameters and standards	
	Intermediate channel	Line-termination interface point Signal format and standards	
	Transmission channel	Span-termination interface point Signal format and standards	
	Optical Tx channel	Photonics interface/reference points Signal format and standards	
	Overhead channels	Interface points and standards Signal content	
Information Signal Processing	Modulation/encoding: Path or line layer	*Analog* Baseband Pulse code Frequency/phase Amplitude	*Digital* Baseband NRZ, RZ, other Frame format Protocol

	Multiplexing:
	    Path or line layer
	    Photon Layer
		Frequency division
		Wavelength division
			Time division

	Span Layer Encoding or Carrier Modulation	Frequency Phase Amplitude	NRZ, RZ Manchester PSK, FSK
Overhead Signal Processing	Network management architecture	High-level MIS interfaces Lower-Level MIS interfaces Alarm, alert, control points Orderwire approach Error-Performance encoding Protection switching Alarm/control signaling Spare channels Define POH, LOH, SOH	
	Overhead signaling	Inband vs. out of band Multiplexing approach Channel sharing? Format/protocol	
Define Functional Elements	Termination equipment	Path Line Span Photonic	
	Orderwire equipment	Voice orderwire Data or spare-channel equipment Alarm sensors Control/monitoring elements	
	Fiber transmission equipment	Transmitter/source Receiver/detector Regenerators Connectors Couplers and WDM Splices Fiber	

Figure 1.11 illustrates the three stages of fiber-optics transmission systems design. Depending on the complexity and type of system or network, one or all of these stages may apply to a particular design application. The three stages include the following.

Stage 1: Services, path, and line-layer design.   The first stage of design is to define the service, path, and line-termination equipment performance parameters required to achieve specified end-to-end performance for baseband services. The analysis at this stage determines and specifies candidate modulation and encoding and multiplexing approaches estimated to achieve desired end-to-end performance.

Stage 2: Transmission subsystem section-layer design.   The first step in this stage of the design is to define the section-layer signal format and overhead

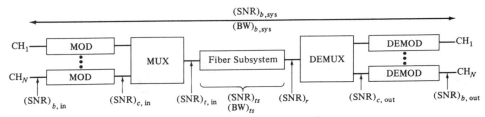

Stage 1:

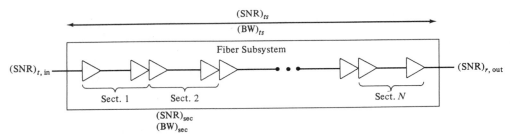

Stage 2:

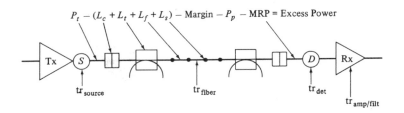

Stage 3:

Figure 1.11   Signal-performance analysis methodology: Stage 1: Systems-design analysis end-to-end transmission parameters. Stage 2: Section design. Stage 3: Photonics design.

requirements and select photonics-layer technology that is estimated to meet the requirements. Candidate components or component types are then chosen and their performance defined to a degree that a first-order analysis of section-layer performance can be determined. One of the objectives here is to determine the number of regenerator sections that is required to achieve distance and performance requirements. Once this is determined, the end-to-end section-layer requirements can be allocated to individual regenerator sections and photonics-layer performance can be determined.

Stage 3: Photonics-layer detailed design.   Once the end-to-end performance is defined for the photonics layer, a detailed optical-link budget analysis is performed to determine whether the photonics-layer design and components chosen achieve the transmission requirements under all operating conditions with all tolerances considered. Bandwidth analysis and an optical-power budget analysis are performed in order to achieve this.

1.7.5 Reliability Analysis

After it is determined that the system will provide the desired signal performance, the architecture and the chosen components are tested from the standpoint of component reliability and overall systems availability.

Reliability is a function of the failure rate of components and functional elements within the system. A high failure rate effects systems outages, maintenance costs, and sparing philosophy. Component choice and operating conditions determine the degree of system reliability.

Availability defines the probability or percentage of the time that each communications channel in the system is available for operation. Availability is a function of component reliability, the degree of redundancy designed into the system, and the time it takes to repair the system when a failure occurs. Systems availability is therefore a function of systems architecture, sparing and maintenance philosophy, as well as component choice.

Chapter 8 treats the availability-analysis approach in detail. As with signal-performance analysis, the process can bring about a change in systems architecture and component choice in order to achieve desired performance.

1.7.6 Cost Analysis

The final stage in the analysis is to determine whether the design meets cost goals. During the process of component selection and functional-element definition, some degree of cost consciousness is generally practiced so that the final product falls within the range desired. Nonetheless, the final cost analysis can alter both architecture and component choice and can even alter the basic performance requirements since cost versus performance is always a trade-off.

Cost modeling and financial analysis is a complete discipline and beyond the scope of this book. Chapter 8, however, provides some guidelines as to the nature of the analysis and cost trade-off process.

REFERENCES

1. Bell Communications Research, Tech. Ref. TR-TSY-000020, issue 3, 1987.

2. John Gower, *Optical Communications Systems,* Prentice Hall International, London, 1984, pp. 135–59.

3. Western Electric Technical Publications, *Telecom Transmission Engineering, Vol. 1: Principles,* Winston-Salem, NC, 1980, chap. 5.

4. J. Richard Jones, previously unpublished data.

5. J. Conradi, F. Kapron, and J. Dyment, "Fiber-Optical Transmission Between 0.8 and 1.4 Mm," *IEEE J. Solid-State Circuits* **SC-13,** 1 (February 1978): 109.

6. Electronic Industries Association, *Generic Specification for Optical Waveguide Fibers,* EIA-4920000-A, 1978.

2

Defining Requirements and Standards

2.1 INTRODUCTION

In order to apply fiber-optics technology to a systems application, it is first necessary to define the requirements in terms of signal performance at the various interfaces as well as such practical characteristics as environmental performance, reliability, maintainability, etc. Many of the parameters generally required were defined in Table 1.1 in Chapter 1. As technology and applications mature, standards are developed in order to aid in this definition process.

A requirements definition must address not only the components of the system, but the interface standards and performance requirements of the signals that are to be transmitted. This chapter, therefore, covers some of the important standards specific to fiber-optic technology as well as some of the more common standards specific to the information signals to be transmitted.

Voice is generally PCM encoded for transmission over fiber, and is therefore time-division multiplexed with other voice or data channels in order to better utilize the broad bandwidth of fiber. By far the largest application of fiber optics for voice transmission is in telephone networks. Voice transmission is, therefore, dealt with in the telephone environment.

Fiber generally finds application for video transmission in three areas: cable TV; broadcast trunks and feeds; and military radar, countermeasures, or other broadband information systems. For ultra-high-quality network-broadcast trunks, where long distances are involved, the video is usually digitized at 45 to 140 Mb/s. Once digitized, the transmission equipment is essentially the same as that for voice. For most other video transmission, some form of analog modulation is used, either direct AM or some form of FM or phase modulation. Since entertainment video is the largest user of fiber, this application area is emphasized to describe video-transmission techniques for fiber.

Fiber also plays a major role in data transmission since it is an inherently low-loss broadband interference-immune medium. It is applied in long-haul transmission systems, local interconnects between computers or data sets, interconnects with computer peripherals, and other data interconnects where security, noise immunity, or broadband characteristics are necessary. Although binary modulation is essentially the same as for digitized voice or video, the transmission products vary widely depending on application. For long-haul or local area trunking, the transmission product and architecture are essentially the same as that for digitized voice or video. For local or metropolitan area computer interconnects, the transmission product and architecture generally take the form of a local-area network (LAN). Where short- to moderate-distance simple interconnects are required, such as an RS232 data link, the approach tends toward a simple point-to-point fiber "modem" set.

2.2 THE EMERGENCE OF FIBER-OPTIC STANDARDS

A number of standardization committees have been established in order to develop standards for fiber-optic components and systems. Until recently, however, standardization within the fiber-optics industry had been de facto due to the fast emergence and rapid changes in the technology. This is largely true in the military where even today there are few "Mil Qualified" parts. Most have been developed and qualified on specific programs to meet a systems requirement. In the commercial area, standardization has focused principally on the long-haul telephone applications where the market demand has been the largest. The rapid growth of the fiber market, the indecision between carriers and the Regional Bell Operating Companies (RBOCs), the influence of foreign suppliers, and the rapid conversion from multimode to single-mode technology in the early 1980s has impacted the orderly development of standards for the telephone industry as well. Only today are the standards organizations beginning to catch up with the industry, and standards are beginning to be formalized for the measurement and performance of fiber components as well as for the implementation of fiber in transmission networks.

The first standards to emerge were fiber-measurement standards and fiber physical standards. These were necessary before much of any large-scale fiber application could take place. Common performance terminology and specifications had to be developed through establishing common measurement approaches. Because the initial market in the United States for fiber was military, DOD STD 1678 was the first widely used measurement standard for fiber optics. Today, organizations including the National Bureau of Standards (NBS), the Electronic Industries Association (EIA), the Institute of Electrical and Electronics Engineers (IEEE), the American Society for Testing and Materials (ASTM), the Society of Automotive Engineers (SAE), the American National Standards Institute (ANSI), the Consultative Committee in International Telegraphy and Telephony (CCITT), Exchange Carriers Standards Association (ECSA), and other government and industry groups listed in this section are also actively involved in the development of fiber standards.

One of the earliest barriers to wide-scale use of fiber was the connector and splice problem. Without a decision on numerical aperture, outer diameter, core size, and other physical tolerances, the connector manufacturers were unwilling to invest in the development of precision connectors. Standardization organizations such as the CCITT (recommendation G651) helped to resolve this problem. The $125 \pm 3\mu m$ outer diameter for high-performance fiber was established. Core diameters of $50 \pm 3\mu m$ for multimode graded-index fiber and 8 μm for single-mode fiber provided a common baseline for connector and splice development.

For many years, most standardization activity had been aimed at components and measurements. With the possible exception of the work on MIL-STD-1773 (fiber-optic databus) by the SAE and the Department of Defense (DOD), little activity was spent on fiber-optic systems standards other than as required for specific programs. Within major user organizations, such as American Telephone and Telegraph (AT&T), systems standards were evolving for internal use that would set the stage for digital product development. One such early standard was AT&T's Tech. Pub. 43806, *Generic Metropolitan Interoffice Lightwave Systems Requirements and Objectives.* More recently, activity has begun on such standards as the fiber distributed-data interface (FDDI) for high-data-rate LANs, developed by ANSI, the synchronous optical network (SONET) standard for synchronous multiplexing in an optical network, and the single-mode optical interface specifications, the last two developed by ECSA. CCITT Study Group XVIII has also begun activity toward developing broadband Integrated Services Digital Network (ISDN) standards that include subscriber interface rates in the 150 to 600 Mb/s range.

2.3 STANDARDS ORGANIZATIONS

Because standardization organizations emerge and change charters often as demand dictates, it is not possible to mention all of those that have contributed to the technology. A few of the key organizations are referenced here.

2.3.1 American National Standards Institute (ANSI)

Committee X3T9 is developing the fiber distributed-data interface (FDDI) standard for high-speed LANs operating in the 100 Mb/s range. FDDI is aimed at high-speed interconnection between mainframes, minicomputers, and their associated peripherals. This standard is covered in more detail in Chapter 7. Other ANSI standards important to fiber, or digital, network design are

T1.101-1987 *Synchronous Interface Standards for Digital Networks*

T1.102-1987 *Digital Hierarchy Electrical Interfaces*

T1.103-1987 *Digital Hierarchy Sync DS-3 Format Spec (SYNTRAN)*

T1.106-1988 *Digital Hierarchy Optical Interface Specs: Single Mode* (see ECSA)

T1.107-1988 *Digital Hierarchy Format Specs (T1X1/87-127)*

2.3.2. Exchange Carriers Standards Association (ECSA)

ECSA is the secretariat for the ANSI and has submitted the following key documents to ANSI for approval:

(a) T1X1/87-126, the proposed draft, *American National Standard for the Fourth Hierarchical Level,* was developed by ECSA to be compatible with the CCITT's 139.264 Mb/s signal as presented in Rec. G.703 and G.755.

(b) T1X1/87-129R1, *Digital Hierarchy Optical Interface Rates and Formats Specification,* or ANSI draft standard T1.105-1988. ECSA T1X1.2 Working Group is developing this standard, also known as the synchronous optical network (SONET) standard, that will establish synchronous compatibility between optical transmission equipments. The building block synchronous transport-signal level 1 (STS-1) is at a rate of 51.840 Mb/s with an optical-carrier level 1 (OC-1) optical counterpart also at 51.840 Mb/s. This standard is to be consistent with the work of the CCITT on network-node interface (NNI) using synchronous multiplexing techniques.

(c) T1X1/87-128R1, *American National Standard for Telecommunications Digital Hierarchy Optical Interface Specifications: Single Mode.* This standard describes the performance characteristics of the single-mode interface to be used with SONET and provides references to other standards that define component performance and "joint" engineering methodologies. It, along with EIA document TSB20, forms some key standards for fiber transmission-systems design. This standard is further described in Chapter 7.

(d) T1M1.2/87-037R2, *Functional Requirements for Optical Terminating Equipment.*

(e) T1X1.4/87-702R2, proposed draft on *Digital Hierarchy Formats.*

2.3.3 Electronic Industries Association (EIA)

The EIA contains the following subcommittee organizations:

> FO-2 Optical Communications Systems
> > FO-2.1 Optical Fiber Telecom Systems
> > FO-2.2 Fiber Optics LAN
> > FO-2.3 Jitter and Wander
> > FO-2.5 Cable Plant Installation
> FO-6 Fiber Optics
> > FO-6.1 Field Tooling and Test
> > FO-6.2 Terminology Definition and Symbology
> > FO-6.3 Interconnection Devices
> > FO-6.4 Test Methods and Instrumentation
> > FO-6.5 Fiber Optics Transducers
> > FO-6.6 Optical Fibers and Materials
> > FO-6.7 Fiber Optics Cables

The EIA has initiated work on over 170 fiber-optic test procedures (FOTP) and standards. Some of the key test procedures are listed in Table 2.1. Other key EIA documents on fiber optics include

RS-440-1978	*Connector Terminology*
RS-458,-459	*Standards for Fiber Classes and Materials*
RS-472-1985	*Generic Spec, Cables*
472A-XX0-86	*Cables for Outside Aerial Use*
472B-XX0-86	*Cables for Underground and Burial Use*
472C-XX0-86	*Cables for Indoor Use*
472D-XX0-86	*Cables for Outside Plant Use*
RS-475-1986	*Generic Spec, Connectors*
RS-492-1987	*Generic Spec, Optical Waveguide*
RS-509-1984	*Generic Spec, Fiber Optic Terminal Devices*
RS-515-1986	*Generic Spec, Fiber and Cable Splices*
TSB-20	*Single Mode Fiber Optic System Transmission Design*

TABLE 2.1   EIA RS-455-XXX, FIBER-OPTIC TEST PROCEDURES

Procedure number FOTP—XXX	Component type	Test description
1	Connectors	Cable flex test
2	All devices	Impact test measurement
3	Connectors	Temperature cycling
4	Connect/component	Temperature life
5	Connectors	Humidity test
6	Connectors	Cable retention test
11	Connectors	Vibration test procedure
12	Connectors	Fluid immersion
13	All devices	Visual and mechanical inspection
14	All devices	Shock test
15	All devices	Altitude immersion
16	Components	Salt spray
17	Cable assemblies	Maintenance aging
18	Component/assembly	Acceleration testing
20	Optical	Optical-transmittance change
21	Connectors	Mating durability
22	System	Ambient-light susceptibility
23	Component seals	Air-leakage testing
25	Cable assemblies	Impact testing
26	Cable	Crush resistance
27	Fibers	Outside-Diameter measurement
28	Fibers	Tensile failure point
29	Fiber	Refractive index profile, transverse method
30	Multimode fiber	Bandwidth Measurement, frequency-domain method
31	Fiber	Tensile proof test method
32	System	Fiber-optic circuit discontinuities
33	Cable	Tensile loading and bending

TABLE 2.1   EIA RS-455-XXX, FIBER-OPTIC TEST PROCEDURES (*cont.*)

Procedure number FOTP—XXX	Component type	Test description
34	Connector	Insertion-loss test
35	Connector	Dust (fine-sand) test
36	Cable assembly	Twist test
37	Cable	Bend test, low and high temperatures
39	Cable	Wicking test
40	Cable	Fluid immersion
41	Cable	Compressive-loading resistance
42	Components	Optical crosstalk
43	Fibers	Output near-field radiation pattern
44	Fibers	Refractive index profile, refracted ray method
45	Fibers	Fiber geometry, microscope method
46	Long GI fibers	Spectral-attenuation measurements
47	Fibers	Output far-field radiation pattern
48	Fibers	On-line diameter measurement
49	Components	Nuclear radiation effects measurement method
50	GI fiber	Light launch for attenuation measurement
51	Multimode fiber	Pulse-distortion measurements
52	Fibers	Temperature dependence of attenuation
53	Long fibers	Attenuation by substitution method
54	Fiber	Bandwidth, mode scrambler requirements
55	Fibers	Coating geometry measurement
56	Fibers	Fungus resistance
57	Fibers	End preparation and examination
58	GI fiber	Core-diameter measurement
59	Fiber/cable	OTDR attenuation measurement
60	Fiber	Length by time-of-flight measurement
61	Components	Nuclear thermal-blast resistance
62	Fiber	Macrobend attenuation
63	Fiber	Torsion test
65	Fiber	Flexure test
66	Fiber	Abrasion resistance of buffer coatings
67	Coated fiber	Self-sticking (blockage) test
68	Fiber	Microbending
69	Fibers	Max and min use temperature
70	Fibers	Advanced aging
71	Components	Temperature shock effects
72	Components	Temperature cycling effects
73	Fibers	Temperature and humidity cycling
74	Fibers	Humidity testing
75	Fibers	Fluid-immersion test
78	SM fiber	Spectral attenuation, cutback method
80	SM fiber	Cutoff wavelength by transmitted power
81	Filled cable	Compound flow (drip) test
82	Filled cable	Fluid-penetration test
83	Cable to connector	Axial compressive loading
84	Cable	Jacket self-adhesion (block) test
85	Cable	Twist test
86	Cable	Jacket-shrinkage test
87	Cable	Knot test

TABLE 2.1   EIA RS-455-XXX, FIBER-OPTIC TEST PROCEDURES (*cont.*)

Procedure number FOTP—XXX	Component type	Test description
88	Cable	Bend test
89	Cable	Jacket elongation and tensile strength
91	Cable	Twist and bend
94	Cable	Stuffing-tube compression
95	Fiber/cable	Absolute optical-power test
96	Cable	Storage temperature and humidity
98	Cable	Freezing test
99	Cable	Gas-flame test
100	Gas blocking cable	Gas-leakage test
101	Cable	Accelerated oxygen aging
102	Cable	Water-pressure cyclng
103	Buffered fiber	Bend test
104	Cable	Cyclic flexing test
107	Terminal	Return loss
109		Reference-point temperature
110	Transmitter	Type verification
111	Receiver	Type verification
112	Power supply	Current
113	Digital terminal	General measurement requirements
114	Terminal	Data-input current
115	Terminal	Control-input current/voltage
116	Terminal	Data-output voltages
117	Transmitter	Optical output power
118	System	Output propagation delay
119	Receiver	Sensitivity and dynamic range
120	Transmitter	Optical output power
121	System	Output propagation delay
122	Receiver	Responsivity and rms noise voltage
123	Receiver	Output voltage and switching times
124	Detector	Dark currents
125	Detector	Numerical aperture
126	Source	Modulation index
160	Cable	Temperature shock test
161	Cable	Temperature cycling
162	Cable	Temperature/humidity cycling
163	Cable	Fungus resistance
164	SM fiber	Mode field diameter, far-field scan
165	SM fiber	Mode field diameter, near-field scan
166	SM fiber	Mode field diameter, transverse offset
167	SM fiber	Mode field diameter, far-field variable aperture
168	Fiber	Chromatic dispersion, spectral group delay
169	Fiber	Chromatic dispersion, phase shift
170	SM fiber	Cutoff wavelength by transmitted power
171	Short fiber	Attenuation by substitution method
172	Firewall connector	Flame resistance
174	SM fiber	Mode field diameter, far-field knife edge scan
175	Fiber	Chromatic dispersion, differential phase shift
176	Fiber	Geometry measurement by Grey-scale analysis
179	Fiber	Inspect cleaved ends by interferometry

2.3.4 Institute of Electrical and Electronics Engineers (IEEE)

The Standards Coordinating Committee on Definitions (SCC 10) is responsible for the development of terms and definitions. Key documents include

ANSI/IEEE STD 100-1984	*IEEE Standard Dictionary of Electrical and Electronics Terms*
IEEE 812-1984	*Definition of Terms Relating to Fiber Optics*

2.3.5 International Electro-Technical Commission (IEC)

The IEC has established the TC46 technical subcommittee on wire and cables, the SC46E subcommittee on fiber optics, and the following working groups:

WG0 Terminology
WG1 Fiber and Cable
WG2 Connectors
WG3 Safety

The IEC has produced the following key documents:

IEC/SC 46E (CO)9 and IEC/SC 46E (sec) 66	*Optical Fibers Measurements Method, Dimensional Tests Sections One and Two*
IEC/SC 46E (SEC) 55	*Generic Spec, Connectors*
IEC/SC 86B (CO) 15	*Revision to Generic Spec for Connectors Regarding Climatic Environmental Tests*
IEC/SC 86B (CO) 16, 17	*Generic Spec for Fiber Optic Branching Devices: Insertion Loss and Return Loss*
IEC 693	*Dimensions of Optical Fibers*
IEC 793-1	*Generic Spec, Optical Fibers*
IEC 794-1	*Generic Spec, Optical Fiber Cables*
IEC 874-1-1986	*Generic Spec, Connectors*
IEC 875-1-1986	*Generic Spec, Branching Devices*
IEC 875-2-1986	*Generic Spec, Branching Devices, Star Coupler*
IEC 876-1-1986	*Generic Spec, Fiber Optic Switches*

2.3.6 Consultative Committee in International Telegraphy and Telephony (CCITT, Geneva)

Under the CCITT, Subgroup SG XV and others are developing fiber-optic transmission standards. Subgroup SG XVIII is developing broadband ISDN standards. Key Documents include:

Rec G651 *Standardizatrion of Fiber Physical Parameters*
Rec G652 *Characteristics of Single Mode Fiber Optics Cable*

2.3.7 Society of Automotive Engineers (SAE)

It was the SAE that began some of the earliest work on military fiber-optic standards. Some early activity includes

> SAE A2-H Subcommittee: Wire and Cable Standards; developed DOD-STD-1678 and DOD-C-85045
>
> SAE A2-K Subcommittee: developed the fiber-optics version of the MIL-STD-1553 data bus standard (MIL-STD-1773)

The SAE has two standards committees working on the fiber-optic data bus:

> AS-2 Committee: High Speed Data Bus standards
> > AS 4074.2 High Speed Ring Bus (HSRB), Tri-Service JI-AWG committee
> > AS 4074.1 Linear Token Passing Multiples Data Bus
>
> AS-3 Committee: DOD-STD-1773 Development and Enhancement
> > STANAG 3910 (NATO) Standard Support

2.3.8 Department of Defense (DOD)

Standards activity is within the Defense Materials Specifications and Standards Office (DMSSO) of the Office of the Undersecretary of Defense for Research and Engineering. There exist Departmental Standards Offices (DEPSO) for parts in the Defense Logistics Agency (DLA) and for systems separately in the Army, Navy, and Air Force. For the Army, the assignee activity is the Army CECOM at Ft. Monmouth, New Jersey. For standard parts, the assignee activity for DLA is the Defense Electronics Supply Center (DESC) in Dayton, Ohio. These groups work with various DOD standards working groups and industry committees and coordinate their activities with the standards and specifications test and evaluation activities within the Tri-Service Fiber Optic Coordinating Structure (TSFOCS). TSFOCS is the R & D arm of the Office of the Undersecretary of Defense for R & D. Within the DESC, Federal Stock Group 60 has been established to cover fiber-optic materials, components assemblies, and accessories. Some key DOD publications include

DOD-STD-1678	*Fiber Optic Test Methods*
DOD-C-85045	*Fiber Optic Cable Standard*
FSC 6625	*Test Equipment*
MIL-STD-1773	*Fiber Optics Mechanization and Aircraft Internal Time Division Command/Response Multiplex Data Bus System*
MIL-STD-188-111	*Subsystem and Engineering Standards for Common Long Haul/Tactical Fiber Optics Communications*
MIL-STD-1863	*Standards for Mechanical Interfaces*
MIL-HDBK-141	*Fiber Optic Design Practices*

2.3.9 North Atlantic Treaty Organization (NATO)

Among other activities, NATO subcommittees have been developing fiber and cable standards for NATO applications. One such committee is NIAG Subgroup 6, which developed fiber-optic interface standards for shipboard application.

2.3.10 National Bureau of Standards (NBS)

The NBS at Boulder develops certain measurement standards for fiber-optic components. One key measurements document is Special Publication 637, *Optical Fiber Characterization*.

2.4 STANDARDS FOR VOICE TRANSMISSION OVER FIBER

The major application for transmission of voice information is the telephone. Although there are many other applications for audio transmission, the telephone is the most predominant, and, therefore, this section focuses on voice communications over telephone networks and the application of fiber therein.

2.4.1 The Telephone Network

Classical telephone networks transmit voice in a totally analog fashion. The telephone handset is basically a four-wire instrument (two transmit and two receive) transmitting a 200- to 3500-Hz speech signal [1]. In order to save copper in the transmission loop, the set combines the signals onto two wires. Two-wire transmission is used in most of the loop. Four wires can be used when the bandwidth exceeds 4 kHz or when distances are long.

Trunking is classically performed using frequency-division multiplexed voice channels transmitted in channel "groups" over coax, microwave, and satellite [2]. Voice networks are gradually evolving to all-digital trunking, where the analog voice channels are converted to pulse-code-modulated (PCM) digital channels at the local exchange carriers central offices (LECCO) or at the interexchange carriers point-of-presence (IECPOP). A simplified functional diagram of this network configuration is shown in Figure 2.1.

Fiber is applied to the digital voice network in the trunking portions, i.e., the interfacility links and the long-haul trunk network. Although fiber in the local loop is already developing at a rapid pace for businesses in major metropolitan areas, it will take many years and a lot of changes in local loop requirements for fiber to replace twisted pair in private residences served by a local exchange carrier. Fiber is replacing twisted pair in locations where multiple voice or data channels can be combined into T-carriers (1.544 Mb/s multiples) through a PBX, PCM channel bank, T-carrier multiplexer, or such.

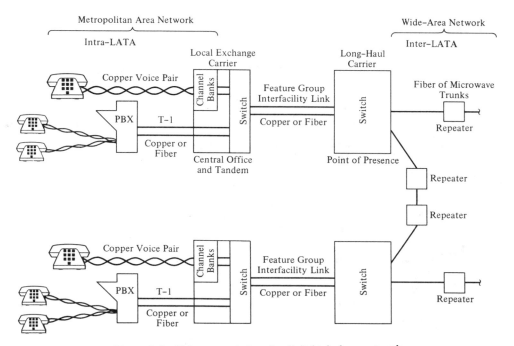

Figure 2.1   Voice transmission: the digital telephone network.

2.4.2 The Voice Signal

The characteristics of a speech signal are complicated by the varying nature of talker volume. For our purposes, we are not interested so much in the nature of speech, but more in the measurement of signal and noise levels relating to voice transmission.

Speech signal is confined within a 200- to 3500-Hz frequency band, with energy peaked at around 800 Hz. Articulation is mostly at the higher frequencies, above 800 Hz. Long-term spectral-energy density peaks at about 400 to 600 Hz, with a spectrum that is 5 dB down at 100 Hz and slopes off from 600 Hz at 30 dB/decade to a value 40 dB down at 10 kHz [3].

Measurement of noise in a voice channel is based more on the degree to which it annoys the listener than on the absolute noise level. For this reason, weighting filters are used to simulate the nature of the human ear. A C message weighting filter is used by AT&T and the Bell Systems and a psophometric message weighting is recommended by CCITT. Both are similar, peak at 1000 Hz, are relatively flat to about 3000 Hz, and are 10 dB down in the band between 300 and 3400 Hz. A simple conversion factor for C message weighting is found by assuming that the filter reduces the white noise in a 3-kHz band by about 2 dB. The psophometric conversion factor is 2.5 dB [3].

Measurements are generally performed using an rms voltmeter compensated for the frequency weighting and the transient response similar to that of the human ear. Measurement terminology is as follows:

(dB)$_{rn}$: Measurements of voice power in (dB)$_{rn}$ are referenced to a noise level of 10–12 watts, or −90 (dB)$_m$ at 1000 Hz. Without weighting, a 1000-Hz tone at −90 (dB)$_m$ gives a 0 (dB)$_{rn}$ reading.

(dB)$_{rnc}$: This is the same measurement as (dB)$_{rn}$ but using a C weighting network.

(dB)$_{rn0}$ and (dB)$_{rnc0}$: These are power measurements referenced to a 0 dB transmission-level point (0 TLP). The 0 TLP is a selected standard measurement reference point in a transmission system.

Volume unit (vu): This is the unit used for expressing speech-signal amplitude. The scale is logarithmic (similar to dB). For a sinusoidal signal delivered to a 600-ohm resistive termination, vu represents power in dBm.

2.4.3 PCM Encoding of Voice

In order to transmit voice over a fiber network, it is generally digitally encoded and combined with many other voice channels (time-division multiplexed) in order for the economies of fiber transmission to work. The most common approach is to encode the voice with 8-bit/sample pulse-code modulation (PCM), resulting in a 64-kb/s digital signal. Each resulting 64-kb/s encoded voice channel is then time-division multiplexed (TDM) and combined with up to N channels using a PCM channel bank.

In the United States, Japan, and Canada, the standard is a 24-channel combination, each at 64-kb/s plus 8 kb/s overhead and synchronization. The output of the channel bank is a 1.544-Mb/s signal, or DS-1. This is illustrated in Figure 2.2.

In Europe, the standard is also 64 kb/s per channel, but with two 64-kb/s overhead channels for signaling and synchronization. The channel bank output is therefore 2.048 Mb/s.

A detailed description of PCM encoding approaches and SNR relationships is found in Chapter 5 and [4].

Pulse-code modulation (PCM): The standard PCM encoder samples the 4-kHz voice signal at the Nyquist rate ($f_s = 8$ kHz) and measures the voltage amplitude of the voice signal in a quantizer of 2^n levels, which converts each measurement into an n-bit binary word. For telephones using a standard 8-bit linear quantizer, the result is a digital output of nf_s or 64 kb/s. Compression algorithms (nonlinear quantizing) can be used to maintain the SNR for both the loud and soft talker. With logarithmic compression ($\mu = 100$), acceptable speech quality can be obtained with 7-bit encoding (56 kb/s). D1D thru D4 channel banks generally use a compression algorithm of $\mu = 255$ instead of the earlier $\mu = 100$ [2]. Logarithmic PCM is the internationally recognized standard for voice (refer to CCITT Rec. G.712). 96 kb/s linear PCM (12-bit) is required for high-fidelity voice. For PCM voice transmission, 10^{-4} BER is generally tolerable and 10^{-6} has negligible effects.

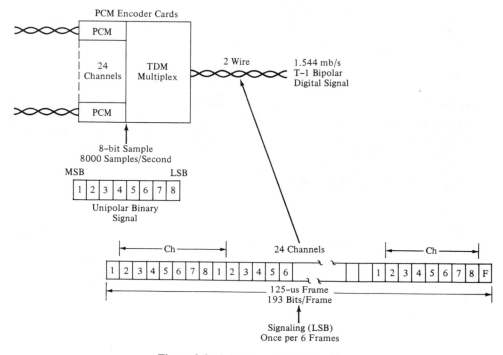

Figure 2.2   PCM channel bank function.

Other modified encoding schemes can be used to achieve similar signal quality at a reduced bit rate, say 32 or 16 kb/s. These approaches follow:

Differential PCM (DPCM): With DPCM, only the change in signal amplitude, from sample to sample, is encoded and transmitted. As a result, the number of bits per sample is less, for the same SNR, and, therefore, the data rate can be reduced. According to [4], the SNR advantage of DPCM over PCM for the same number of encoding bits is 4 to 11 dB. For the same SNR, we can reduce the number of bits by 1 or 2 and, therefore, reduce the data rate to 48 to 56 kb/s.

Delta modulation (DM): In DM, the baseband signal is oversampled and the output is a binary 1 when the signal increases and a binary 0 when the signal goes negative or stays the same. This is essentially a two-state version of DPCM. For voice signals equivalent to 8-bit PCM, DM requires a data rate much higher than 64 kb/s. A 40-kb/s DM has been shown to be equivalent to 5-bit PCM [4].

Adaptive differential PCM (ADPCM): This operates using the same principles as DPCM except that the quantizer continually adjusts to the incoming speech level. It has been shown that 5-bit (40-kb/s) ADPCM provides more preferable quality than 7-bit compressed PCM [4, 5]. Today, ADPCM encoders offer acceptable quality voice at 16 to 32 kb/s, therefore doubling or quadrupling the number of voice channels carried on a DS-1 channel from 48 to 96.

Other techniques that perform complex processing of the signal can achieve speech transmission at such rates as 1.2 to 4.8 kb/s. Predictive coders can operate at 2.4 to 4.8 kb/s and vocoders down to 1.2 kb/s [6].

2.4.4 Voice Trunking, the Multiplexing Hierarchy

In order to trunk voice transmissions from location to location, the DS-1 output from the channel bank is either transmitted directly over a twisted-wire pair or combined with multiple DS-1s and trunked on coax, digital microwave, or fiber. The standard multiplexing hierarchies [7, 8] are given in Table 2.2. A new fourth-level DS4NA standard at 139.264 Mb/s has been proposed by the ECSA T1X1 committee to be compatible with the CCITT standard [9].

TABLE 2.2 PCM MULTIPLEXING HIERARCHIES[a]

Level	USA and Canada	No. channels	Europe	No. channels	Japan	No. channels
1	1.544	24	2.048	30	1.544	24
2	6.312	96	8.448	120	6.312	96
3	44.736	672	34.368	480	32.064	480
4	274.176	4032	139.264	1920	97.728	1440
	139.264[b]	2016				
5	—	—	564.992	7680	400.352	5760

[a] Data Rates are in Mb/s.
[b] Proposed DS4NA.

2.5 DATA TRANSMISSION ON FIBER

This section describes the general data-transmission environment. It focuses on the largest applications for fiber that are either data-trunking or processor-based data networks. The application of fiber in the various portions of the data network is described.

2.5.1 The Data-Communications Environment

The present state of the technology and its evolution has been highly influenced by many years of standardization around early low-bandwidth copper-wire systems. Wideband fiber technology is having a major impact on the way data are transmitted, and although transmission is slowed by these conventional systems standards, totally new systems standards are emerging tailored to the specific capability of fiber. Examples include the 100-Mb/s local area network FDDI standard and the SONET synchronous networking standard.

Figure 2.3 generally illustrates the architecture of a data-communications network using public and private facilities. The role of fiber in this data network today is in the physical medium, as a replacement for the conventional copper-wire medium or as a means of providing entirely new physical and logical data networks based on the unique nature of fiber. Figure 2.4 illustrates the approximate bandwidth

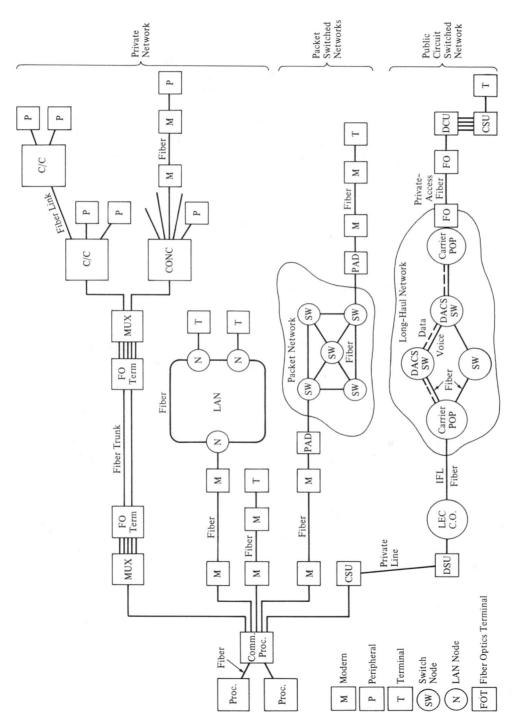

Figure 2.3 Fiber application in data-communications networks.

49

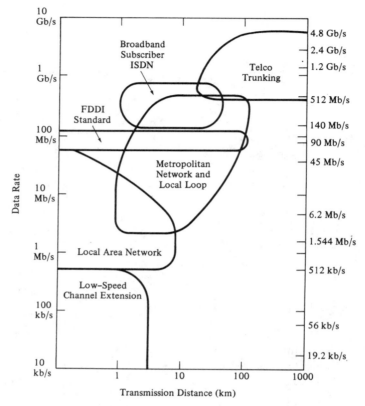

Figure 2.4 Typical product partitioning for the data-communications environment.

requirements of various system classifications within the data-communications network. Key functional elements of the communications network and the role that
fiber might play are described in what follows.

Communications processors. The communications processor, or front-
end processor (FEP), performs the function of communications control to and from
the host processor. The role of fiber here is as a high-speed interface between the
FEP and the host. It provides a means of remoting the FEP over some distance,
whereas the distance with copper is on the order of inches.

Modems. Modems are physical interfacing devices between the data equipment and the communications channel that provide physical and data-link layer functions to transmit the data signal over the physical medium. Fiber replaces the twisted
pair or coax conventionally used for the physical medium. Fiber plays a key role
here since it offers great advantages in longer distances and noise immunity.

Concentrating devices. Concentrators, cluster controllers, and multiplexers combine a number of data input/output (I/O) channels onto one transmission
channel. Multiplexers combine multiple low-speed channels into one high-speed

channel. Cluster controllers and concentrators combine multiple low-speed channels onto a single low-speed channel(s) by sequentially coordinating use of the bandwidth between low-duty-cycle I/O channels. Fiber is a replacement for twisted pair or coax between the concentrator and the peripheral served by it.

Packet networks.   Packet networks work on the principle of converting a low-speed data stream or transaction into a group of high-speed bursts. These bursts, or "packets," are routed to their destination around some available path in the mesh network, where they are reassembled into a slow-speed data channel. Simple multiple access and billing on a per-packet basis are possible. Physically, the network is composed of dedicated lines with packet-switching nodes and access ports (packet assemblers and disassemblers with modems) to form the "packet network." The most common worldwide standard is X.25. Fiber simply replaces copper or microwave as the long-haul or local-loop transmission medium.

Public data networks.   Public data networks are based on the general topology of, and use many of the facilities of, the public telephone network. Public data networks generally involve local-loop service from a local access or local exchange carrier (LEC) and long-haul service using either the switched digital facilities of an interexchange carrier. Switched data circuits can be routed within the interexchange carrier network through digital access crossconnect systems (DACS), through switched 56-kb/s services [10], or through the use of the integrated services digital network (ISDN). ISDN can provide packet-switched service on the D channel or D channel control of data circuits using one or multiples of the B channels. Data circuits can also be established on demand through one of many switched data and value-added networks (VANs). Use of the voice-switched network for data is also possible with dial-up analog modems that access a line through the voice switch and transmit data on an analog carrier. Nonswitched data circuits can be handled as leased private lines. The future of the public network is all-digital transmission with data and voice switching as well as private and virtual private-circuit offerings. The offering of the ISDN permits the combining of voice and data traffic over the same facilities, not only the interexchange carrier facilities (primary rate ISDN), but the local-loop facilities as well (basic rate ISDN). Fiber again is a replacement for microwave, satellite, or copper as the transmission medium in long-haul trunks and local loop. Fiber also plays a role with the larger users in the local-loop environment as a private local-access medium for DS-1 to DS-3 level access between business locations and to the interexchange carrier of choice. Fiber also made possible the SONET standard (see Chapter 7) that permits standard synchronous connection between public and private networks at electrical and optical interfaces. It brings about the transport and switching of large blocks of data at the 139-Mb/s rate or above.

Private networks.   Larger businesses often have enough transmission capacity between locations, or to an interexchange carriers POP, to justify a dedicated private network on the basis of cost and/or reliability and quality as compared to the public local loop. Networks composed totally of private facilities are generally small local-area networks, interfacility trunks, or DS-1 to N × DS-3 local-access networks. They are generally integrated with a nationwide and/or worldwide private network using private and shared bandwidths purchased on public facilities. The private transmission facilities are generally point to point or multiple-access rings or LANs interconnecting data facilities, DS1 switching equipment, and/or PBXs. The application of fiber here is advantageous, since businesses often find the cost of a private network competitive with public tariffs. In addition, a private fiber system offers high performance and reliability.

2.5.2 Data-Communications Standards

For the most part, the fiber transmission engineer uses existing standards established for microwave, satellite, or coax in order to establish end-to-end requirements for a fiber transmission system. In some cases, new standards are emerging that are peculiar to fiber transmission, as summarized in Section 2.2. The standards that some of the transmission standards organizations are addressing today in data communications are summarized in what follows.

International Standards Organization (ISO).   The ISO establishes international standards for over 70 member nations. Jointly with the International Electrotechnical Commission (IEC), they have formed Joint Technical Committee One (JTC-1). Key subcommittees and supporting committees include ANSI's subcommittee SC-13, dealing with the interconnection of equipment, and technical committee TC-97, developing data transmission standards.

American National Standards Institute (ANSI).   ANSI is a Joint Technical Committee secretariat working together with the International Standards Organization and the International Electrotechnical Commission to develop or approve standards for interconnection, including fiber optics interconnection.

ANSI X353 is the technical committee on data-communications standards.
ANSI X3T9 is the ISO part of X3, Information Processing. X3T9 contains the X3T9.5 Subcommittee on Future Interfaces, which developed the FDDI 100-Mb/s LAN standard.

International Telecommunications Union (ITU).   This organization has over 150 member countries and contains three organizations:

Consultative Committee in International Telegraphy and Telephony (CCITT).
Consultative Committee in International Radio (CCIR), which prepares standards for radio communications.
International Frequency Registration Board, which registers radio-frequency assignments.

Some of the ITU standards that are of interest to network designers include the following:

Transmission Systems Standards:
   CCITT Vol. III.1(G)
   CCITT Vol. III.4(H, J)
Digital Networks:
   CCITT Vol. III.3(G)
Data Communications:
   CCITT Vol. VIII.1(V): Data Over Telco Networks
   CCITT Vol. VIII.2 and VIII.3 (X): Data Networks
Television Transmission:
   CCIR Vol. XI, part 1: Television Broadcasting
   CCIR Vol. XII: Long Distance Transmission

Electronic Industries Association (EIA).   The EIA provides regulation, legislation, and standards for U.S. industry.

Defense Communications Agency (DCA).   This U.S. defense agency creates standards for military tactical and strategic communications systems. An example useful to the fiber transmission systems designer is MIL-STD-188-323, the transmission standard for long-haul digital transmission systems.

2.5.3 Transmission Quality Standards

The performance of a digital transmission system is a function of the performance of its components. Performance allocations are assigned to the components based on end-to-end requirements and the number and nature of subsystem elements. In order to standardize on performance, "hypothetical reference circuits" are defined as a "standard" systems configuration for the various transmission methods (terrestrial, satellite, microwave, etc.). The reference circuits are established based on transmission distance, transmission medium, terminal equipment types, multiplexing methods, and other factors that define the mode of transportation and data transmitted. To date, little has been done to establish reference circuits for fiber transmission, so, generally, an established terrestrial standard is substituted for the purposes of systems design. Examples of some of these are

   Television Transmission: CCIR Rec. 567-1; CCIR XV, Vol. XII.
   Digital Transmission: CCIR Rec. 556; CCIR XV, Vol. IX, part 1.
   64-kb/s Transmission: CCITT Rec. G 104, Yellow Book Vol. III. 1.

End-to-end performance over the reference circuit, or any digital transmission system, is defined in terms of the transmitted and received digital pulsed waveform and its effect on the distortion of the data transmitted once decoded. In a digital system, performance is measured by the relative number of bits that were correctly received and decoded. A few of the key performance parameters are defined according to Smith [6]:

Bit-Error Rate (BER):

$$\frac{\text{No. errored bits}}{\text{No. transmitted bits}}$$

in some unit time period.

Error-Free Seconds (EFS): Probability (or percentage) that any one-second interval is error-free.

For statistically independent errors:

$$\%EFS = 100(1 - BER)^{BR} \tag{2-1}$$

For a 1-second interval:

$$\%EFS = 100 \exp\left[-(BER)(BR)\right] \tag{2-2}$$

where BR = bit rate.

Percentage of time that the BER is below a certain threshold:

$$\%T = 100 \sum_{x=0}^{x=N_e} \frac{T_m(BR)(BER)^x}{x!} \exp(-N_e) \tag{2-3}$$

where T_m = error-measurement period
$N_e = T_m (BR)(BER)$ = number of errors in time T_m
BER = long-term average BER
BR = bit rate

The specification of BER must also specify the time over which it is to be measured. For example, the CCITT recommendation for a 64-kb/s PCM voice transmission specifies a maximum 10^{-6} BER using a 1-minute averaging time. As another example, a circuit can be specified as "out of service" if the BER falls below 10^{-3} for more than 10 seconds.

2.6 VIDEO TRANSMISSION

2.6.1 Applications for Video on Fiber

Applications for video transmission on fiber include:

(a) high-quality video trunked from studio to transmitter;
(b) broadcast CATV video;
(c) video trunking within a city or between cities;
(d) baseband video for closed-circuit, security, or minicam links;
(e) high-definition TV (HDTV) trunking and broadcast (fiber to the home).

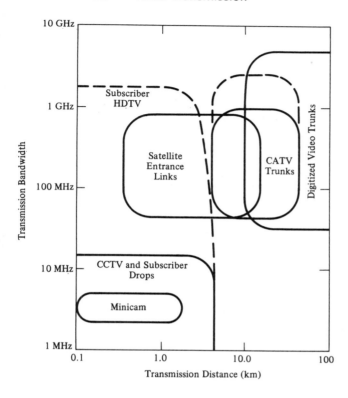

Figure 2.5 Typical product partitioning for the video environment.

Each of these applications requires varying degrees of signal quality and bandwidth. Figure 2.5 indicates the approximate length-versus-bandwidth requirements for these applications.

2.6.2 SNR Relationships

The waveform of a standard video scan line is shown in Figure 2.6. The sync pulse synchronizes the TV scan circuitry. The portion of the picture that is viewed is between "white" and "blanking", therefore, SNR is referenced to this portion. Since there is a fixed relationship between blanking to white and sync to white, the SNR can be referenced either way. The noise is measured as rms noise voltage in a specified bandwidth, generally, 10 kHz to 5 MHz, and usually through some weighting filter that accounts for human viewing characteristics. Noise at higher frequencies is less objectionable than noise at lower frequencies. Various standard definitions of the SNR are given in Table 2.3.

When designing a fiber transmission system for video, it is necessary to relate baseband peak-to-peak SNR to an rms power value. Two conversion factors are involved:

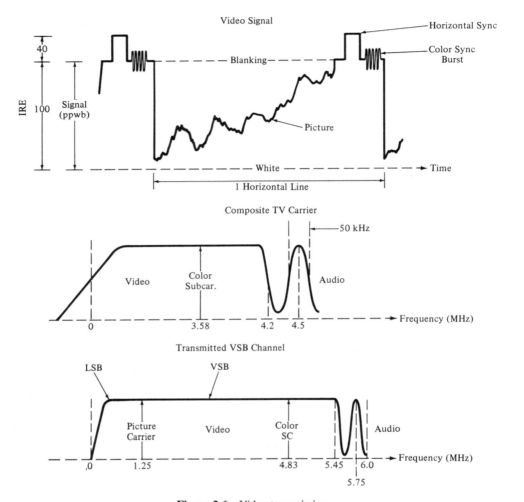

Figure 2.6   Video transmission.

1. the relationship between picture signal level (100 IRE units) and composite-signal sync tip to white (140 IRE units), and
2. the relationship between rms signal power and peak-to-peak signal level.

The conversion from picture to composite is

$$20 \log \frac{140}{100} = 20 \log 1.4 = \boxed{2.92 \text{ dB}}$$

TABLE 2.3 SIGNAL-TO-NOISE RATIO AND WEIGHTING DEFINITIONS FOR VIDEO SIGNALS

Specifying organization	Signal measurement	Noise measurement	Weighting factor (baseband or VSB mod)			Weighting & deemphasis (FM modulation)	
			White noise	Triangular noise	VSB	White	Triangular
National Cable Television Assoc. (NCTA)	RMS power of the VSB signal during sync pulse	RMS noise power measured in a 4-MHz bandwidth in the VSB channel	Not applicable			Not applicable	
Television Allocation Study Organization (TASO)	RMS power of the VSB signal during sync pulse	RMS noise power measured in a 6-MHz bandwidth in the VSB channel	Not applicable			Not applicable	
Consultative Committee in International Radio (CCIR)	Peak-to-peak voltage, reference white to blanking (ppwb)	RMS noise voltage measured in a 4.2-MHz bandwidth in the baseband channel	CCIR OR NCT-7 weighting (M)				
			6.2	10.3	6.7	2.3	12.8
Electronic Industries Assoc. (EIA)	Peak-to-peak voltage, total luminance signal (100 IRE) (ppsw)	Weighted RMS noise voltage measured in 10-kHz to 5-MHz band in baseband channel	EIA RS250B weighting			Not applicable	
			4.2 MHz: 6.8	11.5	4.1		
			5 MHz: 7.4	12.2			
Bell Laboratories (BTL)	Peak-to-peak voltage, sync tip to reference white (ppsw)	RMS noise voltage measured with CCIR or NCT-7 weighting in baseband channel	CCIR or NCT-7 weighting (M)				
			6.2	10.3	6.7	2.3	12.8

References: White and triangular weighting factors for CCIR and BTL are from CCIR Rep. 637-1. White and triangular weighting factors for EIA are from EIA RS-250-B. All VSB figures are from T. M. Straus [16].

The relationship between peak-to-peak (pp) power and rms power is classically 9 dB, which assumes a sinusoidal signal characteristic. When measuring the composite baseband signal in practice, however, the characteristic is more like a combination of a square wave and a sine wave. S. Back [11] derives the relationship as follows:

Sine Wave:

$$pp/RMS = 20 \log 2\sqrt{2} \tag{2-4}$$

Square Wave:

$$pp/RMS = 20 \log 2 \tag{2-5}$$

Average:

$$20 \log \frac{2\sqrt{2} + 2}{2} = 20 \log 2.42 = 7.66 \text{ dB} \tag{2-6}$$

Therefore, the conversion for the (SNR) p-p white to blank/rms noise to the composite (SNR) rms power is

$$(SNR)_{RMS/RMS} = (SNR)_{ppwb/rms} + CF_{composite/picture} - CF_{pp/RMS} \tag{2-7}$$

Classically,

$$(SNR)_{RMS/RMS, comp} = (SNR)_{ppwb/rms} + 2.92 - 9$$
$$= (SNR)_{ppwb/rms} - 6.08 \text{ dB} \tag{2-8}$$

More accurately, according to Back,

$$(SNR)_{RMS/RMS, comp} = (SNR)_{ppwb/rms} + 2.92 - 7.66$$
$$= (SNR)_{ppwb/rms} - 4.74 \text{ dB} \tag{2-9}$$

2.6.3 Video-Transmission Standards

Whenever a video signal is transmitted, it is generally modulated onto a carrier, with the possible exception of short baseband links. When multiple signals are transmitted, such as in television or CATV broadcast, the signals are frequency-division multiplexed in standard 6-MHz wide bands or "channels." Its these channels that are tuned by the TV set or CATV converter. In order to transmit a 4.2-MHz channel, with its audio subcarrier, in a 6-MHz channel, vestigial sideband (VSB) modulation is used. Video transmission standards are therefore referenced to this 6-MHz VSB signal format as illustrated in Figure 2.6.

For trunk transmission over long transmission distances, where high-quality video must be maintained, it is not generally possible to maintain the standard 6-MHz channel since the SNR characteristics of VSB are limited. Here frequency or pulse-code modulation of the baseband signal is used in conjunction with FDM or TDM multiplexing.

Typically, fiber optics does not have the noise, linearity, and dynamic range characteristics to support more than a few channels of VSB transmission, and then only over short distances (<20 km). FM and PCM are the preferred choices for fiber transmission.

Standards organizations have set video-transmission standards for various applications and transmission media. The standards are all referenced to the received-signal quality parameters at baseband video. A summary of some of the key requirements is in Table 2.4.

2.6.4 Vestigial Sideband (VSB) Transmission

In some applications, it is desirable to transmit direct channelized VSB over the fiber. This is usually done for cost reasons and when a limited number of channels (four to six per fiber) is to be transmitted over moderate distances (1 to 20 km) at moderate quality (35 to 50 dB SNR). The advantage of VSB is that the transmission channel bandwidth, $(BW)_c$, is roughly equal to the baseband signal bandwidth, $(BW)_b$. The disadvantage is that there is no $(CNR)_c$ and $(SNR)_b$ improvement, i.e., the transmission SNR requirements are roughly equal to the baseband SNR requirements. The VSB signal, with provisions for a 4.2-MHz video and audio subcarrier, is shown in Figure 2.6.

In order to design using VSB, it is necessary to relate the baseband signal-to-noise ratio for a video channel to the VSB carrier-to-noise ratio, $(CNR)_c$, for that channel. Because of the complex nature of this signal and the measurement techniques applied, there is some degree of controversy over this relationship. S. Bach and J. Nicholas [11] have shown with analysis and measurement that the VSB carrier-to-noise ratio, $(CNR)_c$, is equivalent to the CCIR weighted video SNR(pp/rms) within 0.5 dB. The analysis is summarized as follows:

$$(CNR)_{RMS/RMS} = \frac{\text{carrier power + sideband power}}{\text{rms noise power}} \tag{2-10}$$

$$P_{car} = \frac{V^2}{R} = \frac{(v_c + v_{SB})^2}{R} = \frac{(2v_{SB} + v_{SB})^2}{R}$$

$$= \frac{9v_{SB}^2}{R} = 9P_{SB} \tag{2-11}$$

where v_c = carrier voltage
$\quad P_{car}$ = carrier power
$\quad v_{SB}$ = sideband voltage level = $C_v/2$
$\quad P_{SB}$ = sideband power

Since the video signal is in the upper sideband, then the signal power is

$$P_{SB} = P_{car} - 9.54 \text{ dB} \tag{2-12}$$

The conversion between $(CNR)_{RMS/RMS}$ and $(SNR)_{ppsw/rms,\,w}$ with CCIR weighting is, therefore,

$$(CNR)_{RMS/RMS} = (SNR)_{ppsw/rms,\,w} - 0.24 \text{ dB} \tag{2-13}$$

TABLE 2.4 KEY VIDEO PERFORMANCE STANDARDS FOR TRANSMISSION SYSTEMS

Video specification	Unit	EIA RS250B					NTSC 7	NCTA		
		End-to-end	Long-haul	Satellite	Medium-haul	Short-haul		Terrestrial	MDS	Satellite
Weighted Signal to Noise	dB	54	54	56	60	67	53	52	52	52
Differential Gain	%	10	8	4	5	2	15	10 IRE	10 IRE	8 IRE
Differential Phase	Degrees	3	2.5	1.5	1.3	0.5	5	3.5	3	3
Chrom to Lum Intermod	%	4	4	2	2	1	3 IRE	4 IRE	5 IRE	3 IRE
Luminance Nonlinearity	%	10	8	6	4	2	90 IRE	N.S.	N.S.	N.S.
Chroma-Lum Gain Ineq	100 ± IRE	7	7	4	4	1	3	7.5	N.S.	3
Chroma-Lum Delay Ineq	± ns	60	54	26	33	20	75	40	40	40
Field-Time Waveform Distortion	IRE pp	3	3	3	3	3	4	3	3	4
Line-Time Waveform Distortion	IRE pp	2	1.5	1	1	0.5	4	4	4	4
Short-time Waveform Distortion										
BAR	IRE pp	7	6.5	4	4	4	10	10	N.S.	7
T	100 ± IRE	N.S.	N.S.	N.S.	N.S.	N.S.	6	4	4	4
Long-Time Waveform Distortion										
Bounce	IRE peak	8	8	8	8	8	5	N.S.	N.S.	N.S.
Settling Time	s	3	3	3	3	3	1	N.S.	N.S.	N.S.
Gain Frequency Distortion Multiburst	IRE	N.S.	N.S.	N.S.	N.S.	N.S.	45 – 53	42.5–57.5	41.5–58.5	47–53
Color Burst	ns	N.S.	N.S.	N.S.	N.S.	N.S.	40 ± 4	N.S.	N.S.	N.S.

N.S. = Not Specified

IRE or IRE units is a standard of relative measurement for a video scanline. 100 IRE Units represents the voltage level between blanking and reference white.

calculated as follows:

$+ (SNR)_w$	$(SNR)_{ppsw/rms, w}$
$+ 9.54$	rms P_{car} above rms P_{SB}
$+ 1.16$	VSB modulation index limit of 87.5%
$- 7.66$	pp to rms baseband-signal conversion
$+ 2.92$	140/100 IRE sync to picture level
$\underline{- 6.2}$	weighted to unweighted conversion
$(CNR)_{RMS/RMS}$	

2.6.5 Frequency-Modulation Transmission

FM modulation is a favored approach for transmission of a moderate number (8 to 16) of high-quality (>55 dB SNR) video channels over long distances (>20 km) on fiber. Today, it is lower in cost than PCM and satisfies a need in metropolitan area video trunks for CATV and Broadcast. FM modulation provides a baseband $(SNR)_b$ to transmitted $(CNR)_c$ improvement factor of 10 to 35 dB, depending on the frequency deviation used (1.5 to 10 MHz peak). This compensates for the low dynamic range and $(SNR)_t$ limitations of fiber transmission. The low $(CNR)_t$ requirements also allow a certain amount of nonlinearity (intermodulation) without visible effects. FM modulation requires a lot of bandwidth, however, this is generally a favorable trade-off for SNR when using fiber. From Chapter 5, the relationship for CNR to SNR for an FM modulated signal is

$$(CNR)_{c, RMS/RMS} = (SNR)_{b, RMS/RMS} - 20 \log \frac{\Delta f}{(BW)_b}$$
$$- 10 \log \left\{ \frac{(BW)_c}{(BW)_b} \right\} - 1.76 \text{ dB} \tag{2-14}$$

In order to use this relationship for video transmission, we must convert the video baseband-signal specification, $(SNR)_{pp/rms}$ to a transmitted power specification, $(SNR)_{RMS/RMS}$:

$$(SNR)_{RMS/RMS} = (SNR)_{pp/rms, w} - CF - WF - DF \tag{2-15}$$

where CF is the conversion factor for the conversion between rms picture signal power and peak-to-peak signal level specified (9 dB for sinusoidal and 7.66 dB for ppwb video).

WF is the weighting factor for the reduction in noise power (increase in SNR) due to the weighting filter. Note that "triangular" noise weighting is applicable to FM.

DF is the deemphasis factor for the reduction in noise power (increase in SNR) due to preemphasis and deemphasis.

Therefore, for FM modulation:

$$(CNR)_{c, \text{RMS/RMS}} = [(SNR)_{b, \text{pp/rms}, w} - CF - WF - DF]$$

$$- 20 \log \frac{\Delta f}{(BW)_b} \qquad (2\text{-}16)$$

$$- 10 \log \left\{ \frac{(BW)_c}{(BW)_b} \right\} - 1.76 \text{ dB}$$

For example, assume an FM transmission with preemphasis and a baseband $(SNR)_{\text{ppwb/rms}, w}$ specified with CCIR weighting. The triangular noise weighting factor (WF) is 2.9 dB in a 4.2-MHz bandwidth and the deemphasis factor (DF) is 9.9 dB. The conversion factor (CF) is the conversion from pp signal voltage (white to blanking) to composite (picture plus sync) RMS power:

$$CF = 20 \log \frac{\text{ppwb}}{\text{RMS}} + 20 \log \frac{140}{100 \text{ IRE}}$$

$$= -7.66 \text{ dB} + 2.92 \text{ dB} = -4.74 \text{ dB}$$

With a wide deviation modulator ($\Delta f = 8$ MHz),

$$(BW)_c = 2(\Delta f + (BW)_b) = 2(8 + 4.2) = 24.4 \text{ MHz}$$

$$(CNR)_{c, \text{RMS/RMS}} = [(SNR)_{b, \text{ppwb/rms}, w} - 4.74 - 2.9 - 9.9]$$

$$- 20 \log \frac{8}{4.2} - 10 \log \frac{24.4}{4.2} - 1.76$$

$$= (SNR)_{b, \text{ppwb/rms}} - 32.6 \text{ dB}$$

If only the picture portion was to be FM modulated (sync portion eliminated), then the improvement factor would be reduced by 2.92 dB.

2.6.6 PCM Video Transmission

PCM is the most controllable form of modulation for video transmission over fiber, particularly over long distances where repeaters are required. Its major drawbacks are the cost of encoding and the large amount of bandwidth it utilizes. With PCM, the signal quality at the received end is almost totally a function of the properties of the encoder and not a function of the transmission link if error rates are reasonable ($>10^{-6}$). See Chapter 5, Section 5.3, for a detailed discussion of PCM. The relationship between SNR and the encoding parameters is, from Table 5.4, as follows:

$$(SNR)_{\text{RMS/RMS}} = 6n + 10.8 + 20 \log \frac{V_{S\text{rms}}}{A_q} \qquad (2\text{-}17)$$

where $V_{S\text{rms}}$ = rms signal level
A_q = pp quantizer range
n = number of bits encoding

When PCM encoding a video signal, the baseband signal is encoded generally with separate encoders used for the video and audio portions. Separate encoders are often necessary since the video portion requires >52 dB $(SNR)_{ppwb/rms}$ and, therefore, 7 to 8 bits of encoding at a sampling rate of around 10 MHz, whereas the audio requires 50 to 56 dB $(SNR)_{pk/rms}$ and a sampling rate of around 2.4×15 kHz = 36 kHz. The option is to place the audio on a subcarrier and encode the composite signal, a solution for moderate-quality systems.

When encoding the video signal separately, it is only necessary to encode the picture portion. The sync pulse can be reconstructed at the received decoder end. In this case, the pp white-to-blanking picture signal (V_s) is adjusted in level to match that of the quantizer pp range (A_q). Therefore, $A_q = 2V_s$. The rms signal level, assuming a full-load sinusoid, is, thus, $V_s/\sqrt{2}$. Therefore, Equation (2-17) becomes

$$(SNR)_{RMS/RMS} = 6n + 10.8 + 20 \log \frac{(V_s\sqrt{2})}{2V_s}$$

$$= 6n + 1.8 \text{ dB} \tag{2-18}$$

Converting from $(SNR)_{pp/rms, w}$ to $(SNR)_{RMS/RMS}$, we have

$$(SNR)_{RMS/RMS} = (SNR)_{pp/rms, w} - CF - WF \tag{2-19}$$

where CF = Conversion factor for the conversion of pp signal level to rms signal power

        = 9 dB if sinusoidal and 7.66 dB if ppwb video

    WF = weighting factor for the reduction of SNR; use white-noise weighting for PCM: 6.2 dB for CCIR and 6.8 dB for EIA

Therefore,

$$(SNR)_{pp/rms, w} = 6n + 1.8 + CF + WF \tag{2-20}$$

The bit rate (BR) for PCM is (from Table 5.4)

$$(BR)_c = nf_s \tag{2-21}$$

where n is the number of encoding bits per sample, and f_s is the sampling frequency. For practical filtering, f_s is selected as a 2.35 to 2.4 multiple of $(BW)_b$. Often with video, a 10.74 MHz sampling rate is selected (multiple of color burst frequency) to reduce interference products; therefore,

$$(BR)_c = n(10.74 \text{ MHz}) \tag{2-22}$$

For example, assume a 65 dB $(SNR)_{ppwb/rms, w}$ measured with CCIR white-noise weighting in a 4.2 MHz $(BW)_b$. If we assume that the sync pulse is not encoded, then the PCM encoding parameters would be as follows:

$$65 \text{ dB SNR} = 6n + 1.8 + 7.66 + 6.2$$

$$n = \frac{65 - 15.7}{6} = 8 \text{ bits}$$

$$(BR)_c = 8(10.74) = 86 \text{ Mb/s}$$

Figure 2.7 illustrates the encoding parameters versus $(SNR)_{ppwb/rms}$ for unweighted measurements.

$$SNR = \frac{\text{Peak white to blanking}}{\text{rms noise in baseband } \overline{BW}}$$

$$= 6n + 1.8 + 7.66$$

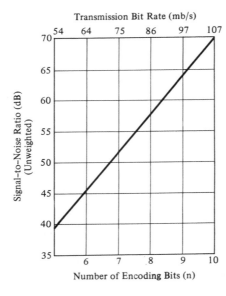

Transmission Bit Rate (mb/s)

Figure 2.7 Relationship between number of binary bits (n) and the signal-to-noise ratio (p-p white-to-blanking/rms) for linear quantizing PCM encoded video. Conditions are set for a sampling rate of 10.74 MHz, a video baseband bandwidth of 4.2 MHz, and a transmission rate $= n \times f_s$.

Much of the standardization activity centered on PCM video for the subscriber loop is centering on the range of 135 to 150 Mb/s for National Television System Committee (NTSC) standard video [12]. The standard for broadband ISDN, a potential vehicle for video, is 139.264 Mb/s [13]. These rates permit a single channel of video (at about 100 Mb/s) simultaneous with other information such as FM radio, two channels of video with compression, or a lower signal quality.

2.6.7 High-Definition Television (HDTV)

Improvements in television picture quality are occurring in three areas: (1) improved NTSC, (2) enhanced-quality TV (EQTV) with a 525-line format, and (3) high-definition television (HDTV).

Improved NTSC is achieved through video preprocessing at the studio and postprocessing at the TV Receiver. Transmission requirements remain relatively the same as for NTSC. HDTV originates at the studio camera with 1250 lines and 20 MHz of bandwidth. This is compressed (through interlacing) to 625 lines and 8.4 MHz baseband bandwidth in the Philips approach [14], for example. The display is then reconstructed at the set to 1250 lines. According to E. H. Hara, Department of Communications, Canada, HDTV may require about 30 MHz of bandwidth and a digitized bit rate of between 400 to 700 Mb/s [13]. L. S. Smoot, Bell Communications Research [15], indicates that the digital Codecs for HDTV will operate at either 135, 180, or 270 Mb/s encoding, with 135 Mb/s being the most expensive and 270 Mb/s the most likely solution. He indicates 135 Mb/s as the most likely Codec bit rate for EQTV although it can be achieved with 90 Mb/s.

REFERENCES

1. Western Electric Technical Publications, *Telecommunications Transmission Engineering, Vol. 2: Facilities,* Winston-Salem, NC, 1984.

2. Western Electric Technical Publications, *Telecommunications Transmission Engineering, Vol. 3: Networks and Services,* Winston-Salem, NC, 1980.

3. Western Electric Technical Publications, *Telecommunications Transmission Engineering, Vol. 1: Principles,* Winston-Salem, NC, 1980, chaps. 12 and 28.

4. S. Haykin, *Communications Systems,* Wiley, New York, 1978, pp. 407–463.

5. P. Noll, "A Comparative Study of Various Schemes for Speech Encoding," *Bell System Technical Jornal,* **54** (November 1975): 1597–1614.

6. D. R. Smith, *Digital Transmission Systems,* Van Nostrand Reinhold, New York, 1985, pp. 63–123.

7. American National Standards Institute, *American National Standards for DS1, DSc, DS2, and DS3 Levels of the Digital Hierarchy,* ANSI BSR T1.102-1987, New York, 1987.

8. R. L. Freeman, *Telecommunication Transmission Handbook,* Wiley, New York, 1981, p. 449.

9. American National Standards Institute, *Proposed Draft for the American National Standard for the DS4NA Fourth Hierarchical Level,* ECSA T1X1.87-126, New York, 1987.

10. US Sprint, switched 56-kb/s announcement at the Telecommunications Association (TCA) show, San Diego, 1988.

11. S. Bach, "Analysis of Vestigial Sideband C/N and Weighted Video S/N," paper presented at the Telecommunications Association, 1984.

12. American National Standards Institute, *Draft for Digital Hierarchy Optical Interface Rates and Formats Specification,* ECSA T1X1/87-129R1, New York, 1988.

13. E. H. Hara, "An Approach to Broadband ISDN," paper presented at the *Litewave Journal* Symposium, "Fiber in the Subscriber Loop," 1986.

14. A. Toth, "Prospects for Advanced Television and Its Potential Impact on the Fiber Optics Loop," paper presented at the Fiber Optic Network Seminar, Palo Alto, CA, 1986.

15. L. S. Smoot, "Broadband ISDN Access," paper presented at the *Litewave Journal* Symposium, "Fiber in the Subscriber Loop," 1986.

16. T. M. Straus, "The Relationship between the NCTIA, EIA, and CCIR Definitions of Signal to Noise Ratio," IEEE Transactions on Broadcast, Vol. BC-20, No. 3, Sept. 1974.

3

Fiber-Optic Component Specification

3.1 INTRODUCTION

Once the requirements for the system are defined, and the basic systems architecture developed, the next step is to define the characteristics of each functional element within that architecture. This is an iterative process whereby one estimates the characteristics, performs the design analysis, and refines the characteristics until the proper results are achieved. The first activity is, therefore, to select or specify candidate equipment and components that, when integrated as a system, are estimated to meet the end-to-end requirements. In a complete systems design, all systems elements must be defined. This chapter, however, focuses on the components unique to fiber optics, those contained within the photonics layer.

It is generally best to start by specifying or selecting components that achieve a certain performance result from the link-budget analysis. The selected candidates then are screened, or the component specification detailed, to account for other systems-design criteria such as (a) environmental stability, (b) cost, (c) reliability, (d) physical requirements, (e) product maturity, (f) support requirements, and other practical parameters.

This chapter provides a brief summary of the available component-technology choices, and approaches component specification from the standpoint of the photonics-layer-signal performance. The methodology is to use link-power and bandwidth-budget parameters as the design reference. These parameters were described briefly in Chapter 1 and are described in depth in Chapter 6.

3.1.1 Application-Driven Selection Criteria

It is best to begin a design with some idea of what type of components and which mix of components are candidates for a specific application. Table 3.1 summarizes typical component selection for various application catagories. Chapter 7 provides additional guidance as to how the technology is typically applied. Figure 3.1 illustrates length-versus-bandwidth trade-offs for various mixes of fiber, source, detector, and wavelength choices.

3.1.2 Selection Methodology

The logical process of component selection and specification depends largely on the particulars of the design and what the unknowns are. In order to generalize, however, the designer normally follows the following logical sequence:

1. Select the basic technology by using design aids, experience, judgment, and supplier specifications. Select:

single-mode versus multimode;

wavelength (820, 1300, 1550 nm, etc.);

fiber type: core size, attenuation range, bandwidth/dispersion;

laser versus LED source;

detector/receiver technology (APD, PIN/FET, etc.).

2. Select a specific receiver design and determine its sensitivity (MDP) in dBm at the specific operating bandwidth or data rate. The determination of MRP is accomplished through computation from receiver-component parameters or from design tables and curves supplied by the manufacturer.

3. Select the specific optical-source and transmitter designs, and determine source-coupled (transmitted) power (P_t) in dBm. The transmitter is selected to match the fiber technology, the wavelength, and the signal characteristics (analog, digital, etc.).

4. If the system is to use wavelength-division multiplexing (WDM) or is architectured as a passive optical LAN where couplers are used, select the coupler types and specify port-to-port "tap" insertion loss (L_T) in dB. WDM coupler selection is generally based on fiber technology, the wavelength of the transmission "window," the channel-wavelength separation, signal-isolation requirements, and other practical considerations. Coupler insertion loss, isolation, and other requirements are traded off primarily with source and detector technology. In the LAN case, coupler selection and the resultant port-to-port insertion losses generally dominate the optical-loss budget. The trade-offs are generally coupler design versus fiber technology, wavelength, and transmitter/receiver performance.

5. Select the connector type and specify unit insertion loss (l_c) in dB. Connector selection is generally based on fiber technology and physical parameters, and is

TABLE 3.1 TYPICAL COMPONENT-TECHNOLOGY SELECTION BY APPLICATION

			Metropolitan area network		Local area network	
	Wide-area trunking	Interfacility	CATV/Video	Home distrib.	Data LAN	Data link
Repeater spacing (typical) (km)	30–50	3–20	10–20	0.1–1	1–100	0.03–3
Information	Data and PCM Voice/Video	Data and PCM voice	1–100 ch. 6-MHz video	1–4 ch. TV and data, phone	Data or data/voice	RS232 and other low-speed data
Transmission rate (per fiber)	135–2400 Mb/s	45–560 Mb/s	6–800 MHz 45–1.2 Gb/s	6–20 MHz >139 Mb/s	10–100 Mb/s	DC to 20 Mb/s
Link-loss margin (optical power) (dB)	20–30	20–30	10–20	3–10	17–24	10–40
Wavelength (μm)	1.3 and 1.55	0.85 and 1.3	1.3 and 1.55	0.85 and 1.3	0.85 and 1.3	0.85
Fiber	Single-mode	Multimode GI, Single-mode	Single-mode	Multimode SI, Single-mode (future)	Multimode, 50–100 μm core	Multimode, 50–100 μm core
Source	Laser diode	LED, laser diode	Laser diode	LED	LED	LED
Detector	PIN/FET, APD	PIN/FET	PIN/FET, APD (Dig. only)	PIN	PIN	PIN
Spliced method	Fusion or bonded	Fusion or mechanical	Fusion or bonded	Mechanical	None	None
Connector	Biconic, FC w/pigtail	Biconic, FC w/pigtail	Biconic, FC w/pigtail	Plastic, SMA, field install	SMA or MIC, package mount	Low-cost SMA, fiber to device
Couplers	2 ch. WDM	WDM, bidirect	Multich. WDM	Bidirectional	Multiport tee, star	Not usually
Wavelength MUX	For expansion, infrequent use	Rarely	Sometimes used for capacity	Bidirectional single-fiber	No	Not usually

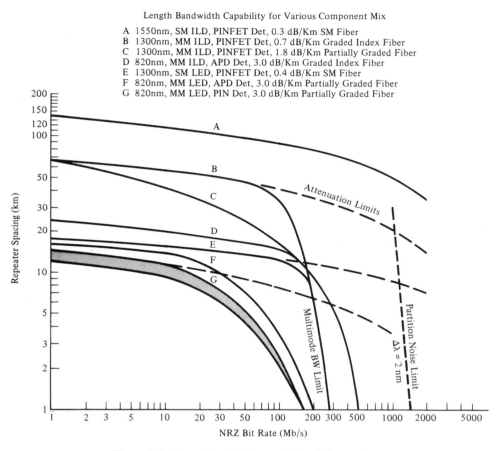

Length Bandwidth Capability for Various Component Mix

A 1550nm, SM ILD, PINFET Det, 0.3 dB/Km SM Fiber
B 1300nm, MM ILD, PINFET Det, 0.7 dB/Km Graded Index Fiber
C 1300nm, MM ILD, PINFET Det, 1.8 dB/Km Partially Graded Fiber
D 820nm, MM ILD, APD Det, 3.0 dB/Km Graded Index Fiber
E 1300nm, SM LED, PINFET Det, 0.4 dB/Km SM Fiber
F 820nm, MM LED, APD Det, 3.0 dB/Km Partially Graded Fiber
G 820nm, MM LED, PIN Det, 3.0 dB/Km Partially Graded Fiber

Figure 3.1   Length bandwidth capability of fiber product.

often dictated by the supplier of the transmitter and receiver. In coupler-based systems, WDM or LAN, connector selection goes hand in hand with coupler selection in determining coupler insertion loss.

6. Select the specific fiber and fiber-cable type and specify unit-length attenuation, l_f, in dB/km, and bandwidth, $(BW)_f$, in MHz-km, or dispersion, t_{df}, in ns/km. Fiber parameters are generally selected such that the transmitter and receiver can operate adequately over the required distance between repeaters, after connector and splice losses are taken into account. Cable-design selection is generally based on fiber-count requirements, installation environment, and construction methods.

7. Select the splicing technique and specify splice unit insertion loss (l_s) in dB. Splice-technology selection is generally based on the insertion-loss requirements, plant-construction constraints and costs, and concerns about long-term reliability. Insertion-loss trade-offs center around fiber type, splicing method, and loss-measurement method.

In practice, this sequence can be modified where certain elements are fixed or predetermined. The methodology is also iterative in that as selections are made and tested, they are subsequently altered as necessary to optimize performance.

3.2 RECEIVER AND DETECTOR SPECIFICATION

Receiver and detector selection is a function of the required receiver sensitivity, or the minimum required received optical power (MRP), for a specified signal performance at the output of the optical receiver. Component choice often begins with the receiver, since there is a limited number of choices that can be made, and receiver performance defines the basic noise limits on photonics performance.

3.2.1 Selection and Design Methodology

Generally, selection proceeds as an integral part of the signal-performance analysis (Chapter 6) as follows:

(a) Select the operating center wavelength. This is usually based on fiber attenuation and distance requirements, not on receiver performance. If it is not a high-performance application and the wavelength can be at the option of receiver design, a generally short wavelength (800 to 900 nm) is chosen to take advantage of the stability, high performance, and low cost of silicon devices. A long wavelength (1300 or 1550 nm) is always required for single-mode high-performance trunking.

(b) Select the transmission signal format (analog, digital, FM/FDM, NRZ, RZ, Manchester, etc.). If the encoding or signaling format is optional, select to optimize performance at the receiver (Chapters 4 and 5). For digital signals, this is generally a scrambled NRZ.

(c) Determine the performance range of the receiver from the preliminary performance analysis (Chapter 6, step 2d) in terms of bandwidth, or bit rate, and minimum required received power (MRP).

(d) Select a detector type, PIN (no gain) or APD (with avalanche gain), depending on wavelength region, available device technology, cost, and required receiver sensitivity.

(e) Receiver preamplifier design is generally selected along with detector choice since it is the combination of detector and preamplifier that determines receiver sensitivity. Choices include not only the design function (integrating or trans-impedance), but the front-end amplifying component as well (bipolar transistor or FET).

(f) Determine the fiber-to-detector coupling means and associated loss and select the detector, and coupling "pigtail" fiber if applicable, of a size and type necessary to optimize coupled power while preserving speed. The smaller the detector, the less capacitance it has and the faster and more sensitive it is. In cases where the detector is manufactured with an integral fiber "pigtail," coupling loss is already accounted for in the detector-responsivity or receiver-sensitivity specifications as the case may be.

(g) If the receiver is not a purchased device, but is to be developed or analytically modeled from individual component parameters, select and list the key parameters required for the detector and the preamplifier. For the detector, these parameters generally include:

responsivity: r (amperes/watt)
gain: G (if an APD)
capacitance: C_d (picofarads)
rise time: t_r (nanoseconds)
leakage (dark) currents: I_{du} and I_{dm} (nanoamperes)
bias voltage (volts)

The design parameters for the preamplifier include:

input resistance: R_i (ohms)
input capacitance: C_a (picofarads)
transconductance of the FET: g_m (mhos)
channel-noise factor for the FET
corner frequency of the FET transistor: f_c (mHz)
Bipolar transistor h_{fe}

(h) Having selected the design, compute the MDP or MRP from receiver-design curves supplied by the manufacturer or from the analytical formulas supplied in this book in Chapter 4, Sections 4.9 and 4.10.

(i) If detector coupling loss (L_d) is not included in receiver-sensitivity or detector-responsivity specifications, it must be added to compute the minimum required received optical power (MRP) at the optical aperature to the receiver.

(j) MRP is used in the link-power-budget analysis to refine the design of the photonics layer.

3.2.2 MRP Specification

Minimum required received power (MRP) is a measure of receiver sensitivity defined for a specific SNR or BER and bandwidth or bit rate at the receiver output. MRP implies the power within the optical aperature of the receiver, not necessarily the power detected. The minimum required detected power (MDP) is the received power less any coupling losses (L_d) between the detector and any coupling device or "pigtail." This is shown in Figure 3.2 and related as follows:

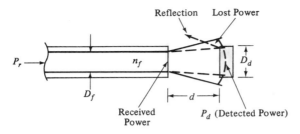

Figure 3.2 Relationship of detected power to received power.

$$\text{MRP (dBm)} = \text{MDP (dBm)} + L_d \text{ (dB)} \qquad (3\text{-}1)$$

The value of MDP depends on:

(a) the choice of detector and receiver preamplifier: PIN, APD, FET preamp, bipolar preamp, integrating or transimpedance design;

(b) the wavelength of operation and, therefore, the device material: silicon, germanium or GaAs;

(c) the required received SNR based on modulation, other transmission parameters;

(d) receiver bandwidth required to pass the signal;

(e) source noise, modal noise, and source nonlinearities that decrease SNR;

(f) temperature, which determines resistive noise and effects the parameters of active components.

Chapter 4 provides an in-depth analysis of MDP, incorporating all of these parameters. Figures 3.3(a) and 3.3(b) represent plots derived from the MDP relationships in Chapter 4, which can be used as design aids in receiver/detector and wavelength selection.

Since MRP includes the fiber-to-detector coupling loss (L_d), it depends on the following additional factors:

(a) coupling mechanism: connector aperature or fiber pigtail;

(b) coupling geometries between the immediate coupling aperature and the detector, including relative diameters, distance, and numerical aperature; and

(c) Fresnel reflections.

Detector-to-fiber coupling loss, incorporating these factors, is discussed in Section 3.2.10. Optical-receiver or packaged-detector specifications generally incorporate the detector coupling loss in their performance parameters. Noise equivalent power, responsivity, and received power are generally referenced to an external connector interface. Detector quantum-efficiency specifications generally include the effects of coupling loss when factory coupled. For noncoupled devices, quantum-efficiency numbers generally include the effects of surface reflection at a specific wavelength when antireflection coated.

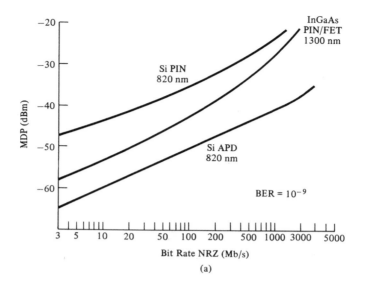

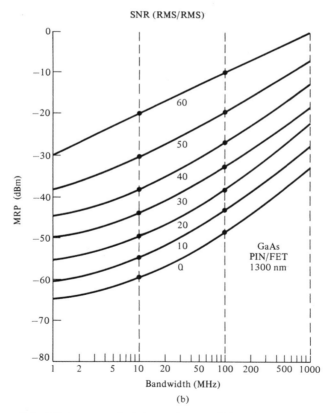

Figure 3.3 Minimum required detected power (MDP) for optical receivers. (a) Digital transmission, and (b) analog transmission.

MRP is often specified in terms of nominal and worst-case values, in an effort to account for component tolerance and coupling-loss variation. Although it is generally not practical to determine the statistical nature of all parameters contributing to MRP, the statistical relationship for MRP can be written:

Mean:

$$MRP = [MDP + P_{p,T} + P_{p,t} + P_{p,n}] + L_d \qquad (3\text{-}2)$$

Standard Deviation:

$$\sigma_{MRP} = (\sigma_{MDP}^2 + \sigma_{L_d}^2 + \sigma_T^2)^{1/2} \qquad (3\text{-}3)$$

where MDP is the mean under nominal conditions

$P_{p,T}$, $P_{p,t}$, and $P_{p,n}$ are the deratings (power penalties) in the mean MDP due to temperature, time and transmitted noise, respectively

σ_{MDP} and σ_{L_d} are the standard deviations for the receiver and coupling loss, under nominal conditions

σ_T is the increase in standard deviation from nominal due to temperature

MRP is the minimum required received power

In practice, unless the data are available, it is often easier to simply recalculate the receiver MDP for each operating extreme, rather than attempting to isolate and calculate all of these variables.

3.2.3 Detector Specification*

The photodetector is generally selected and specified, together with the preamplifier, to provide the highest signal-to-noise ratio and response-time characteristics. High SNR implies a detector with high responsivity, low leakage-current noise, and low capacitance. Fast response implies small size.

The choice is between PIN detectors that exhibit responsivities (r) in the range of 0.5 to 0.85 amperes of photocurrent per watt of incident optical power, and avalanche photodetectors (APD) that have an internal current-gain mechanism (G) that further multiplies responsivity by a factor of 10 to 250, depending on structure and materials. Table 3.2 summarizes some of the typical characteristics of commercial devices.

The detector material depends on the wavelength specified for the system and, therefore, the materials that absorb light in that wavelength. Detector characteristics are different at different wavelengths; therefore, the choice of systems operating wavelength involves a trade-off of detector performance.

In order to understand these trade-offs, it is necessary to define some of the operating parameters of a photodetector [1, 2, 4]:

---

*The author wishes to acknowledge Dr. Tran Muoi for the general information upon which this section was developed.

TABLE 3.2 TYPICAL PERFORMANCE OF DETECTOR PRODUCT*

Parameter	Unit	Silicon		Germanium		InGaAS	
		PIN	APD	PIN	APD	PIN	APD
Wavelength range	nm	400–1100		800–1800		900–1700	
Peak	nm	900	830	1550	1300	1300 (1550)	1300 (1550)
Responsivity							
Chip	R	0.6	77–130	0.65–0.7	3–28	0.63–0.8 (0.75–0.97)	60–70
Coupled	R	0.35–0.55	50–120	0.5–0.65	2.5–25	0.5–0.7 (0.6–0.8)	10–30
Quantum efficiency	%	65–90	77	50–55	55–75	60–70	60–70
Gain	G	1	150–250	1	5–40	1	10–30
Excess noise factor	x	—	0.3–0.5	—	0.95–1	—	0.7
K_{eff}	k	—	0.02–0.08	—	0.7–1	—	0.3–0.5
Bias voltage	$-V$	45–100	220	6–10	20–35	5	<30
Dark current							
Nonmult	nA	1–10	0.1–1.0	50–500	10–500	1–20	1–5
Mult	nA	—		—		—	1–5
Capacitance	pF	1.2–3	1.3–2	2–5	2–5	0.5–2	0.5
Rise time	ns	0.5–1	0.1–2	0.1–0.5	0.5–0.8	0.06–0.5	0.1–0.5

* The author wishes to acknowledge Dr. Tran Muoi for some of the information upon which this table was developed.

*(a) Photocurrent (i_p): This the current generated when optical power (P_d) is absorbed by the detector:

$$i_p = rGP_d \tag{3-4}$$

The relationship is linear between received power and photocurrent.

*(b) Quantum Efficiency (η) [4]: Consists of internal and external quantum efficiency.

Internal QE (η_I):

$$\eta_I = \frac{\text{average number of electrons emitted}}{\text{average number of incident photons}} = \frac{I_p/e}{P_d\lambda/hc} \tag{3-5}$$

where e is the electron charge $= 1.6 \ 10^{-19}$ J

h is Planck's constant $= 6.63 \ 10^{-34}$ J-s

c is the velocity of light $= 3 \ 10^8$ m/s

λ is the wavelength, m

$$\eta_I = \exp(-\alpha w_s) - \exp[-\alpha(w_d + w_s)] \tag{3-6}$$

where α is the absorption coefficient of the detector material in that region, w_s is the thickness of the surface $p+$ region, w_d is the thickness of the depletion i region.

External QE (η_E) Includes Surface Reflection:

$$\eta_E = \eta_I \left[1 - \left(\frac{n_d - n_o}{n_d + n_o} \right)^2 \right] \tag{3-7}$$

where n_d and n_o are the refractive indexes of the detector and surrounding medium, respectively. With an antireflection coat, η_E is close to η_I.

*(c) Responsivity (r): Responsivity is the ratio of the photocurrent generated for a unit of optical power:

$$r = \eta(e\lambda/hc) \ \text{A/W} \tag{3-8}$$

In this text, responsivity whose symbol is a capital R is used to describe the responsivity of an APD with gain accounted for; $R = rG$.

Figure 3.4 illustrates the relationship between responsivity and wavelength typical of PIN and avalanche photodetectors (APD) of different materials. The differences in materials, relative to the PIN case, are due to the limitations on maximum gain for longer-wavelength devices.

(d) Gain (G): This is avalanche gain of electron-hole pairs within avalanche detectors. Maximum gain for APDs varies from about 10 to 250, depending on material (see Table 3.2).* $G = 1$ for PIN detectors.

---

* The author wishes to acknowledge Dr. Tran Muoi for the general unpublished information upon which this section was developed.

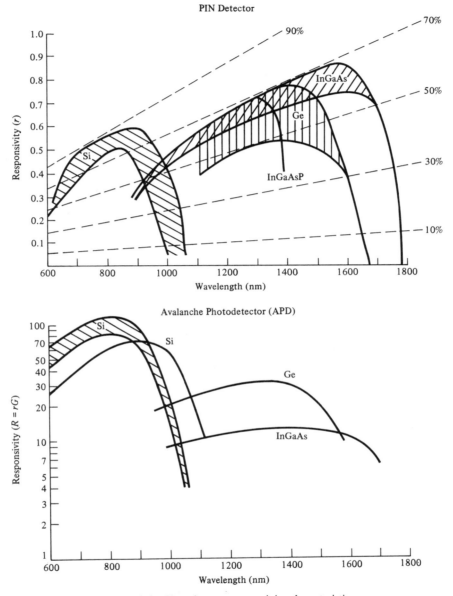

Figure 3.4 Photodetector responsivity characteristics.

(e) Excess-noise factor (F): This is a factor in avalanche photodetectors (APD) that, due to the randomness of the multiplication process, increases the multiplication of noise at a greater rate than signal [3]. Factor F is a complex function of ionization rate (k) and gain (G), as discussed in Section 4.8 of Chapter 4. The plot of F versus gain for various detector types is illustrated there as well. Often, F is expressed as the approximation $F = G^x$; however, this is only roughly accurate for all values of gain. The relative values of G and F for various materials are included in Table 3.2.

(f) Noise equivalent power (NEP): This is the radiant optical power required to produce an rms detector current equal to the rms noise current for a given noise bandwidth and given modulation frequency [1]. Detectivity = 1/NEP. The relationship [2] is

$$\text{NEP} = \frac{hc}{e\lambda}\frac{I_p}{\eta} \tag{3-9}$$

where I_p is the photocurrent. If it is dominated by dark-current noise (I_d), then

$$I_p = [2eI_d(\text{BW})_n]^{1/2} \tag{3-10}$$

where $(\text{BW})_n$ is the noise bandwidth, usually equated to 1 Hz for this measurement.

3.2.4 SNR Considerations

Selection of detectors is generally based on the most practical selection, other parameters considered, that will provide the lowest noise contribution and fastest response. Table 3.3 summarizes and compares the noise sources associated with PIN and APD detectors. Their derivation is discussed in Chapter 4. In both cases, quantum noise and leakage-current noise are present. Quantum noise is a function of optical power, i.e., the more incident power, the more noise. This effect creates a trade-off between PIN and APD detectors as a function of optical power or SNR.

From Section 4.10 in Chapter 4, the rms SNR of a signal at the output of the preamplifier, assuming a sinusoidal signal, is

(a) for a PIN detector:

$$\text{SNR} = \frac{\frac{1}{2}(M_o r P_o)^2}{2e(rP_o + I_{du})(\text{BW})_n + \langle i_{nA}^2\rangle} \tag{3-11}$$

TABLE 3.3 NOISE SOURCES IN PHOTODETECTORS

Noise source	PIN photodetectors	Avalanche photodetectors
Quantum $\langle I_{nQ}^2 \rangle =$	$2erP_d(\text{BW})_n$	$2erG^2FP_d(\text{BW})_n$
Leakage current $\langle I_{nL}^2 \rangle =$	$2eI_{du}(\text{BW})_n$	$2e(G^2FI_{dm} + I_{du})(\text{BW})_n$
Total $\langle I_{nD}^2 \rangle =$	$2e(rP_d + I_{du})(\text{BW})_n$	$2e[(rP_d + I_{dm})G^2F + I_{du}](\text{BW})_n$

where e = electric charge = 1.6×10^{-19} coulomb
  r = responsivity (A/W)
  G = avalanche gain
  F = excess noise factor = $kG + (1 - k)(2 - 1/G)$
  P_d = detected optical power in Watts
  I_{du} = unmultiplied dark current in Amps
  I_{dm} = multiplied dark current in Amps

(b) for an APD detector:

$$\text{SNR} = \frac{\frac{1}{2}(M_o r G P_o)^2}{2e[G^2 F(r P_o + I_{dm}) + I_{du}](\text{BW})_n + \langle i_{nA}^2 \rangle} \tag{3-12}$$

With the PIN detector not having gain, the noise terms that dominate are leakage current (I_{du}) and preamplifier-noise current (i_{nA}). This makes SNR roughly a squared function of received optical power. For the APD, on the other hand, gain dominates, making quantum noise the dominant noise term. APD noise, therefore, becomes an approximate direct function of received optical power. These relationships are plotted for gains, responsivities, and leakage currents representative of silicon and GaAs detector devices in Figure 3.5. It is evident that for high-incident optical power or high $(\text{SNR})_e$, there is a crossover in performance between the PIN and the APD devices. Since the APD receiver is generally more complex due to APD stability and the APD more costly, the trade-off can be an important one.

With digital transmission, the SNR is fixed by the BER. For a 10^{-9} BER, the $(\text{SNR})_{\text{RMS/RMS}}$ is on the order of 15.6 dB, from Section 4.5 in Chapter 4. From Figure 3.5 we can see that at a SNR of 15.6 dB, the APD has a significant advantage in MDP.

In the case of the APD, both quantum noise and bulk leakage current (I_{dm}) are multiplied by gain and excess-noise factor (F). With surface leakage (I_{du}) fixed and quantum noise and bulk leakage noise increasing with gain at a greater rate than signal, there is a point at which SNR degrades with further increases in gain. This "optimum-gain" point is shown in Figure 3.6.

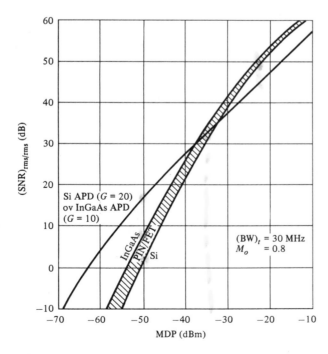

Figure 3.5 Signal-to-noise ratio versus detected power for PIN and avalanche devices.

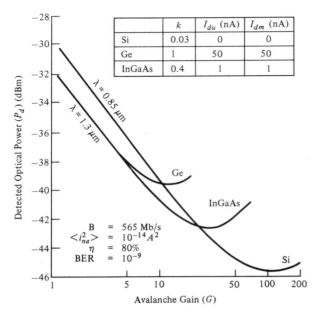

	k	I_{du} (nA)	I_{dm} (nA)
Si	0.03	0	0
Ge	1	50	50
InGaAs	0.4	1	1

Figure 3.6 Required optical power for a 565-Mb/s receiver using APDs as a function of avalanche gain. The author wishes to acknowledge Dr. Tran Muoi, unpublished, [4] for submitting this previously unpublished figure to this publication.

The trade-off between PIN and APD also involves the practical limitations on a bias supply and the need for compensation due to the effect of temperature on the transfer function. Silicon APDs require 100 to 250 VDC reverse bias; germanium, 20 to 50 VDC, and GaAs, 10 to 30 VDC. This is in contrast to PIN detectors, which require between 10 to 50 VDC. The trade-off here is the cost and complexity of the bias supply.

Another effect that involves the complexity of the bias supply and the receiver circuitry is the variation in APD gain with temperature. Figure 3.7 shows this effect for a silicon APD. For a fixed bias voltage, gain decreases significantly as a function

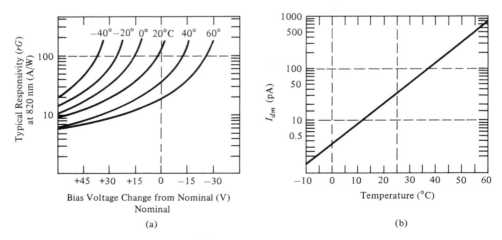

Figure 3.7 Temperature sensitivity of an avalanche photodetector. (a) Gain and (b) dark current.

of temperature. If the bias is not increased to compensate, the signal amplitude becomes a function of temperature. Figure 3.8 shows an automatic-gain-control (AGC) function that can be included in APD receivers to compensate for temperature effects.

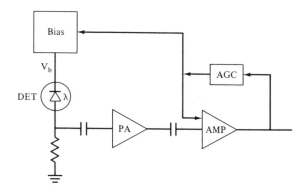

Figure 3.8 Receiver design for an avalanche photodetector illustrating automatic gain control to compensate for thermal drift and received power variations.

3.2.5 Detector Response

There exist two limitations to detector rise time:

1. carrier diffusion and transit time across the surface and depletion regions of the diode;
2. diode capacitance that, in conjunction with load and amplifier input resistance, contributes to the RC time-constant limitation of the receiver. By ignoring the contribution of capacitance, the detector response time, as shown in Figure 3.9, is effected by the response time of the surface (T_s) and the response time of the depletion region (T_d) [4, 5].

$$N_s T_s = [1 - \exp(-\alpha w_s)] \frac{0.95 w_s^2}{D_n} \tag{3-13}$$

$$N_d T_d = [1 - \exp(-\alpha w_d)] \frac{w_d}{v_s} \tag{3-14}$$

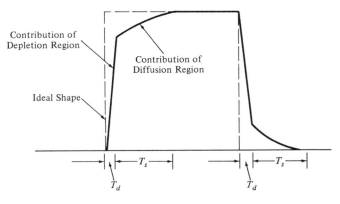

Figure 3.9 Photodetector response characteristic.

where N_s and N_d correspond to the fraction of the light absorbed in the surface and depletion regions, respectively. Terms w_s and w_d correspond to the width of the surface and depletion regions, respectively. D_n is about 30 cm²/s, and v_s, the carrier saturation velocity, is about 10^{-7} cm/s at room temperature. In the depletion region for silicon, the transit time is about 10 ps/μm. Outside this region, it is more like 4000 ps/μm.

The trade-off is between quantum efficiency and speed. The thicker the depletion region, the more light is absorbed. Higher-speed devices require thinner depletion regions and thus have lower quantum efficiencies.

Avalanche devices have an additional delay characteristic, or gain-bandwidth limitation: at low gain, the response time is a function of the ratio of ionization rates; for high-gain values, bandwidth varies inversely to G [4].

Diode capacitance [5] is a function of active area A as well as depletion region width w_d:

$$C_d = \frac{A (E_o E_r)}{w_d} \tag{3-15}$$

where $E_o E_r$ is the product of the permittivity of space and the dielectric constant of the material, which is approximately 1 pF/cm for silicon.

For this reason, active areas are very small for fiber devices (less than 100 μm in diameter), and, generally, the devices come terminated with a fiber stub, or "pigtail," to permit small size with low coupling loss.

Detectors are reverse-biased devices, and the response time and other characteristics depend on the value of the bias voltage. Large (two or three multiples) improvements in response time and capacitance are experienced with increasing bias voltage. Figure 3.10 shows the trend.

3.2.6 Preamplifier-Design Selection

Selection of a preamplifier design depends on the detector chosen, the bandwidth required, the dynamic range necessary, and such practical considerations as cost.*

The key element in preamp design is the amplifying device. The choice is generally between a bipolar transistor or an FET transistor. The bipolar transistor is adequate for low-to-moderate bandwidth designs (up to 50 MHz) or designs employing APDs where receiver noise is dominated by detector gain. A GaAs FET is generally the best-performing device for low-noise high-speed applications, but it is expensive and has an increase in noise density with frequency.

The choice of design also depends on the detector. If an APD is chosen, receiver noise can be dominated by quantum noise in the detector due to avalanche gain. The preamplifier, therefore, does not have to be ultra-low-noise. The design can emphasize bandwidth and low cost at the expense of some noise performance. On the other hand, if a PIN detector or low-gain APD is chosen, the dominant noise is preamplifier-noise current. Design must emphasize low noise in this case, gener-

---

*Acknowledgement and thanks to Tran Muoi of PCO for his contribution of the basic receiver-noise equations in this section.

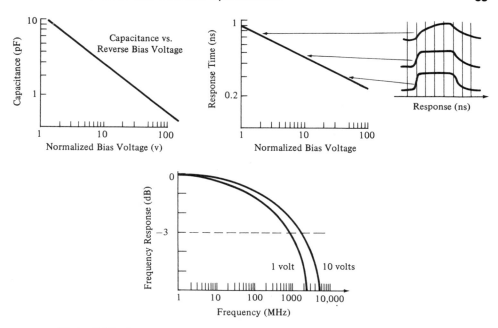

Figure 3.10 Improvement in response time and capacitance with increased bias voltage. The curves are not specific to any one device, but represent the order of magnitude of change that can be observed in practical devices.

ally requiring more expensive devices and possibly complex compensation for front-end bandwidth limitations.

Figure 3.11 shows the three most common preamplifier design choices. The noise performance of each design is a function of thermal noise generated within dominant resistive elements at the input (R_i), or in the feedback loop, and of noise generated within the gates and elements of the front-end transistor (i_{nA}):

$$\langle i_{nA}^2 \rangle = 4kT \, (\text{BW})_n / R_i + \langle i_{nA}^2 \rangle \tag{3-16}$$

where k = Boltzmann's constant = 1.38×10^{-23} J-K^{-1}

T = temperature,

$\langle i_{nA} \rangle$ = amplifier-transistor noise current in amps

The objective is to make R_i as large as possible to reduce thermal noise, but to provide a means for reducing the effect that a large R_i has on frequency response. This is accomplished in different ways for each of the three designs. Figure 3.12, derived from [4] and the relationships in Sections 4.8 and 4.9, compares the relative sensitivity of each design for some typical cases. Chapter 4 provides the analytical relationships for this figure.* Table 3.4 provides a summary of other performance parameters.

*The author wishes to acknowledge Dr. Tran Muoi for the general information upon which this section was developed.

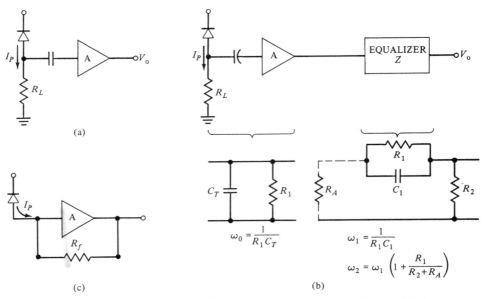

Figure 3.11 Receiver preamplifier design. (a) Voltage amplifier, (b) high-impedance integrating, and (c) transimpedance.

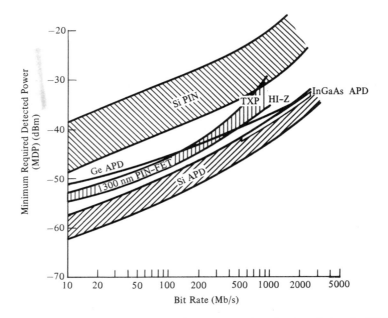

Figure 3.12 Typical receiver sensitivity representative of various designs for 10^{-9} BER [4].

TABLE 3.4 RECEIVER-DESIGN CHARACTERISTICS

Parameter	Voltage	High-impedance integrating	Transimpedance
Receiver bandwidth	$\dfrac{1}{2\pi R_i C_t}$	$\dfrac{G_{eq}}{2\pi R_i C_t}$	$\dfrac{A}{2\pi R_f(C_t + AC_f)}$
Input impedance	R_i	R_i	$\dfrac{R_f}{A}$
Output voltage	$i_p R_i A$	$\dfrac{i_p R_i A}{G_{eq} - 1}$	$i_p R_f$
Dynamic range	$(DR)_{amp}$	$\dfrac{(DR)_{amp}}{G_{eq} - 1}$	$(DR)_{amp}$
PIN	30 dB	15–20 dB	30 dB
APD	30–40 dB	20–35 dB	35–40 dB

where $C_t = C_d + C_a$

      C_f = stray capacitance of amplifier in farads

      G_{eq} = equalizer extension factor

         = $1 + \dfrac{R_1}{R_2 + R_A}$

$(DR)_{amp}$ = dynamic range of the preamplifier, which is a function of load resistance. For a high-impedance amplifier, DR decreases as $1/(BR)^3$ for an FET and $1/BR$ to $1/(BR)^2$ for a bipolar transistor

      R_i = total input resistance in ohms

      R_f = total feedback resistance in ohms

      i_p = photocurrent in Amps

      C_t = total input capacitance in farads

      C_d = detector capacitance in farads

      C_a = amplifier input capacitance in farads

Voltage amplifier.   This is the simplest design. The detector photocurrent flows through a resistor, producing a signal voltage that is subsequently amplified. The output voltage range is a direct function of resistor value and amplifier gain. If a relatively low-noise amplifier is used, noise performance is dominated by thermal noise, which is a direct function of resistor value as shown in Equation (3-16). The higher the resistance, the lower the noise current. The frequency response, on the other hand, is an inverse function of resistor value in combination with detector and amplifier input capacitance. The result is that a low-noise design (high input resistance) implies a poor frequency response.

High-impedance integrating front end.   This design functions as the voltage amplifier does and then compensates for the frequency response with an equalizer network. In order to achieve low noise, the input resistor is made large. The result is a reduced frequency response, therefore, resulting in an integration of the input signal. The first stage is then followed by a differentiating network to restore response. The equalizer is essentially a frequency-dependent gain function (G_{eq}). This is one of the most low-noise designs. The problem is that the equalizer network can be rather complex and the dynamic range limited due to practical limitations of the equalizer amplifier chain.

Transimpedance amplifier. The transimpedance design uses feedback to reduce input impedance. This permits fast response due to the low effective input RC time constant and low thermal noise since R_f can be made large. The result is that the RC time-constant limitation is multiplied by amplifier gain. The signal output is a function of the size of the feedback resistance. The transimpedance amplifier has a wide dynamic range, but is limited in noise performance or frequency response by the gain-bandwidth limits of practical transistor devices.

3.2.7 Effect of Signal Distortion

For digital receivers, pulse distortion (percentage of pulse energy in the bit period) and extinction ratio (relation of energy in a high state to a low state) have a direct effect on receiver sensitivity. In an analog transmission system, the optical modulation index M_o has a direct effect on the receiver signal-to-noise ratio. The effect of M_o is handled directly in the analytical relationship for receiver MDP; however, the other effects are dealt with as power penalties when the link-budget analysis is performed.

3.2.8 Temperature Effects

Temperature has an effect on the following parameters, which in turn effect MRP:

(a) detector leakage current I_d increases with temperature;

(b) thermal noise in resistive elements increases as a direct function of temperature;

(c) APD gain changes significantly as a function of temperature (Figure 3.7) and must be compensated with automatic gain control;

(d) other detector and active element parameters such as response time, gain, capacitances, etc. can be effected;

(e) temperature can cause drift in bandwidth-determining components, thereby increasing signal distortion, intersymbol interference, or noise bandwidth;

(f) temperature can cause references to shift in gain control, bias, and threshold-detection circuits; these shifts can have a more significant effect on MDP than that of the previous contributors if receiver design is not of high quality;

(g) temperature can effect the physical fiber-to-device coupling, but the effect is usually negligible.

If data are not available to quantify thermal drift, it would be advisable to leave approximately +0.05 dB/°C MDP variation as a margin for detector/preamplifier sensitivity change. If the MDP is being determined from an analytical model, as given in Chapter 4, the total effect of temperature can best be determined by recalculating at each temperature extreme with the appropriate component parameter values.

Sensitivity change as a result of temperature variation can be related as a power penalty as well. In cases where dark current dominates the receiver sensitivity, MDP decreases approximately as the square root of I_d and the power penalty P_p is

$$P_p \text{ (dB)} = 5 \log \frac{I_{d2}}{I_{d1}} \tag{3-17}$$

In cases where thermal or amplifier noise dominates, the approximate relationship is

$$P_p \text{ (dB)} = 5 \log \frac{T_2}{T_1} \tag{3-18}$$

where T is the temperature in degrees Kelvin.

3.2.9 Effects of Aging

In the optical receiver, there are no dominant age-dependent components with the exception of power regulators. The detector itself, being reverse-biased, is essentially a passive component. Aging effects most likely come from absorbed humidity, contamination, or other such effects that cause drift in reference or bandwidth-determining components. The effects should be minimal with proper design and environmental protection. Receiver failure due to optical component failure is generally due to excess optical power on the detector. The photocurrent combined with the bias voltage across the detector determine the required thermal power to be dissipated at the detector. If excessive, the detector will be damaged. Where such conditions can exist, for example, in a military nuclear environment, current limiting is designed into the detector-bias source to prevent burnout.

3.2.10 Detector Coupling Loss

As indicated in Section 3.2.2, the coupling loss between the fiber, or fiber pigtail, and the detector effects the value of MRP. Figure 3.2 shows the loss mechanism. The analytical relations for the two cases, "butt" coupling and coupling with some separation, follow.

Butt coupling.   If the fiber is in contact with the detector, then the loss mechanisms consist of the difference in detector diameter D_d and fiber core diameter D_f and the fiber-to-detector reflection R_{fd} due to the difference in fiber refractive index n_f and detector refractive index n_d.

$$L_d \text{ (dB)} = 20 \log \frac{D_d}{D_f} + 10 \log (1 - R_{fd})$$

$$R_{fd} = \frac{(n_d - n_f)^2}{(n_d + n_f)^2} \tag{3-19}$$

The refractive index of fused silica fiber is about 1.46. For silicon, it is about 3.4, and for InP, about 3.8. Make sure that the reflection loss is not double counted if it is already included in the external quantum efficiency of the detector.

Coupling with separation. If the fiber is separated from the detector slightly, as can be the case with a connected detector with no pigtail, then the loss mechanisms must include the effect of the emission cone of the fiber and the added reflection of the fiber-to-air (R_{fa}) and air-to-detector (R_{ad}) interface. This reflection can be reduced by using an index-matching compound or epoxy if it is practical to do so. The loss mechanisms include the larger diameter of the irradiation caused by the emission half angle (θ) and the separation distance (d), and the added reflection due to the refractive index of the air or other surrounding medium (n_0). For air, n_0 is 1.

$$L_d = 20 \log \frac{D_d}{2d \tan \theta + D_f} + 10 \log (1 - R_{fa}) + 10(1 - R_{ad})$$

$$R_{fa} = \frac{(n_f - n_0)^2}{(n_f + n_0)^2} \tag{3-20}$$

$$R_{ad} = \frac{(n_d - n_0)^2}{(n_d + n_0)^2}$$

The air-to-detector reflectance can be reduced to a minimal amount with a $\frac{1}{4}$-wavelength antireflection coating. Again, make sure that the detector reflectance is not double counted if included in the responsivity.

3.3 SOURCE AND TRANSMITTER SELECTION*

The selection of a transmitter design is based on signal bandwidth and format. For analog transmission, the focus is on the linearity of the source and of the drive electronics in order to achieve the lowest in-band harmonic distortion products for the highest transmitted optical power. For digital transmission, sources are selected for speed and low extinction ratio, and transmitter drive characteristics are matched to the source in order to optimize response.

From a link-power-budget standpoint, the key parameter is the optical power coupled into the fiber, which we call transmitted power (P_t). Other source parameters effecting the link-power-budget calculations are total harmonic distortion or signal-to-distortion ratio (SDR), extinction ratio (ER), modulation index (M_o), and source noise, since these either impose limitations on end-to-end link signal-to-noise performance or limit the amount of available optical power that can be used for the signal.

From the standpoint of the link-bandwidth budget, the key parameter is the source rise time $t_{r,\,source}$ for digital transmission and source bandwidth $(BW)_s$ for analog transmission. The term $t_{r,\,source}$ is used to roughly describe fall time as well.

3.3.1 Optical-Source Specification

For fiber-optics transmission, there exist three types of semiconductor sources: light-emitting diodes (LEDs), injection-laser diodes (ILDs), and superradiant diodes

---

*Acknowledgment and thanks to C. J. Hwang of General Optronics for his contributions to and support of this section.

(SRDs). They are all P–N junction devices. The output-power-versus-drive-current characteristics for each device is shown in Figure 3.13 and some typical source parameters are provided in Table 3.5.

The devices are all heterostructures whereby the forward drive current injects holes and electrons into the active region with conduction band (quasi-Fermi) energies E_e and E_h that are a function of applied voltage V and electron charge e [10]:

$$E_e - E_h = eV \qquad (3\text{-}21)$$

These carriers eventually recombine after some carrier lifetime in a combination of radiative and nonradiative fashion. The radiative recombination generates a photon with energy equal to the energy difference across the band for the recombining holes and electrons:

$$h\upsilon = E_2 - E_1 \qquad (3\text{-}22)$$

where υ is the frequency, and h is Planck's constant. The recombination process is described by three processes: spontaneous emission, stimulated emission, and absorption. The recombination of carriers can be independent and spontaneous and the resultant radiation can either be absorbed or stimulate another electron-hole pair to combine, creating radiation in phase with the original stimulating radiation. Under high-current injection conditions, a state of population inversion occurs, where $E_2 - E_1 < E_e - E_h$, and the stimulated rate becomes higher than the absorption rate, resulting in an optical-gain mechanism in the active region.

The LED.   For the LED, current injection is low and spontaneous emission dominates, and, therefore, light output is almost proportional to input current. Two basic device structures exist, the surface emitter and the edge emitter. These are represented in Figure 3.16. The surface-emitter emission area is small in order to mate to the fiber and has a Lambertian beam pattern, therefore, emitting its power in a 180-degree angle as $P = P_o \cos \theta$. Variations of the surface emitter, such as the

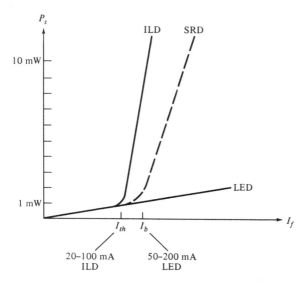

Figure 3.13 Typical transfer characteristics for LEDs, SRDs, and laser diodes.

TABLE 3.5 PERFORMANCE TYPICAL OF OPTICAL SOURCES

	LEDs			Lasers			
	Class	800–850	1300	Class	800–850	1300	1500
Wavelength		800–850	1300		800–850	1300	1500
Material		GaAlAs	GaInAsP		GaAlAs	GaInAsP	GaInAsP
Spectrum width (nm)		30–60	50–150	MM	1–2	2–5	2–10
Mode spacing (nm)				MM	0.3	0.9	0.13
Line width SM (MHz)				SM FP		150	150
				DFB		10–30	10–30
				EC		1–10	1–10
				EG		0.002–1	0.002–1
Structure					PI, CSP, BH	BH	BH
Output power (mW)		0.5–4.0	0.4–0.6	CSP	20–50		
				BH	2–8	1.5–8	1.5–8
Coupled power (mW)							
100-μm core — Surface		0.1–1.5		MM		0.5–7.0	
100-μm core — Edge		0.3–0.45	0.04–0.075				
50-μm core — Surface		0.01–0.05	0.015–0.035	MM	0.5–0.25	0.4–3.0	0.5
50-μm core — Edge		0.05–0.13	0.03–0.06	SM	1.5–3		
Single-mode — Edge			0.003–0.03	MM		0.25–1	0.25–0.8
Extinction ratio					25:1	25:1	25:1
Drive current (ma)		50–150	100–150	CSP	40–80		
				BH	10–40	25–130	
Rise time (ns) — Surface		4–14					
Rise time (ns) — Edge		2–10	2.5–10	BH	0.3–1	0.3–0.7	0.3–0.7

TABLE 3.5 PERFORMANCE TYPICAL OF OPTICAL SOURCES (*cont.*)

Parameter	Type	LEDs		Type	Lasers		
Modulation Frequency (GHz)		0.08–0.15	0.1–0.3	BH	2–3	2–3	2–3
Temperature Drift							
Wavelength nm/°C		0.3	0.6		0.15–0.2	0.3	0.9
Power, %/°C		−0.2, −0.5	−0.9				
Threshold current, %/°C					0.8	1.6–2	3
Linearity							
2nd harmonic		−30 to −40 dB @ $M_o = 0.5$			−40 to −55 dB		
3rd harmonic		−35 to −40 dB @ $M_o = 0.5$			−50 to −70 dB		
Beam width (half)							
Parallel	Surface	120–180°		CSP	5°		
Perpendicular	Surface	120–180°		CSP	10–25°	10–30°	10–30°
Parallel	Edge	180°		BH	10–25°		
Perpendicular	Edge	30–70°		BH	20–35°	30–40°	30–40°
Lifetime (Million hours)		1–10	50–1000		1–10	0.5–50	0.5–50

MM = Multimode
SM = Singlemode
FP = Fabry Perot
DFB = Distributed Feedback
EC = External Cavity
EG = External Grading
CSP = Channel Substrate Planar
BH = Buried Heterostructure
PI = Proton Implantation

etched well and Burrus [14] have been developed in order to reduce absorption of light and bring the fiber closer to the active region. The edge emitter uses a stripe-geometry active region similar to the laser, however, the emission facets are sawed rather than cleaved so that the mirrored cavity is not created. The emission pattern in the vertical direction is Gaussian with an emission angle determined by the width of the active region and the refractive-index differences in the structure. In the parallel direction, the pattern is Lambertian.

As compared to a laser, the LED has a relatively large emission angle, slower response, lower output power, and a broader spectral width, and is therefore used in lower-speed applications (1 to 100 MHz) with large-core or 50-μm core multimode fibers where distances are short (1 to 10 km).

The laser diode.   In the ILD, injection levels above threshold are high such that gain is greater than loss, and the surface is cleaved to form a mirrored cavity, thus reflecting about 30% of the light back into the cavity to create more stimulated emissions. Below the threshold current, the ILD acts as an LED because spontaneous emission dominates.

The ILD has a narrow emission angle, fast response, high output power, and a very narrow spectral width, and is generally used for high-speed (50 MHz to 10 GHz) applications with 50-μm core multimode or single-mode fibers where distances are more demanding (10 to 100 km).

The SRD.   The SRD performs somewhere between an LED and an ILD. It has an injection level such that the gain is greater than the loss, but does not have the mirrored cavity; therefore, it produces a mixture of spontaneous and stimulated emissions. It is used in applications where the multimode properties are necessary to reduce speckle modal noise in multimode fiber, but where high coupling efficiency, high output power, and narrow spectral width are required.

3.3.2 Transmitter Design

Given the required signal to be transmitted and the selection of the optical source, the transmitter is designed to

(a) gain adjust and filter the input signal in order to match to the source dynamic range and eliminate out-of-band signals;

(b) convert the analog or digital signal voltage to a source drive current;

(c) control bias and signal drive-current levels as appropriate to drive the source within the desired operating range;

(d) provide optical feedback as a means of controlling bias current as temperature and signal patterns change or as the source degrades with time; and

(e) where necessary, provide thermal control by mounting the source on a thermoelectric cooler to stabilize thermal drift with case temperature and to improve source lifetime.

Figures 3.14 and 3.15 show basic transmitter designs for various signal conditions and source choices. For LED transmitters (Figure 3.14), generally little or no bias-current compensation is used to stabilize power variations with temperature or time. In the case of long-wavelength LEDs, where environmental stability is sometimes a problem with the more sophisticated transmitters, thermoelectric cooling and bias control are added, similar to that of the laser drivers in Figure 3.15.

LED transmitters for analog transmission are simply linear voltage-to-current converters. Feedback can be added to the driver to achieve further drive-current linearity. Sometimes nonlinear matching networks are added to the feedback loop so that the drive-current transfer function compensates for nonlinearities in the source transfer function. With high-speed transmitters (20 MHz and above), it is more difficult to design a high-speed linear current source, so an amplifier, such as a 50-ohm RF drive amplifier, is often used, which is then matched for impedance and nonlinearity to the LED.

When an LED is chosen and the transmitted signal is digital, the transmitter is simply a current switch. Current is supplied to the LED in the "on" state at a value necessary to achieve the required peak optical power $P_{t,\text{pk}}$. The "off" power state is achieved by switching the current to the LED "off." Figure 3.15(b) illustrates a simple realization with a shunt-current path provided by a transistor switch. This is adequate for slow designs (10 Mb/s). Faster designs usually require differential current switches.

When the device chosen is a laser, then temperature compensation and bias-current compensation are required, as shown in Figure 3.15, in order to achieve stability and device reliability with time and temperature. The source is generally mounted on a thermoelectric cooler along with a temperature sensor and a back-facet optical detector in order to sense changes in operating temperature and output optical power.

The back-facet detector uses the optical power emanating from the back of the laser to monitor mean or average optical power and therefore adjusts bias or threshold levels to maintain the proper operating range. With digital transmission, there is often a desire to make minor adjustments to bias as a result of changing signal characteristics in order to optimize the extinction ratio or the laser turn-on delay (refer to Chapter 6). The input signal is integrated and the resulting error signals are then summed into the bias-compensation network.

3.3.3 Coupled (Transmitted) Power (P_t)

Transmitted coupled power P_t is generally expressed in terms of average or rms coupled power. An average coupled-power specification generally assumes a power coupled into the aperature of the mating fiber or fiber "pigtail," with the source either biased at the midpoint of its intended operating range or, in the case of a digital transmitter, a 50% duty cycle or pseudorandom square-wave signal is input.

Transmitted power P_t is a function of source power P_s and source-to-fiber coupling loss L_i. Source power is the total power emitted by the source within the

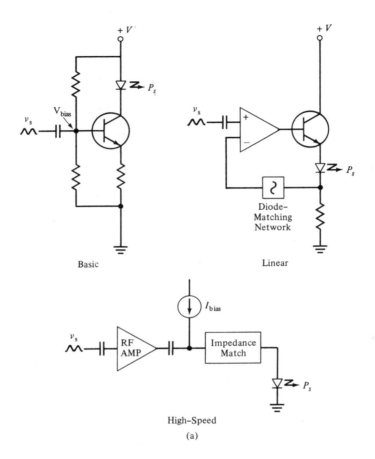

Basic

Linear

High–Speed

(a)

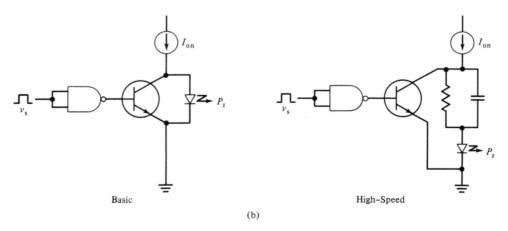

Basic

High–Speed

(b)

Figure 3.14 Light-emitting diode (LED) transmitter-design configurations. (a) Analog transmission and (b) digital transmission.

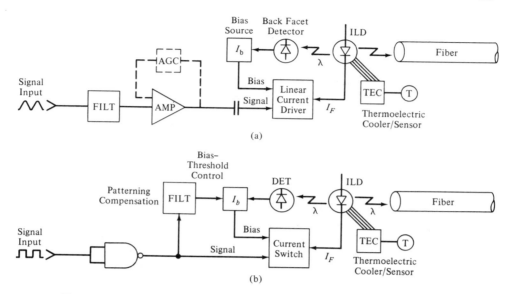

Figure 3.15 Laser-diode transmitter-design configuration. (a) Analog transmission and (b) digital transmission.

source-specific emission cone. Injection coupling loss is the loss of optical power experienced when coupling a fiber to the source. The relationship is shown in Figure 3.16 and can be summarized as follows:

$$P_t \text{ (dBm)} = P_s \text{ (dBm)} - L_i \text{ (dB)} \tag{3-23}$$

$$L_i \text{ (dB)} = 10 \log \frac{P_t}{P_s} = 10 \log \eta_c \tag{3-24}$$

where η_c is the coupling efficiency.

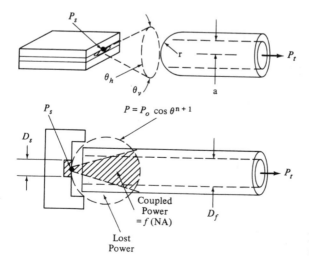

Figure 3.16 Relationship between source power and coupled power.

Generally, the source is supplied precoupled to a "pigtail" fiber or is specified as coupled into a specific fiber aperature. In this case, L_i is accounted for in the values given for P_t.

Source power.   Source power P_s is dependent on the device type: LED, SRD, or ILD (Figure 3.13); design of active region for speed versus power; device material and structure; forward drive current; signal format; modulation index and extinction ratio; and case temperature and age with respect to operating life. Table 3.5 compares various device classifications.

Coupling loss.   The coupling efficiency between the device and the fiber is dependent on the spatial properties of source emission, the numerical aperature, modal properties, and core size of the fiber, and the relative geometric relationship of the source and fiber. The relationship is shown in Figure 3.16. In addition to basic coupling efficiency, L_i is dependent on such practical considerations as coupling accuracy and device tolerances; mechanical drift with time and temperature; and source spatial-mode patterns that can change with time, temperature, and bias conditions.

For an LED, the output-power emission pattern is defined relative to peak power P_o by [15]:

$$P = P_o \cos \theta^n \tag{3-25}$$

and for a Lambertian source, where $n = 1$:

$$P = P_o \cos \theta \tag{3-26}$$

For a butt-coupled LED, the power coupled into a fiber of acceptance angle $\theta_f = \sin^{-1} NA$ and core diameter D_f is, from Figure 3.16:

$$P_t = P_s(1 - R_{sf})\frac{D_f}{D_s}[1 - (\cos \theta_f)^{n+1}] \tag{3-27}$$

where R_{sf} is the reflectance at the source-to-fiber interface, D_s is the source-emission-area diameter, and n is the exponent determining the shape of the emission. If NA is very small, then this can be approximated as [15]:

$$P_t = P_s(1 - R_{sf})\frac{D_f}{D_s}\frac{n+1}{2}(NA)^2 \tag{3-28}$$

If the source is Lambertian ($n = 1$), then the following holds:

$$P_t = P_s(1 - R_{sf})\frac{D_f}{D_s}(NA)^2 \tag{3-29}$$

Note that all three formulas assume $D_f < D_s$. If $D_f > D_s$, make the ratio $D_f/D_s = 1$.

If the source is a laser with semidivergent angles of θ_{dv} and θ_{dh} (vertical and horizontal, respectively) at the $1/e$ points, then the relationship becomes [10]

$$P_t = P_s(1 - R_{sf}) \, \text{erf}\left(\frac{\tan \theta_a}{\tan \theta_{dv}}\right) \text{erf}\left(\frac{\tan \theta_a}{\tan \theta_{dh}}\right) \tag{3-30}$$

where θ_a is the maximum acceptance half angle as dictated by fiber position and core radius (a), and if the fiber is lensed, the radius of curvature (r). In Figure 3.17, C. J. Hwang has plotted the $\text{erf}(\tan \theta_a / \tan \theta_d)$ as a function of the divergent angle θ_d for various a/r using typical core and cladding refractive indexes of 1.454 and 1.444, respectively. For a Lambertian source, for example, $\theta_d = \cos^{-1} 1/e = 69°$. From Figure 3.17, for an a/r of 0.84, erf ( ) = 27%. Therefore, if we assume an air-to-fiber reflectance of 3.5%, we get:

$$P_t = P_s(1 - 0.035)(0.27)(0.27) = 7\%$$

This is an improvement of about a factor of 2 over a nonlensed approach.

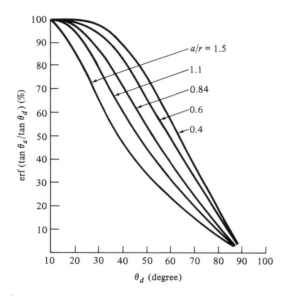

Figure 3.17 Lensed-fiber coupling-efficiency relationships. The author wishes to acknowledge and thank Dr. C. J. Hwang of General Optronics [10] for providing this diagram in support of this section.

3.3.4 Source Response

Figure 3.18 shows some typical response characteristics for optical sources. Table 3.5 also provides some response-time values for typical product.

There are two primary mechanisms within the source that contribute to limiting source response: (1) carrier lifetime (τ) and (2) capacitance-dominated internal impedance. If the injection-current modulation varies at a rate faster than carrier lifetime, obviously the light output will not respond to the modulation frequency.

In LEDs, carrier lifetime can be expressed as [8]

$$\frac{1}{\tau} = \frac{1}{\tau_{rr}} + \frac{1}{\tau_{nr}} + \frac{2v}{d} \qquad (3\text{-}31)$$

where τ_{rr} = radiative lifetimes
$\quad \tau_{nr}$ = nonradiative lifetimes
$\quad d/2v$ = ratio of diffusion length to recombination velocity (range of 50 ns)

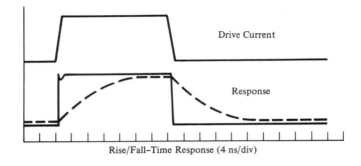

Drive Current

Response

Rise/Fall–Time Response (4 ns/div)

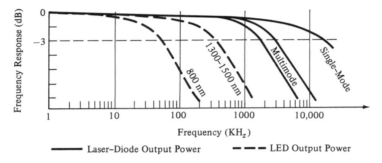

Figure 3.18 Typical response characteristics for LEDs and laser diodes.

The response is therefore a function of carrier lifetime:

$$P_w = \frac{P_0}{(1 + w^2\tau^2)^{1/2}} \tag{3-32}$$

The 3-dB optical bandwidth is approximately [8.9]

$$(BW)_s = \frac{\sqrt{3}}{2\pi t} = \frac{\sqrt{3}}{2\pi\eta_i\tau_{rr}} \tag{3-33}$$

where η_i is the internal quantum efficiency $= \tau/\tau_r = 1/(1 + \tau_r/\tau_{nr})$.

Radiative lifetime τ_{rr} can run between 1 to 100 ns, inversely proportional to doping concentration. The trade-off is usually between source output power and speed. Higher-speed devices generally have smaller active regions and higher dopant concentrations and thus less power output, as illustrated in Figure 3.19.

A model of the LED internal impedance is shown in Figure 3.20 along with a compensation network. Linke [11] indicates that the internal capacitance associated with junction capacitance C_j and parasitic capacitances C_p, together with junction resistance R_j and series resistance R_s, limit source response. Linke indicates that the shape of the roll-off curve for many lasers fits that of a simple RC network. C_j and R_j vary with forward current and have values of about 60 pF and 1 ohm, respectively. Total R_s is estimated at between 5 to 10 ohms.

According to Baird [12], it is possible to increase the rise time of LEDs by a factor of 2 to 4 by using current-peaking circuitry similar to that shown in Figure

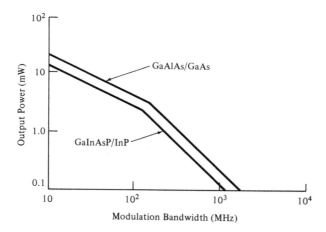

Figure 3.19 Power versus bandwidth limits of LEDs. The author wishes to acknowledge and thank Dr. C. J. Hwang of General Optronics [10] for providing this diagram in support of this section.

3.20. By focusing on the internal mechanisms, we see that C_j is a slowly varying function of forward current I_f. It, and C_p as well, can be compensated somewhat by driving additional current through the device when turning it on and off, as shown by the $I_f(t)$ waveform. Baird gives the 10 to 90% rise time (t_r) for an LED as follows:

$$t_r = \left(\tau + \frac{4kTC_s}{eI_f} \right) \ln 9 \qquad (3\text{-}34)$$

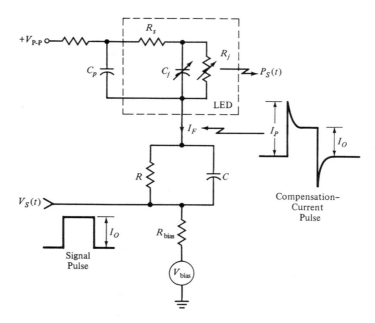

Figure 3.20   LED speed-up circuitry.

where I_f = forward current
τ = carrier lifetime
C_s = total source capacitance at I_f

With the circuitry shown in Figure 3.20, the rise time becomes

$$t_r = \left(\frac{4kTC_s}{eI_p}\right) + \left(\tau - \frac{4kTC_s}{eI_p}\right) \ln \left(\frac{1 - I_o/10I_p}{1 - 9I_o/10I_p}\right) \qquad (3\text{-}35)$$

For no current peaking, $I_p = I_o$, then $t_r = 2.2\tau$. For a current peak of $I_p = 2I_o$, for example, t_r is approximately 0.55τ for small values of C_s.

3.3.5 Source-Wavelength Characteristics

Optical sources have a spectral characteristic defined by a center wavelength, spectral width, and a number of spectral modes or "lines" in the case of the laser. Figure 3.21 illustrates the typical spectral characteristics of LEDs and ILDs.

The spectral properties of the source affect the following design decisions:

(a) the device mean wavelength must match the detector-sensitivity characteristics;
(b) the mean wavelength, spectral width, and drift characteristics must match the selectivity characteristics of wavelength-selective couplers and filters;
(c) the mean wavelength determines fiber attenuation;
(d) the mean wavelength and spectral width determine fiber-material dispersion.

The selection of a source for most noncoherent transmission systems initially focuses on the effect that center wavelength has on fiber attenuation and the effect spectral width has on fiber-material dispersion or bandwidth.

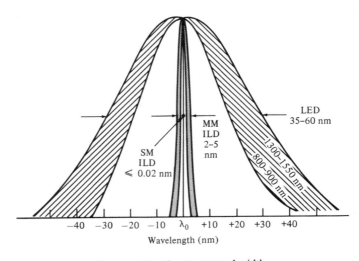

Figure 3.21   Source spectral width.

LEDs.   The spectrum of an LED is essentially a band of incoherent spectral sources centered around some mean that is specified as device center wavelength. The full-width half maximum (FWHM) spectral width is quite large, 3 to 6% of the mean wavelength within the 850-nm range and 6 to 12% at longer wavelengths.

Multimode laser-diode spectrum.   For lasers (ILDs), the device can operate within a single primary spectral line or a few discrete lines, depending on design. These spectral lines are dictated by the number of longitudinal oscillation modes permitted by the cavity structure of the ILD as determined by the length of the ILD cavity and the effective refractive index. The cleaved laser chip itself forms a Fabry-Perot "mirrored" cavity since the refractive index of the material relative to air provides about a 30% reflectivity. The approximate mode spacings in multimode Fabry-Perot ILDs (MM FP) for various wavelength ranges are given in Table 3.5* with a natural width of about 5 to 10 MHz with the natural cleaved mirror. The table also indicates the spectral width for various design classes. Typical telecommunications laser product specifies a spectral width between 3.5 and 30 n, with 5 n being most common for high-speed transmission systems.

Single-mode laser-diode spectrum.   Single-mode lasers also have a finite spectral width that depends on how well the modes can be defined by the design of the active region. CSP (Channeled Substrate Planar) lasers exhibit a single primary mode, but suffer from unpredictable longitudinal mode hopping. Buried heterostructure (BH) devices offer the degree of transverse mode stability required for single-mode fiber coupling. As indicated in Table 3.5, depending on design, the spectral width of the primary mode can vary between a few nanometers to less than a megahertz. For coherent optical communications, spectral confinement in the megahertz region is required because the optical carrier itself is amplitude, phase, or frequency modulated and treated as a coherent carrier. Kobrinski and Brackett [16] provide line-width performance characteristics for the various narrow-line laser structures listed in Table 3.5. The external cavity structures are formed by placing a reflecting cavity or a grating external to a BH laser to select and narrow a primary longitudinal mode.

Spatial properties.   Whereas the longitudinal modes determine the spectral properties, the transverse mode determine the near-field (on the laser surface) and far-field patterns (Fourier transform of near field). There are two sets of transverse modes, vertical and lateral. The patterns in the vertical direction are determined by the thickness of the active layer and the refractive index difference between it and the confinement layers. It is nearly always Gaussian in shape. In the lateral direction, the pattern is structure dependent and varies from Gaussian to somewhat rectangular [10]. In order to couple effectively to single-mode fibers, the transverse mode has to be stable. The buried heterostructure provides this transverse mode stability.

---

*The author wishes to acknowledge and thank Dr. C. J. Hwang of General Optronics [10] for providing some of the information for this table.

3.3.6 Temperature Effects

Output power and the center wavelength of semiconductor sources vary significantly as a function of temperature. Figures 3.22 and 3.23 show the tendency. Degradation is greater for devices operating at the longer wavelengths.

For LEDs, the power variation is almost a linear function with temperature.

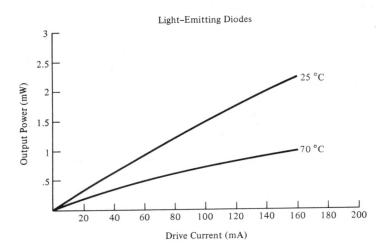

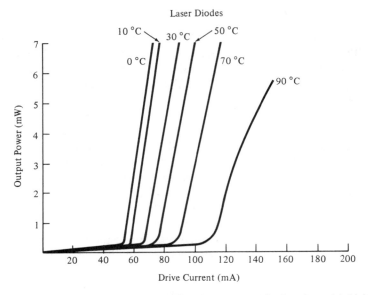

Figure 3.22 Temperature effects on optical-source transfer functions. (a) Light-emitting diodes and (b) laser diodes.

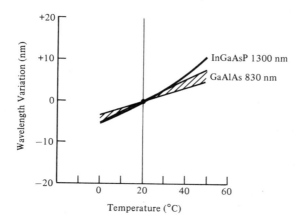

Figure 3.23 Laser-diode variation in center wavelength with temperature.

For commercial devices, the temperature coefficients are observed to be on the order of

−0.2 to −0.5%/°C for 850-nm LEDs

−1%/°C for 1300-nm LEDs

In order to compensate, automatic gain control (AGC) is required at the receiver or temperature-tracking and bias/gain feedback is provided at the transmitter.

With lasers, although some reduction in slope efficiency is experienced, the primary concern is drift in the lasing threshold. The temperature coefficient for threshold current is expressed by the relationship*:

$$I_{\text{TH}}(T) = I_o \exp (T/T_o) \tag{3-36}$$

where T_o is a characteristic temperature, and I_o is a constant particular to the device.

The transmitter bias-control circuitry must track the threshold drift or the laser will shut down or be destroyed. This is generally done with optical feedback, which maintains a constant average power output. Compensating by increasing bias or signal current, however, can drastically shorten life as more drive current produces more heat and thus a further need for drive current. For this reason, most laser transmitters and nearly all long-wavelength transmitters incorporate integral solid-state coolers with the devices to stabilize them. The transmitted power, with such cooling and active feedback, is stable to the limits of the error and sensitivity of this circuitry, and the mechanical stability of the fiber and bias-detector coupling. This can be on the order of about 0.2%/°C.

An added reason for temperature control is that changes in slope efficiency can effect signal power even if bias control is maintained. The wavelength shift can also effect the fiber attenuation and have a major effect on link performance if wavelength-sensitive couplers are involved.

*The author wishes to acknowledge and thank Dr. C. J. Hwang of General Optronics [10] for providing these materials in support of this section.

3.3.7 Degradation Over Time

There are a number of failure mechanisms in semiconductor sources. Some cause rapid catastrophic failure and others cause degradation in performance over time. The gradual variations over time must be accommodated in the systems design. With time, the LED and laser transfer characteristics act very similar to the effects with temperature. In computing the link-power budget, a 1- to 3-dB power margin (depending on your definition of end of life) should be provided to account for these effects.

For a laser, the threshold increases over time and often the slope efficiency decreases slightly as with the temperature curves in Figure 3.22. In addition, the modal properties may change, effecting the coupling. For fully integrated transmitters, threshold and average power are stabilized by the control circuitry, leaving only the smaller, more unpredictable variations. Another problem with laser aging might be increased nonlinearity for analog signaling and increased extinction ratio in binary transmission.

LED and laser lifetime is generally determined from the extrapolation of statistics of accelerated aging tests performed at elevated temperatures in the 70-to-100°C range.

The "end of life" is generally defined as a reduction of 1 dB in power to 3 dB, depending on convention. Room-temperature lifetimes on the order of 1 to 50 million hours for lasers [17, 18] and to over 1 billion hours for LEDs [10] are predicted by extrapolation using the Arrhenius law.

For semiconductor sources, the output intensity variation from initial value I_o with time, $I(t)$, is a function of some degradation constant D_o, activation energy E_a, temperature T, and time t as described by the formula [10, 18]:

$$I(T) = I_o \exp\left[-D_o \exp\left(-\frac{E_a}{kT}\right)t\right] \qquad (3\text{-}37)$$

The extrapolation of lifetime at room temperature from high-temperature tests uses the Arrhenius law [18] that relates the degradation rate D_2 at a temperature T_2 to the measured degradation rate D_1 at test temperature T_1 and activation energy E_a:

$$D_2 = D_1 \exp\left[\frac{-E_a}{k}\left(\frac{1}{T_2} - \frac{1}{T_1}\right)\right] \qquad (3\text{-}38)$$

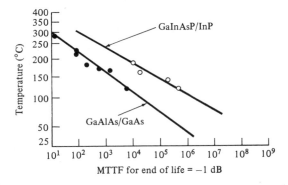

Figure 3.24 LED lifetimes extrapolated for high-temperature tests. The author wishes to acknowledge and thank Dr. C. J. Hwang of General Optronics [10] for providing this diagram in support of this section.

For example, degradations for GaAlAs run in the range of 0.09%/kh at 30°C to 0.8%/kh at 60°C. E_a ranges from 0.56 to 0.65 eV for GaAlAs/GaAs and 0.85 to 1.0 eV for GaInAsP/GaAs [10]. Lifetimes of 10^{+6} to 10^{+7} hours for short-wavelength LEDs and $3 \times 10^{+8}$ to $5 \times 10^{+9}$ hours for long-wavelength devices have been extrapolated in this fashion. Figure 3.24 shows results for two classes of LED.

3.4 FIBER SELECTION

This section discusses the selection of fiber performance in the cabled state.*

3.4.1 Selection of Fiber Type

Fiber specification is based on the optimization of optical power coupled at the transmiting end as well as optical power available to the detector at the receiving end, while preserving the necessary bandwidth to pass the modulating signal with minimal distortion. Performance typical of cabled fiber product is summarized in Table 3.6. The classification is per ANSI/EIA-4920000-A-1987 [20]. (The data are compiled from product literature and information provided by David Charlton [19]. Figure 3.25 shows some of the standard and more common fiber core and cladding dimensions.

The following parameters must be considered in the trade-off analysis associated with fiber specification:

(a) Attenuation: At the chosen wavelength, attenuation must be low enough to transmit the required optical power end to end, allowing margin for connector loss, splice loss, coupling loss, and systems operational considerations. This is generally a function of the operating wavelength and class of fiber.

(b) Source coupling loss: The core size, numerical aperature, and modal properties must be such so as to minimize coupling loss at the source. The larger the core and the numerical aperature, and the greater the number of propagating modes, the more light is coupled. Unfortunately, these conditions are somewhat in opposition to those for high bandwidth and low attenuation.

(c) Detector coupling loss: The core size and numerical aperature must be small enough so that most of the light exiting the fiber falls on the detector. This condition must be traded against source coupling loss and against detector performance, i.e., smaller detectors have faster response and lower noise contribution.

(d) Splice loss: The tolerances of core concentricity, diameter, and ovality should be high enough to minimize joining loss. A large core size and a large numerical aperature reduce the effect of poor tolerances. These conditions are in opposition to those required for wide bandwidth and low loss, however.

(e) Connector loss: As with splicing loss, the tolerances on the core parameters, and the core with respect to cladding, must be high. The diameter of the outer cladding must also be carefully controlled since this is the reference surface for con-

*The author wishes to thank David Charlton of Corning Glass Works [19] for information provided in support of this section.

TABLE 3.6 TYPICAL PERFORMANCE CHARACTERISTICS OF CABLED FIBER

| General class | Multimode | | | | | | | Single mode | | |
EIA class	IA and IB				IC	II	III	IVA	IVB	IVC
Index descriptor	Graded and Quasigraded				Step	Step	Step	Dispersion unshifted	Dispersion shifted	Dispersion flat
Core material	Glass				Glass	Glass	Plastic	Glass	Glass	Glass
Cladding material	Glass				Glass	Plastic	Plastic	Glass	Glass	Glass
Profile (g)	1–3 graded, 3–10 quasigraded				>10	>10	>10	N.A.	N.A.	N.A.
Core dia. (μm)	50	62.5	85	100	50–100	200–600	484–980	8.7–10	7–8.7	7–8.7
Clad dia. (μm)	125	125	125	125–140	125–140	230–650	500–1000	125	125	125
Tolerance:										
Core dia.	±3 μm	±3 μm	±3 μm	±4 μm	±8 μm	±10 μm		±8%	±8%	±8%
Concentricity	<6%	<6%	<6%	<6%	<6%	<10%		<1 μm	<1 μm	<1 μm
Clad dia.	±2 μm	±3 μm	±3 μm	±4 μm	±10 μm	±10 μm		±2 μm	±2 μm	±2 μm
Attenuation (dB/km)										
@ 570 nm							70			
@ 650 nm							130–160			
@ 850 nm	2.6–3.5	3.0–4.1	3.0–4.1	3.0–7.0	4.0–6.0	3.0–8.0				
@ 1310 nm	0.7–1.6	0.8–1.8	1.0–1.8	1.5–5.0				0.4–0.7	0.25–0.3	0.4–0.5
@ 1550 nm									0.1	0.25–0.3
@ 2–5 μm predicted										
Numerical aperture	0.19–0.25	0.24–0.3	0.27–0.31	0.21–0.3	0.15–0.3	0.27–0.37	0.47			
Material dispersion (ps/nm/km)										
@ 850 nm	100–120	100–120	100–120	100–120	100–120	100–120		N.A.		
@ 1300 nm	0.9–3.5	3.0–10	3.0–10	3.0–10	3.0–10			0.9–4.0		3.5
@ 1550 nm								20	3.5	3.5
(BW)$_0$ (MHz-km)										
@ 850 nm	200–600	150–350	150–500	20–500	10–60	9–25	0.5	10^{+5}	10^{+5}	10^{+5}
@ 1300 nm	400–1500	300–1000	300–1000	20–400						

N.A.: not applicable

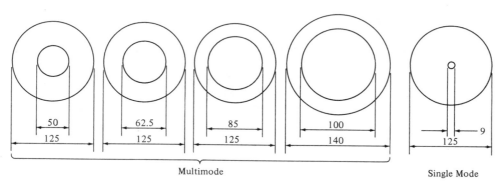

Figure 3.25   Fiber dimensional standards. (Note: All dimensions are in micrometers.)

nector coupling. As with splices, a large core and a large numerical aperature reduce the effect of poor tolerance but limits bandwidth and attenuation performance.

(f) Waveguide bandwidth and dispersion: The dispersion performance and, therefore, the profile (step index, graded index, single mode) must be selected in order that the modulated signal is passed over the complete span length with minimal distortion. The more modes that the fiber allows to propagate, and the larger the profile g (closer to step index), the greater the dispersion. Single-mode fiber is the highest-bandwidth lowest-dispersion choice. Unfortunately, it has the smallest core size and numerical aperature, which limits its coupling efficiency at the source and increases the demands on connector and splice tolerances.

(g) Material dispersion: Wavelength, source spectral width, and fiber-material dispersion, at that wavelength, must be selected to provide adequate signal bandwidth over the complete span length. The zero-dispersion point is around a wavelength of 1300 nm for fused silica; however, the fiber can be designed to move this point over the range from 1330 to 1550 nm. There is, therefore, a trade-off between wavelength and bandwidth as well as between wavelength and attenuation.

(h) Performance in coupler-based systems: The performance of various coupler types can vary widely depending on the fiber technology for which they were developed. For example, some excellent insertion loss (less than 0.5 dB) and backscatter isolation (-50 to -60 dB) performances have been achieved for fused single-mode couplers [21] that are not possible for multimode. In a LAN application, these considerations can drive the choice of fiber technologies. For multiple-wavelength coupling, the fiber performance (bandwidth and attenuation) must be considered at all operating wavelengths in order to optimize the design.

(i) Excess cabling loss: Fiber attenuation increases due to microbending if it is under pressure from cabling materials or installation stress. Although this is generally minimized by proper cable design, the effect is also reduced by increasing fiber numerical aperature (NA). A trade-off exists because modal dispersion increases with NA as well as the material dispersion. Material dispersion increases because the added Ge used to increase the NA also moves the zero-dispersion point to longer wavelengths, more toward the OH absorption peak [19].

3.4.2 Fiber Attenuation

Attenuation in cabled fiber is caused by various mechanisms:

1. Rayleigh scattering and scattering due to concentration fluctuations in the dopant oxide are both functions of $1/\lambda^4$. Rayleigh scattering is a function of isothermal compressability B, glass transition temperature T, refractive index n, and wavelength λ [15]:

$$\sigma_s = 8\pi^3(n^2 - 1)\frac{kTB}{3\lambda^4} \tag{3-39}$$

Both scattering mechanisms are inverse functions of the numerical aperature (the lower numerical-aperature fibers have lower scattering losses). In high-silica glasses, where refractive indexes are low, Rayleigh scatter accounts for 75% of the scattering loss.

2. Other mechanisms include absorption due to impurities introduced in the glass-making process; ultraviolet (UV) resonances associated with the structure crystal atoms; infrared (IR) resonance tail associated with the dopant oxide bonds (Ge-O is 11 μm, P-O is 8 μm, B-O is 7.3 μm, and Si-O is 9 μm); and the OH absorption bands at 725, 825, 875, 950, 1130, 1240, and 1390 nm. The OH absorption peaks are overtones of the fundamental OH vibration at 2730 nm. OH is introduced by water production in the glass-making process. These peaks are indicated in Figure 3.26 at values typical of production-grade fiber.

3. Microbending losses can vary with two parameters: temperature and installation stresses on the cable or pressure on the buffered fiber. These losses are a strong function of numerical aperature and how well the modes are confined by the waveguide design [22]. Keck gives the relationship as [15]

$$L_f = N\langle h^2\rangle\frac{a^4}{D^6\lambda^3}\left(\frac{E_b}{E_f}\right)^{3/2} \tag{3-40}$$

where N is the number of bumps per unit length of average height h, D is the fiber outer diameter, a is the core radius, and E_f and E_b are the elastic modulus of the fiber and the surrounding cable/fiber buffering, respectively. Gardner [23] measured as much as a 3-to-1 increase in dB/km loss due to micobending for a decrease in numerical aperature of from 0.158 to 0.12.

4. Macrobending effects the lesser-confined modes in the fiber waveguide. Evanescent fields reach into the cladding and, therefore, can be effected by distortion of the cladding as well as the core. The increase in attenuation is proportional to exp $(-R/R_c)$, where R is the bending radius, and R_c is the critical bend radius for high loss [15]. $R_c = a/(NA)^2$, where a is the core radius, and NA is the fiber numerical aperature [2].

5. Nuclear radiation can cause transient and permanent absorption effects (as discussed in Chapter 1, Section 1.3).

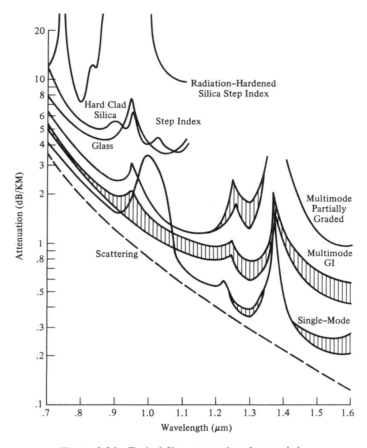

Figure 3.26 Typical fiber-attenuation characteristics.

6. Absorption of hydrogen into the glass can increase fiber attenuation (Chapter 1, Section 1.3).

Attenuation is almost always specified in terms of an attenuation coefficient α in decibels per kilometer. We often use the notation l_f for the unit loss (dB/km) of the fiber to be consistent with other unit-loss parameters. Attenuation for a length of fiber D is, therefore,

$$L_f = Dl_f \tag{3-41}$$

Accounting for measurement tolerances and variations with time, temperature, and wavelength we have:

Mean Loss:

$$L_f = D_f[l_f + l'(t) + l'(T) + l'(w)] \tag{3-42}$$

Standard Deviation:

$$\sigma_f^2 = D_f[\sigma_l^2 + \sigma_{l,T}^2 + \sigma_m^2] \tag{3-43}$$

where $D_f = D_s(1 + u) + D_r$ = maximum fiber length
    D_s = span route distance
    u = fraction uncertainty between route distance and true cable and fiber lengths
    D_r = added length assumed for repair
    l_f = mean loss coefficient under nominal measurement conditions
$l'(t), l'(T)$ = derating of mean loss coefficient with time of exposure and temperature
    $l'(w)$ = change in mean loss coefficient based on operating wavelength different than that where attenuation was measured
    σ_l = standard deviation of cable loss from mean value due to normal production and acceptance tolerances
    $\sigma_{t,T}$ = standard deviation of the derating criteria
    σ_m = measurement uncertainty

Wavelength dependence.   Fiber attenuation is a strong function of wavelength. Total fiber attenuation in cable, under normal conditions (free of radiation or excessive bend) can be represented as a function of wavelength:

$$L_f(\lambda) = \frac{S}{\lambda^4} + Af(\lambda) + \frac{B}{(NA)^6} \qquad (3\text{-}44)$$

where constants S, A, and B are established for a particular fiber design; NA is the numerical aperature. S represents the constant for scattering (an approximate function of NA squared), A is the constant for impurities absorption, and B is the constant for microbending. Figure 3.26 shows typical attenuation curves for fiber product of different classes.

Temperature dependence.   Fiber is inherently stable with temperature. When buffering and cable materials are added, however, the different expansion coefficients cause pressures on the fiber as temperature changes. These pressures in turn create microbending losses by distorting the waveguide. The effects are generally negligible except at cold temperatures (below -20 to $-40°C$) where the plastic coating materials change state. Figure 3.27(a) shows the typical effect of primary fiber buffering and Figure 3.27(b) shows typical effects with cabled fiber for the case where a secondary (harder) buffering has been added to the fiber, such as Hytrel® or nylon, and for the case where the primary buffered fiber floats in a gel within a loose tube or channel [24].

Another cause of attenuation change with temperature is the variation of source wavelength with temperature. Since fiber attenuation is wavelength-dependent, the effect is straightforward.

Aging.   Fiber does not change attenuation with age unless exposed to nuclear radiation, or to strong hydrogen environments, prolonged water immersion or steam that can result in hydroxyl (OH) migration into the core [25, 26]. In the normal environment, cabled fiber is only effected by physical stresses that, like temperature, cause microbending. Often, the cable improves as installation stresses settle out. In

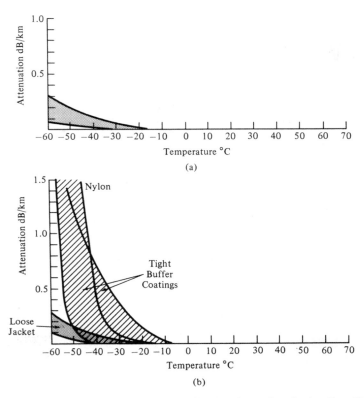

Figure 3.27 The effects of temperature on fiber loss due to the microbending effects of coating and cable buffering materials. (a) Fiber with soft coating and (b) cabled fiber.

areas where there is much rock or frost heave, localized stress and loss can result. Since these stresses are generally localized and unpredictable, it is difficult to quantify a margin for aging.

3.4.3 Fiber Bandwidth and Dispersion

Dispersion is an effect whereby optical modes propagate down the fiber at different rates and arrive at the receiving end at different times. This has the effect of spreading the signal energy in time. As shown in Figure 3.28, if the signal is a digital pulse, it arrives reduced in amplitude and spread in time. If the signal is analog modulated, then the dispersion effect appears as a bandwidth limitation. Dispersion is dictated by the two mechanisms shown in Figure 3.28: multimode dispersion (intermodal) and material dispersion (intramodal).

Multimode dispersion, also called waveguide dispersion, results from the various modes, in multimode fiber, propagating at different rates in the fiber. This is more easily seen as the rays that make up a single optical-signal element, traveling down different path lengths within the fiber.

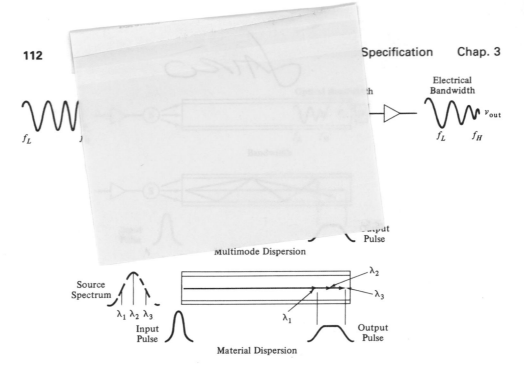

Figure 3.28   Fiber bandwidth and dispersion definitions.

Material dispersion is caused by different wavelengths of light emitted by the source traveling at different velocities down a fiber as the refractive index n varies with wavelength:

$$\text{velocity} = \frac{c}{n} \qquad (3\text{-}45)$$

Since optical sources have a finite spectral width, multiple wavelengths are launched into the fiber even though they may be confined to a single propagating mode. As the mode propagates down the fiber, the various wavelengths separate in time, causing a dispersion effect.

Multimode dispersion is present in all multimode fibers, but not in single-mode. Material dispersion is present in both multimode and single-mode fibers. The dispersion mechanisms and relationships for the three major fiber categories are shown in Figure 3.29 and are explained in what follows.

Multimode dispersion, t_{dw}, and bandwidth, $(BW)_w$, are generally specified on a per-unit-length basis:

$$t_{dw} \text{ (ns/km)}$$

$$(BW)_w \text{ (MHz-km)}$$

Material dispersion (t_{dm}) is related in terms of dispersion per nanometer of spectral width per unit length:

$$t_{dm} \text{ (ps/nm/km)}$$

If a Gaussian pulse shape is assumed, then bandwidth can theoretically be related to dispersion using the following relationship from Section 4.4.2:

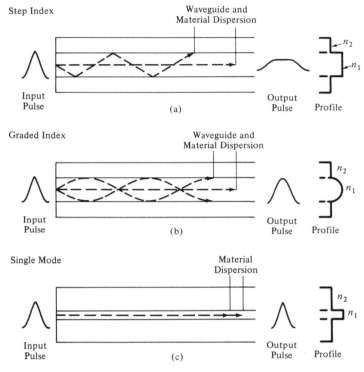

Figure 3.29   Diffusion mechanisms in fiber.

$$BW_{elect} = 312/t_d \text{ MHz per ns } t_{d,FWHM} \qquad (3\text{-}46)$$

$$= 133/t_d \text{ MHz per ns } t_{d,RMS} \qquad (3\text{-}47)$$

Other relationships are given in Chapter 4.

Multimode dispersion (t_{dw}).   Multimode dispersion is a result of waveguide geometry and refractive indexes such that multiple-ray paths can propagate in the fiber. Under this condition, certain rays reach the output end before others, therefore spreading the optical pulse or signal in time. As is shown in Figure 3.29, the effect is the most pronounced in step-index fiber and, as such, fiber bandwidth is limited. With single-mode fiber, multimode dispersion is zero since only one mode (one ray path) exists. Graded index is a special case whereby an attempt is made to equalize the propagation speeds by reducing the refractive index toward the cladding so as to speed up the rays traveling in the longer paths (higher-order modes) thus equalizing the transit time regardless of mode number.

The effect of multimode dispersion is reduced in multimode fibers by grading the refractive index of the core. The relationship is approximately of the form [27]

$$t_{d,rms} \approx \frac{DN\,\Delta}{2c} \frac{g}{g+1}\left(\frac{g+2}{3g+2}\right)^{1/2}\left(\frac{g-2-e}{g+2}\right)^2 \qquad (3\text{-}48)$$

where g represents the profile exponent, N is the refractive index, D is the length, and c is the speed of light.

$$n^2(r) = n^2(0)\left[1 - 2\Delta\left(\frac{r}{a}\right)^g\right]$$

$$e = \frac{-2n}{N}\frac{\lambda}{\Delta}\frac{d\,\Delta}{d\lambda}$$

If the refractive index profile g can be adjusted such that $g = e + 2$, then the material-dispersion term goes to zero at a particular wavelength. This is the theory behind graded-index fiber. The material dispersion is reduced to within what practical production tolerances and source spectral width permit at a particular design wavelength. Figure 3.30 shows the results of profile optimization at two wavelengths, 850 and 1300 nanometers, and a third profile designed to obtain a more broadband "dual-window" performance. Most high-performance graded-index fibers today have this broadband characteristic and can be used at two wavelengths.

As fiber length increases, particularly as fibers are spliced together to create longer spans, the amount of dispersion per unit length appears to decrease in some fibers. This gives the effect of improved performance over that calculated from a lin-

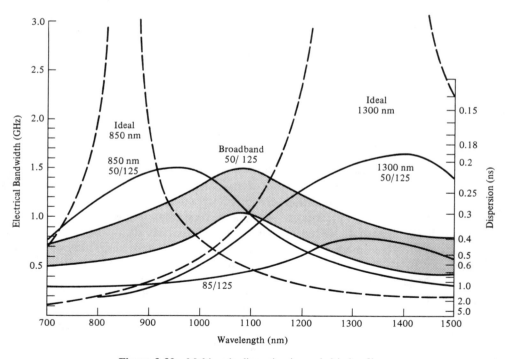

Figure 3.30 Multimode dispersion in graded-index fiber.

ear extrapolation of per-unit-length specifications, as shown in Figures 3.31 and 3.32. This is due to the bandwidth-limiting higher-order modes being scattered, attenuated, and converted to lower order modes. The effect is called concatenation. The quantifying factor is an exponent (γ) called the concatenation factor:

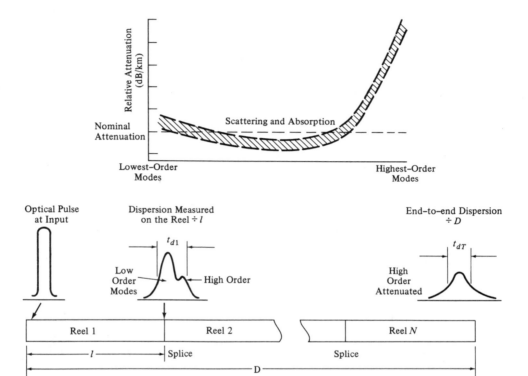

Figure 3.31   Mode mixing in concatenated fiber.

$$(\text{BW})_{fw} = \frac{(\text{BW})_w \ (\text{MHz-km})}{D^\gamma \ (\text{km})} \qquad (3\text{-}49)$$

$$t_{dw,\,\text{tot}} = t_{dw} \ (\text{ns/km}) \times D^\gamma \ (\text{km}) \qquad (3\text{-}50)$$

where the concatenation factor is typically:

0.5 to 0.6 for step-index fiber

0.7 to 0.9 for graded-index fiber at 850 nm

0.8 to 1.0 for graded-index fiber at 1300 nm

A fixed concatenation factor cannot be relied on, particularly with low-attenuation graded-index fibers. The conservative designer uses $\gamma = 1.0$ unless specific data on the fiber is reliable.

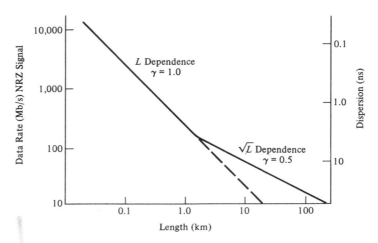

Figure 3.32   The effect of fiber length and concatenation on multimode dispersion.

Multimode dispersion can be effected by microbending pressures and tension on the fiber. The reeled cable or cable under tension may show better performance before than after installation. Measurements should always be made in the loose state.

Material dispersion (t_{dm}).   Material dispersion is caused by different wavelengths of light traveling down the fiber at different speeds. Because even a single excited mode can be composed of a finite spectrum of light, the effect occurs in single-mode fiber as well as multimode. Dispersion is a function of wavelength (λ), change in refractive index with respect to wavelength ($dn/d\lambda$), and transmission distance (D) [27]:

$$t_{dm,\,\mathrm{rms}} = D \frac{\lambda}{c} \frac{d^2 n}{d\lambda^2} \, d\lambda \tag{3-51}$$

The amount of material dispersion created in silicon fibers [28] is shown in Figure 3.33. A natural null (zero dispersion) occurs at 1300 nm. By modifying the cladding design in single-mode fibers, this curve can be "shifted" to a null at 1550 instead of 1300 nm or, with further refinement, optimized at both wavelengths, as shown in Figure 3.34.

Material dispersion (specified in picoseconds/nanometer/kilometer) is a function of source wavelength (t_{dm}), spectral width ($\Delta\lambda$), and propagation distance (D):

$$t_{dm,\,\mathrm{tot}} = t_{dm} \; (\mathrm{ps/nm/km}) \times \lambda \; (\mathrm{nm}) \times D \; (\mathrm{km}) \tag{3-52}$$

Bandwidth derived from reel measurements.   Sometimes it is desired to calculate total link dispersion or bandwidth from the data taken from factory mea-

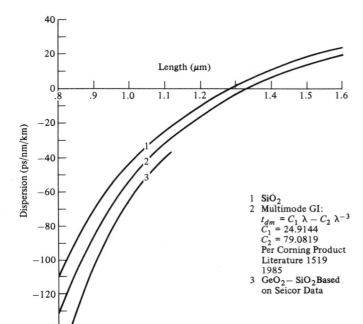

Figure 3.33 Material-dispersion range for typical fiber products.

1 SiO_2
2 Multimode GI:
$t_{dm} = C_1 \lambda - C_2 \lambda^{-3}$
$C_1 = 24.9144$
$C_2 = 79.0819$
Per Corning Product Literature 1519 1985
3 $GeO_2 - SiO_2$ Based on Seicor Data

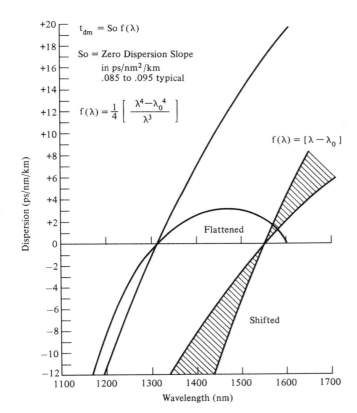

$t_{dm} = So\ f(\lambda)$

So = Zero Dispersion Slope in $ps/nm^2/km$
.085 to .095 typical

$f(\lambda) = \frac{1}{4} \left[\dfrac{\lambda^4 - \lambda_0^4}{\lambda^3} \right]$

$f(\lambda) = [\lambda - \lambda_0]$

Flattened

Shifted

Figure 3.34 Material dispersion in single-mode fiber.* (Based on Corning Product literature)

117

surements on reeled fibers. In order to do so, care must be taken that the measured data is meaningful for fibers in the loose state. Assuming Gaussian received pulse shapes, the relationships are as follows:

Total Dispersion:

$$\sigma_t = (\sigma_{t1}^2 + \sigma_{t2}^2 + \cdots + \sigma_{tn}^2)^{1/2} \qquad (3\text{-}53)$$

Total Bandwidth:

$$(\text{BW})_f = \cfrac{1}{\left(\cfrac{1}{(\text{BW})_1^2} + \cfrac{1}{(\text{BW})_2^2} + \cdots + \cfrac{1}{(\text{BW})_n^2}\right)^{1/2}} \qquad (3\text{-}54)$$

where $(\text{BW})_1$ through $(\text{BW})_n$ are the individual bandwidth measurements for individual reels 1 through n.

This relationship assumes that the concatenation factor is 1. If the concatenation factor is less than 1, then the effects of distance must be factored in. If we assume that the reels are short enough that the concatenation effect has negligible effect on the original reel bandwidth measurements, then we can account for concatenation by multiplying the previous linear assumption for $(\text{BW})_f$ by D/D', where D' is the distance factor adjusted for concatenation (γ). In a simpler form, this becomes

$$(\text{BW})_{f,\,\text{concat}} = (\text{BW})_{f,\,\text{linear}} \times D^{1-\gamma} \qquad (3\text{-}55)$$

3.5 CONNECTOR SELECTION*

3.5.1 Selection Trade-offs

Connectors are required whenever the fiber must be joined and periodically disconnected. This generally occurs at terminal equipment, optical cross-connect panels, and couplers. There are a number of connector classifications, each with their own set of important performance considerations:

(a) Fiber to fiber: These connectors couple fiber ends together at patch panels or the "pigtail" ports of terminal devices. The important characteristics are the ability for the connector to mate core to core with a high degree of lateral and angular precision and the ability to bring the two fiber ends close together without contact but with minimal reflective or displacement loss. They also must be able to withstand many mating cycles with minimal degradation or variation, and they must be readily cleanable. Generally, these connectors are tailored for the fiber to be used, often are factory installed on "pigtail" fibers for highest precision, and employ the

*The author wishes to acknowledge and thank Dennis Knecht of Seicor Corporation for his contribution of the diagrams and loss relationships to this section.

most creative optical and mechanical design solutions to minimize loss. Designs vary greatly, depending on whether they are to be used in fixed installations, military tactical installations, large-core fiber systems or single-mode systems, and depending on whether they are to be installed at the factory or in the field.

(b) Fiber to source: When the optical source is contained within the connector housing, the fiber is mated directly to the source. These are generally found in the lower-cost terminal systems employing LEDs and large core multimode fibers. Applications include LANs and short-haul data modems. Of principal concern is that the fiber be brought close to the source emission surface (or focal point in some designs) and that the core be centered in the emission cone. Trade-offs are generally between connector cost, source compexity/cost, and coupling precision.

(c) Fiber to detector: The span fiber is mated to the detector either through a glass window or a large-core waveguide stub (if the connector is integral with the detector) or to a larger-core multimode fiber pigtail (if the detector is separate from the connector). Since the diameter of this detector coupling aperature can generally be larger than the span fiber core, the coupling tolerances are less critical.

(d) Multiport coupler connector: Connectors are generally used at multiport couplers (STAR, TAP, or WDM couplers) in order to permit reconfiguration of the network or for maintenance. Since most couplers are fabricated using a fiber waveguide with the same characteristics as the coupling fiber, the tolerances are just as critical as with fiber-to-fiber connectors. This depends to some degree on the design and mode-conversion properties of the coupler, however.

Selection of a connector design, therefore, depends largely on its application. When purchasing terminal equipment, connector choice is not always involved, since the connectors are usually selected by the manufacturer of the terminal and coupling equipment. When designing at the component or subsystem level, however, detector choice becomes a critical element. Table 3.7 lists the characteristics of some common connector designs. The parameters generally considered include the following.

(a) Coupling loss: Other parameters considered, design for minimum coupling loss. This parameter is discussed in more detail in the next section. Note that the loss often specified for the connector only includes extrinsic losses associated with mechanical connector tolerance. Intrinsic losses related to fiber tolerance must also be considered.

(b) Degradation and repeatability: Insertion loss changes in value from connection to connection due to mechanical tolerances creating different core alignments each time. Insertion loss also degrades with mating cycles due to contamination and wear on connector surfaces. The connector must be capable of cleaning between mating cycles without being damaged. A degradation or variance of 0.5 dB is generally considered reasonable, although often 1 dB or more is experienced even with the best connectors.

TABLE 3.7 CONNECTOR PERFORMANCE

Multimode

Fiber size	50/125		62.5/125		85/125		100/140		Repeatability dB	Temperature range °C
Biconic	0.6–0.8	0.85–1.5	0.6–0.8	0.85–1.5	0.6–0.8	0.85–1.2	0.4–0.7	0.85	0.3–1	−40 to +80
NEC—D4	0.4–0.5	0.7–1	0.4–0.5	0.7–1	0.4–0.5	0.7–1	0.5	0.7–1	0.1–0.3	−40 to +80
NTT—FC	0.4–0.8	0.5–1	0.3–0.8	0.6–1	0.3–0.8	0.6–1	0.5–0.7	0.7–1	0.05–0.3	−40 to +85
SMA	0.6–1.5	1–3	0.6–1	1–2	0.6–1	1.2	0.5–0.8	1–2	0.1–0.5	−60 to +125
ST	0.2–0.8	0.5–1.5	0.2–0.6	0.5–1.5	0.2–0.6	0.5–1.5	0.2–0.5	0.5–1.2	0.1–0.2	−40 to +70
Expanded beam	1.5	2	1.5	2	1.5	2	1.5	2	0.2	−30 to +80
Thermoplastic	1.5–1.7	2	1.5–1.2	2	1–1.5	2	0.8–1	1.5–2	0.05–1	−40 to +70
Mini-BNC	0.6	1	0.5	1	0.5	1	0.4	1	0.1–0.3	−20 to +80
Military	1.2	1.7	1.2	1.7	1.2	1.7	0.9	1.7	0.2	−65 to +200

Single mode

Single mode	Insertion loss Mean	Insertion loss Maximum	Repeatability dB	Temperature range °C
Biconic	0.5–0.7	0.75–1.3	0.1–0.3	−20 to +60
NEC—D4	0.5–0.8	0.6–1.2	0.02–0.3	−40 to +80
NTT–FC	0.15–0.8	0.3–1.2	0.05–0.5	−40 to +85
NTT PC	0.3–0.5	0.5–1.0	0.1	−40 to +85
Expanded Beam	0.3–0.6		0.2	−40 to +80

Data compiled from suppliers' data sheets and *Lightwave* Magazine special report, July 1986, page 28.

(c) Connector cost: The precision mating properties are generally traded against cost. The lowest-cost connectors are generally low-tolerance large-core fiber connectors made from plastic or stainless steel. Small-core and single-mode connectors that are purchaced factory mounted on "pigtails" are also reasonable in cost.

(d) Physical properties and mounting configuration: These must match the application. Sometimes practical limitations of the connector dictate the configuration. For example, it is generally more economical and practical to purchase a precision connector with a factory-terminated "pigtail" fiber and splice it onto the cable plant than to try to use a field-terminated connector.

(e) Environmental stability: Insertion loss changes with temperature variation due to the differences in expansion coefficients of the different glass, metal, and plastic materials. Vibration, stress, dust, corrosion, and other extreme conditions can also effect performance. The connector chosen should be qualified to the installation conditions.

3.5.2 Connector Insertion Loss

Connector insertion-loss mechanisms are shown in Figure 3.35 and include the following:

1. Intrinsic fiber-loss mechanisms: These are tolerance mismatches intrinsic to the fiber such as core size, numerical aperature, core concentricity and profile;
2. Extrinsic loss mechanisms: These are associated with fiber-end preparation and connector tolerances such as lateral offset, angular misalignment, fiber-end separation, reflections, and surface roughness.

Figures 3.36 through 3.42 show these loss mechanisms. Loss is presented here as a positive number in dB. (The diagrams and the following relationships were contributed by D. Knecht [29].)

*Core-area mismatch (Figure 3.36).   For multimode fiber, we use a simple geometric relationship assuming modes are evenly distributed across the core area (conservative). D_s is the diameter of the sending core and D_r is the diameter of the receiving core.

$$L_c \text{ (dB)} = 20 \log \frac{D_s}{D_r} \tag{3-56}$$

In single-mode systems, we are concerned about mode-field mismatch and, therefore, mode-field radius. Mode-field radius w is a function of operating wavelength λ, cutoff wavelength λ_c, and core radius a. The power-transfer coefficient is [29, 30]

$$L_c = 20 \log \frac{1 + (w_s/w_r)^2}{2(w_s/w_r)} \tag{3-57}$$

*The author wishes to acknowledge Dennis Knecht for providing the formulas in this section.

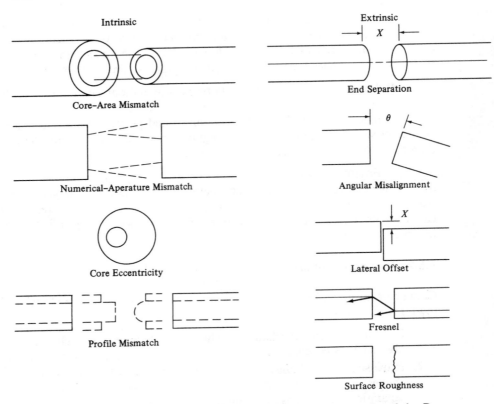

Figure 3.35 Coupling-loss mechanisms. The author wishes to acknowledge Dennis Knecht [29] for submitting this figure to the publication.

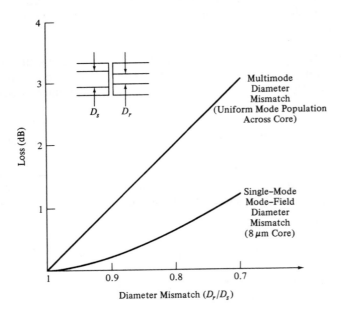

Figure 3.36 Coupling loss due to core-area or mode-field diameter mismatch.

where the mode-field radius is [29, 31]

$$w = a\left[0.65 + 0.434\left(\frac{\lambda}{\lambda_c}\right)^{3/2} + 0.0149\left(\frac{\lambda}{\lambda_c}\right)^6\right]$$

Numerical-aperature mismatch (Figure 3.37).

$$L_c \text{ (dB)} = 10 \log \frac{(NA)_s}{(NA)_r} \tag{3-58}$$

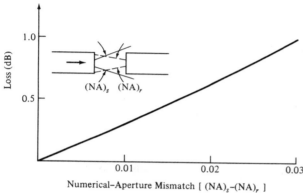

Figure 3.37 Coupling loss as a function of NA mismatch.

*Lateral offset and core concentricity (Figure 3.38).

In multimode fiber, the loss due to lateral offset is a function of the amount of offset (d) and the core diameter of the fiber (D_f):

$$L_c \text{ (dB)} = 10 \log \left[1 - \frac{2}{\pi}\frac{d}{D_f}\left(1 - \frac{d^2}{D_f^2}\right)^{1/2} - \frac{2}{\pi}\sin^{-1}\frac{d}{D_f}\right] \tag{3-59}$$

For single-mode fiber, the relation is a function of lateral offset (d) and average mode-field radius (w_o) as approximated by [29, 32]

$$L_c \text{ (dB)} = 4.34\left(\frac{d}{w_o}\right)^2 \tag{3-60}$$

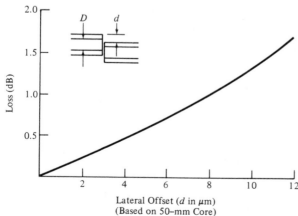

Figure 3.38 Coupling loss to lateral offset or core/clad concentricity.

*Profile mismatch (Figure 3.39).   Assuming that the power transfer is a direct function of the number of modes coupled in a multimode fiber, and, therefore, a ratio of the grading profiles (g), from Equation (1-3) the loss is

$$L_c \text{ (dB)} = 10 \log \frac{g_s(g_r + 2)}{g_r(g_s + 2)} \tag{3-61}$$

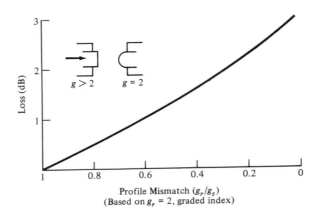

Figure 3.39   Coupling loss attributed to profile mismatch.

*End separation (Figure 3.40).   The end-separation loss in multimode is a function of separation distance (d), core diameter (D_f), fiber NA, and the refractive index of the gap medium (n_0) [29]:

$$L_c \text{ (dB)} = 20 \log \frac{D_f/2}{D_f/2 + d \tan [\sin^{-1}(NA/n_0)]} \tag{3-62}$$

In single-mode fibers, the loss is a function of mode-field radius (w), fiber refractive index (n_f), and separation (d). The transmission factor is [29, 30]

$$T = \frac{\{4Z^2 + (w_s/w_r)^2\}}{[Z^2 + (w_s^2 + w_r^2)/4w_r^2]^2 + Z^2(w_r/w_s)^2} \tag{3-63}$$

where $Z = d\lambda/2\pi n_f w_s w_r$.

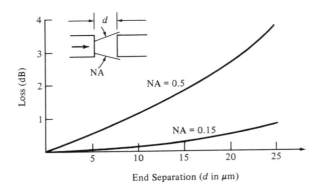

Figure 3.40   Coupling loss as a function of end separation.

*Angular misalignment (Figure 3.41).   Angular misalignment is a function of numerical aperature: the smaller the NA, the smaller the effect on loss. In multimode fibers, the relationship is [29]

$$L_c \text{ (dB)} = 10 \log \left[\frac{\cos \theta}{\pi} \left\{ \frac{\pi}{2} - P(1 - P^2)^{1/2} - \sin^{-1} P \right. \right.$$

$$\left. \left. + q\left(r(1 - r^2)^{1/2} + \sin^{-1} r + \frac{\pi}{2} \right) \right\} \right] \qquad (3\text{-}64)$$

where $P = \dfrac{\cos \theta_c (1 - \cos \theta)}{\sin \theta_c \sin \theta}$

$\quad\quad q = \dfrac{\cos^3 \theta_c}{(\cos^2 \theta_c - \sin^2 \theta)^{3/2}}$

$\quad\quad r = \cos^2 \theta_c (1 - \cos \theta) - \dfrac{\sin^2 \theta}{\sin \theta_c \cos \theta_c \sin \theta}$

$\quad\quad \theta_c = \sin^{-1} \dfrac{\text{NA}}{n_f}$

In single-mode fibers, the relation is a function of fiber refractive index (n_f), mode-field radius (w), wavelength (λ). and the angle of tilt (θ). The transmission factor is [29]

$$T = \left(\frac{2w_s w_r}{w_s^2 + w_r^2} \right)^2 \exp - \frac{2(\pi n_f w_s w_r \theta)^2}{(w_s^2 + w_r^2)\lambda^2} \qquad (3\text{-}65)$$

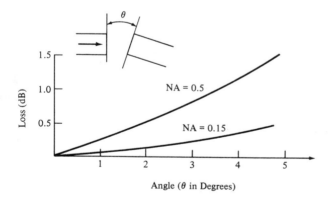

Loss (dB)

Angle (θ in Degrees)

Figure 3.41   Coupling loss as a function of angular misalignment.

Fresnel reflection loss (Figure 3.42).   In a perfect connector, the joining loss reaches the limit dictated by the Fresnel reflection loss at the glass-to-air interface.

$$L_c \text{ (dB)} = 10 \log \left[1 - \left(\frac{n_f - n_0}{n_f + n_0} \right)^2 \right] \times 2 \qquad (3\text{-}66)$$

For a fiber having a refractive index (n_f) of 1.47 interfacing with air having a refractive index (n_0) of 1, the loss is about 0.32 dB for the two surfaces. If the fibers physically touch, this loss component disappears. This is not practical since the two glass

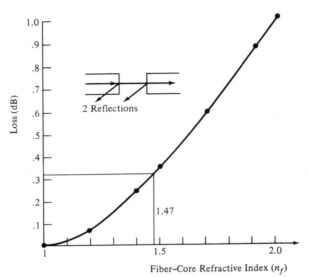

Figure 3.42   Coupling loss due to Fresnel reflection.

surfaces will scratch or fracture each other. Generally, some separation is required to overcome damage. For single-mode fiber, a separation of around 5 μm is required as well, in order to reduce the effects of reflection creating coherent wavefront constructive and destructive interference. This causes an odd loss oscillation phenomenon with submicron movements.

3.5.3 Link-Budget-Loss Relationships

Because of all the many variables, connector loss is truly a statistical function, as shown in Figure 3.43. From connection to connection, tolerances align differently, giving different insertion-loss results. Connector loss should be specified statistically and treated in the same manner. Unfortunately, many connector manufacturers specify a typical or a maximum value. The typical can be treated as a mean, but the maximum has vague meaning. It should be defined statistically, such as 2 standard deviations. In the link-power budget, connector loss (L_c) should be represented as follows:

Mean Loss:

$$L_c = N_c[l_c + l'(t)] \tag{3-67}$$

Standard Deviation:

$$\sqrt{N_c}[\sigma_c^2 + \sigma(t)^2 + \sigma(T)^2]^{1/2} \tag{3-68}$$

where l_c = mean loss per connector when new

           $l'(t)$ = permanent degradation from the mean due to mating-cycle wear over time

   $\sigma_c, \sigma(t), \sigma(T)$ = per-connector standard deviations related to tolerance, matings, and temperature

           N_c = number of connectors in series

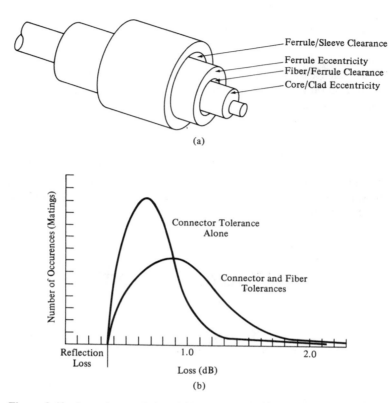

(a)

(b)

Figure 3.43 Loss characteristics of fiber-optic connectors. (A) Loss mechanisms in a ferrule-in-sleeve design, and (b) statistical insertion-loss characteristics from multiple matings.

Specified connector loss values must be considered with care since the specifications often exclude anticipated fiber-related tolerances and, therefore, provide optimistic values. Where it is suspected that fiber tolerances are not included, a close approximation is to statistically sum the intrinsic loss (l_i) expected of a fiber joint into the extrinsic connector loss (l_e) specified.

$$L_c = N_c(l_e + l_i) \pm \sqrt{N_c(\sigma_e^2 + \sigma_i^2)} \tag{3-69}$$

3.5.4 Temperature Effects

Since connector loss depends on the mechanical alignment of devices composed of many dissimilar materials, some variation with temperature change can be expected. For high-quality designs, this is typically held to within a 0.5-dB variation for a ±50°C temperature change. Since the direction of the variation is usually indeterminant (positive or negative), it is represented in the previous relationships as an increase in standard deviation rather than a shift in mean value.

3.5.5 Aging and Mating Cycles

Connection loss not only varies randomly with mating cycles due to tolerance effects, but exhibits a gradual permanent increase (mean shift) due to contaminants collecting on the mating surfaces, scratching of the optical surfaces, and wear of mechanical components.

Connectors should be carefully cleaned before each mating and should not be opened to dirty environments. Wear is difficult to avoid, however, so the connectors should be designed and procured with the anticipated life (in terms of mating cycles) in mind. For a reasonable quality design, the mean variation with use should be specified to be no greater than 0.5 degradation. Figure 3.44 shows the loss as might be observed with 1000 matings of a ferrule-in-sleeve type connector mating 50-μm multimode fiber.

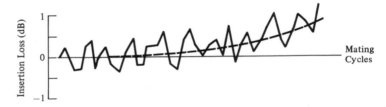

Figure 3.44   The effect of mating cycles and wear on optical-connector insertion loss.

3.6 SPLICE SELECTION

3.6.1 Splice Application

Splices are required in a system whenever two fibers are to be permanently joined. Some of the most common splicing applications include the following.

(a) Outside plant splice: Whenever two reels of cable are to be joined to form a long continuous fiber span, the fibers must be spliced in order to provide a continuous optical waveguide. This is generally performed using fusion or bonded joining methods in a protected environment with some environmental control. The spliced fiber is then coated and protected and placed in an environmentally sealed enclosure. The enclosure is either pole mounted, direct buried, or placed in a manhole or vault. Of consideration here is minimum joining loss, field spliceability, long-term stability, and reliability, and the cost of the equipment to perform the splice and measure it.

(b) Pigtail splice: In fixed installations, generally, the connectors are not field installed to the trunk cable. Instead, pigtail fibers with factory-installed connectors are spliced to the cable in a rack-mounted splice enclosure. The trade-off here is generally between the cost and labor required to splice the connector with pigtail versus the cost and labor required of a field-installable connector, and the relative performance of each. Generally, the spliced pigtail is superior on all counts.

3.6.2 Technology Trade-Offs

In general, splicing technology can be divided into three categories:

1. Fusion Splicing: In fusion splicing, fibers are permanently joined by first cleaving the fibers to obtain a precision optical surface, aligning the fibers, and then fusing the glass together. This is accomplished in a precision manner such that a continuous waveguide is formed with little distortion and no gap where the cores are joined. In order to achieve accurate alignment, the fiber is visually aligned under a microscope and, in the case of precision single-mode equipment, actively aligned by locally launching and detecting maximum light transfer. Splicing loss is dependent on the ability of the alignment mechanism to align the cores, the tolerance of the core physical parameters, the cleave angle made on the bare fibers before fusion as well, the control of the glass-melting conditions, the stability of the alignment during fusion, and the cleanliness of the environment.

2. Mechanical splicing: We classify this technology as a splice whereby the fiber is joined in a housing that contains slots, alignment rods, elastomerics, or some other mechanism that achieves and maintains fiber outer-surface alignment. As with the fusion method, fibers are first precision cleaved to obtain a good optical surface. This form of splicing generally has the highest loss since alignment of cores is dependent on fiber concentricity and outer-diameter tolerance control as well as the ability of the splice housing to control fiber alignment. However, once the fibers are aligned, a permanent bond is achieved with thermally or ultraviolet cured optically compatible epoxy. This bond also eliminates the air gap.

3. Aligned and bonded splicing: With this form of splicing, fibers are mounted into a sleeve (usually glass), the cores are actively aligned (by measuring the transmission of launched light), and then the sleeves are bonded together using an optically compatible epoxy. Generally, the fibers are polished after being mounted in the sleeves to obtain a precision optical surface. This approach was principally developed by AT&T under the names "Bonded" and "Rotary" splicing. Splice loss is the lowest of any approach since cores are actively aligned for minimum loss, angular alignment is better controlled, and there is negligible movement during curing. This approach is, however, one of the most time consuming and labor intensive due to the polishing process and the epoxy curing time.

The following issues are generally considered when evaluating the trade-offs between the three technologies:

(a) Loss: For lower wavelength, large-core and multimode fibers, splice loss is a consideration but not critical. Mechanical and fusion splicing are generally adequate. For ultra-low-loss single-mode applications, however, splicing loss becomes a major contributor to the link-power budget. Mechanical splicing tolerances and stability become major limitations. Fusion splicing is adequate if active alignment mechanisms are used to minimize loss. In extremely low-loss applications where fusion splicing loss or tolerances are unacceptable (particularly at 1550 nm), actively aligned and bonded splicing is the chosen technology.

(b) Cost of equipment: Splicing equipment can be expensive. Equipment is required to cleave the fiber (or polish it), to align the two fiber ends, to form a permanent bond, and to measure loss. The lowest-cost equipment is for mechanical splicing. It generally consists of a cleaver, an alignment fixture, and an epoxy-curing oven or lamp. Measurement is performed after the fact using an OTDR or power meter. Fusion splicing requires a precision cleaver and a machine with a microscope, an electronically controlled fusing and alignment mechanism, and (for best results) a local injection/detection system for active core alignment. With active alignment, the OTDR is often omitted until the final end-to-end measurements. For the active alignment and bonded approach, a machine or fixture with local injection/detection is required to aid in alignment of the fiber cores. A polishing fixture and epoxy-curing station is also required. Often equipment cost dictates the choice of technique when loss is not critical.

(c) Cost of splice materials: The cost of incidental splice materials is generally in the case of fusion, since each splice only requires some protective coating and sleeve. The bonded approach requires relatively low-cost materials as well. In the case of the mechanical splice, however, the sleeve can become a significant cost item. The equipment and labor costs associated with fusion and active alignment and bonding have to be compared against the higher cost and lesser performance of the mechanical sleeve.

(d) Labor intensity: The cost of splicing labor often outweighs many of the other cost criteria. Mechanical splicing is often perceived as the least time consuming, however, with well-trained crews, fusion is often as fast. Cable, fiber, and enclosure preparation are generally the most time-consuming part of the operation and is common to all procedures. Approaches that require fiber polishing and slow epoxy-curing time become less desirable from a labor standpoint.

(e) Reliability and stability: Fusion splicing is generally perceived as the most reliable and stable since the glass is permanently fused. Epoxies are generally suspect due to questions of curing control and differences in thermal expansion coefficients. In reality, however, there is little data to show that one technique is more reliable and stable than another in the long term. Fusion splices are equally subject to glass-surface damage due to handling, which can result in fiber failure over time. Fusion splices, as well as mechanical, if not properly secured, can vary in loss with temperature due to local microbending.

The proper selection and use of splice enclosures is just as important as the selection and application of the splicing technology. The enclosure, and its internal splice organizer, serves a number of important functions:

(a) cable sheath termination;
(b) cable sheath grounding;
(c) fiber organization;
(d) splice coating and protection from dirt;
(e) protection from water, dirt, and environment;

(f) maintenance of proper fiber bend radius and protection from crushing;

(g) storage of unit tubes in a way that prevents kinks and crushing;

(h) storage and separation of fiber service loops.

The enclosure must be environmentally sealed. A preferred design for underground application is to have an outer housing as well that is filled with a water sealant. It is necessary that the enclosure be reenterable, since repairs or resplicing is often required in practice. The enclosure also must contain a splice organizer that holds and protects the spliced fibers and maintains the proper bend redius. The organizer should permit reentry and resplicing as well. One of the most common causes of splice failure is a poorly selected enclosure that has permitted fiber crushing, excessive bends, unit tube kinking, and entry of dirt or water. It is prudent to spend money on a well-designed enclosure and organizer.

3.6.3 Splice Loss (L_s)

Loss mechanisms are similar to those of connectors with the exception that the intrinsic losses can dominate with extrinsic factors greatly reduced, depending on the splicing approach. Refer to Section 3.5.2 for the loss relationships.

Splice loss, as with connector loss, is also a statistical parameter. Since there are generally more than two splices in any span, splices should always be treated statistically. The most common problem with splicing loss is that inexperienced designers invariably specify a maximum splice loss at a value roughly equal to where the 1-sigma value should be, and then blame the lack of achievement on the splicing technique instead of normal fiber or measurement-tolerance statistics. An inordinant amount of time is then spent resplicing and remeasuring until the desired loss is measured or time runs out. The amount of fiber handling also weakens the fiber and reduces systems reliability. This is the wrong way to design a system or to specify splice loss.

A proper approach is to design the system based on a mean loss with margin for a 2- or 3-sigma variation where the standard deviation is summed statistically. During splicing, the splicers should be required to meet a mean loss (say 0.15 dB) and challenged to remeasure and remake splices a few times when the 1-sigma value is not met (say, three remakes to try to achieve 0.25 dB or less). If not achieved after a few tries, leave it alone unless excessive (say, above 0.5 dB). In this way, the extrinsic variables are reduced, tooling and technique can be checked, but unrealistic rework is avoided.

By recognizing the statistical nature of splice loss, total loss for a span (L_s) is calculated as follows.

Mean:

$$L_s = (N_s + N_r)(l_s + l'T) \tag{3-70}$$

Standard Deviation:

$$[(N_s + N_r)(\sigma_s^2 + \sigma_{sT}^2)]^{1/2} \tag{3-71}$$

where l_s = mean loss per splice at room temperature
$l'T$ = change in mean loss at temperature extremes
σ_s = standard deviation per splice due to tolerances
σ_{sT} = standard deviation of temperature effect
N_s = number of splices in series
N_r = added margin (number of splices) for repair

Worst-case loss for systems design is defined as the total mean loss plus two to three times the total 1-sigma value.

The following loss values are typical to those achieved in practice for various splice technologies, but can vary substantially as fiber and splice technology improves, or from device to device. In order to achieve the lower loss values, fiber end-cleave angle must be below 1 degree. Results from typical mechanical cleavers give a mean cleave angle of about 1 to 1.3 degrees with a standard deviation of from 0.6 to 1.13 degrees [33].

1. Actively aligned and bonded splicing: When cores are actively aligned for maximum light coupling and bonded in place, mean losses of around 0.03 to 0.04 dB with a 0.03-dB standard deviation for single mode can be achieved [29].

2. Fusion splicing: When fiber is actively aligned for minimum loss then fused in place, mean losses of around 0.05 to 0.07 dB (single mode and multimode) with a 0.03- to 0.04-dB standard deviation (single mode) and 0.09 dB (multimode) are possible assuming identical fibers [34, 35]. Fiber batch-to-batch tolerance must be added to this in a practical environment, which could double these numbers. If fiber is optically aligned under a fusion-splicing microscope, then a mean of around 0.15 to 0.2 dB with a standard deviation of 0.15 has been regularly observed in field measurements on dissimilar fibers [36].

3. Mechanical splicing: When fiber is aligned by the outer cladding, mean losses of around 0.11 to 0.13 dB with a 0.04- to 0.1-dB standard deviation can be obtained for single mode and multimode by rotating the fibers for minimum loss. With random insertion into the splice, losses of about 0.15 to 0.3 dB and standard deviations on the order of 0.15 dB are obtained [37, 38].

Figure 3.45(a) shows the preceding in the form of what might be observed as splicing histograms for the three approaches. Figure 3.45(b) shows the effect that OTDR measurement error can have on the results [39, 40]. It is necessary to note this error when performing field measurements. The error is due in part to measurement resolution of the OTDR and largely to the fact that each reel that is spliced may have different loss characteristics. These differences translate to different amounts of backscatter at the splice point, masking the true splice loss. Some splices actually appear to have gain due to this effect. OTDR measurements of splices typically show an average of 0.2-dB difference (and as much as 0.5 dB) when measured from opposite directions [40]. The OTDR can, therefore, only be used for continuity checks and approximate loss measurements.

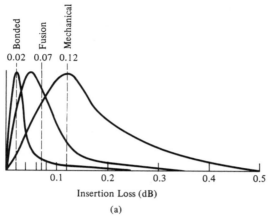

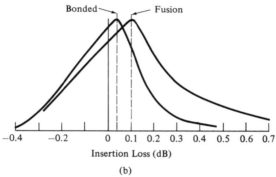

Figure 3.45 Splicing histograms. Histograms represent the results of splicing identical single-mode fiber. The loss measurements were also made with an OTDR that contributes to the spread in measured loss. Three splicing methods are represented, fusion, bonded, and mechanical. (a) True measurement and (b) measurement with an OTDR.

3.6.4 Environmental Loss Variations

Once a splice is properly fused or bonded in place, there should be no variation with temperature. Fusion splices are inherently stable. Fibers that are bonded in place with epoxies usually are stable as well unless the design, technique, or materials used are not proper. With a mechanical splice, if epoxy bonding is not incorporated, temperature can effect loss. Variations of ±0.2 dB over a −55 to +85 °C cycling range were observed with the 4-rod mechanical splices [38] and about 0.14- to 0.17-dB mean change (0.08 to 0.12 standard deviation) with the elastomeric over a +70 to −40°C cycling [37].

The major culprit with temperature is the manner in which the completed splice is protected and secured in the enclosure. If proper fiber or buffer tube bend radius is not observed, or if a fiber is pinched or buffer tube kinked, large unpredictable changes with temperature are observed. Also, if the spliced area is not properly coated with a soft material, like RTV, or the sleeve fits tightly over the spliced area, microbending losses with temperature occur. A well-designed organizer and careful technique are mandatory requirements. When good design and installation practices are observed, margin is not required for splice-loss variation with temperature.

As with temperature, there are no significant loss mechanisms that consistently degrade with time. If good design and installation practice are used, no margin is required for aging. Often, microbending loss due to improper stowage in the enclosure, or due to pressure on the cable, is mistaken for degradation of the splice over time. Aging tests with UV-cured epoxy over 4 years at temperatures of 50 to 70°C indicated no change to optical properties [38]. Elastomeric splice data [37] on accelerated 13-month tests, designed to simulate 30 years of life, indicated changes of from 0 to about 0.1 dB but mostly loss improvement.

Vibration cannot effect a fused or bonded splice unless there is pressure on the fiber splice causing microbending. With a mechanical splice, however, vibration can cause some splice-loss perturbations. Variations of up to 0.2 dB were observed with a 4-rod splice design with 10-Hz 10-G vibration cycling [38].

3.7 COUPLER SELECTION

Couplers are required whenever optical power is to be injected into a fiber from more than one source, or extracted from a fiber to more than one destination. The effect that the coupler has on the link-budget analysis is primarily the addition of insertion loss to the link and perhaps some reflection and crosstalk noise components.

3.7.1 Coupler Application

The selection of a coupler first centers on the application. Figure 3.46 shows four basic coupler configurations. Table 3.8 classifies these as to design and summarizes the typical product performance for each design.

1. T coupler: It is either a 3-port coupler or a 2×2 coupler. It splits power in one direction from one fiber to two. It can be used as a power splitter of various splitting ratios, a power combiner, or a LAN terminal optical input/output coupler. A T coupler can be wavelength-independent (WIC) or a single wavelength-dependent coupler (WDC).

2. Star coupler: This is an $N \times N$ coupler that couples power from one of N fiber input ports to all N output fibers. It can be used as a multiport power splitter or combiner, or as a multiport star coupler in a LAN. Ideally, this is a wavelength-independent device.

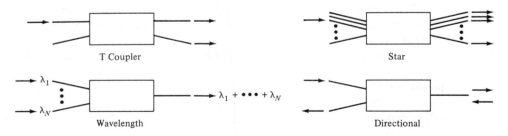

Figure 3.46 Coupler design.

TABLE 3.8 TYPICAL PERFORMANCE OF FIBER COUPLERS

Design class	No. of ports	Coupling ratio	C.R. tolerance (±%)	Excess Loss (dB)	Uniformity (dB)	Isolation directivity (−dB)	Polarization sensitivity (%)
2 × 2 T							
Single mode	2	0.5	5–15	0.07–1.0	0.1–0.2	−40 to −55	0.1–6
WIC or WDC	2	0.25	or	0.07–1.0	0.1–0.2	−40 to −55	0.1–6
	2	0.1	0.03 dB/nm	0.07–1.0	0.1–0.2	−40 to −55	0.1–6
2 × 2 T							
Multimode	2	0.5		<1	0.5	−40	
	2	0.25		1–2	0.5	−40	
	2	0.1		1–2	0.5	−40	
	2	0.0625		1–2	0.5	−40	
N × N WIC	3	0.33	10	<0.5	0.1–0.2		1
Star coupler	4	0.25		0.5–2.0	0.5–0.6		
	8	0.125		<2.5	0.6		
	16	0.0625		<6.0	0.6		
	32	0.03125		<8.0	0.6		
Directional 2 × 2 WIC	2	1		0.5–1.0	5–10	−40 to −50	<5

Design class	No. of ports	Wavelength spacing (nm)	Wavelength channels (μm)	Excess loss (dB)	Wavelength sensitivity (± %)	Isolation directivity (−dB)	Polarization sensitivity (%)
Wavelength Multiplexing							
2 × 2 WDM Wavelength dependent	2	0–1	1.3, 1.5	0.5	0.1 to 0.4%/nm	−16 to −28, −59 to −60 w/filter	<2.5%
N × N WDM Single-mode	3	200–200	1.2, 1.3, 1.5	4.0		−25	
	6	20–30	1.3 or 1.5	2–3		−25	
	8	20–30	1.3 or 1.5	2–3		−25	
N × N WDM Multimode	3	100–200	1.2, 1.3, 1.5 or 1.1, 1.2, 1.3	3.5		−25	

3. Wavelength-division MUX/DEMUX coupler: This can be a 2 × 2 wavelength-dependent coupler (WDC) or an $N \times N$ wavelength-division multiplex (WDM). The multiplexer couples power from multiple sources, operating at different wavelengths, onto a single fiber. The demultiplexer separates the signals at the receiver end onto different fibers depending on wavelength. The WDM coupler multiplexes different signals on a single fiber optically rather than electronically. The 2 × 2 WDC coupler is also often used as a directional coupler for transmission of two signals, using two different wavelengths, in opposite directions on a single fiber. The power-splitting loss of a T or star coupler is eliminated by using wavelength-selective couplers.

4. Directional couplers: This is a 2 × 2 WID coupler that routes the optical power to a specific port depending on the direction of propagation. Recent single-mode coupler products permit directionality with a single wavelength.

3.7.2 Performance Parameters

Coupling ratio.

$$ \text{CR} = \frac{P_{oc}}{P_{o1} + \cdots + P_{oN}} \tag{3-72} $$

Excess loss (dB).   Excess loss (L_e) is due to scattering, absorption, and imperfections within the coupler.

$$ L_e = 10 \log \frac{P_{o(\text{tot})}}{P_{i(\text{tot})}} \tag{3-73} $$

Directivity (dB).   Directivity is related to the amount of power reflected back by the coupler into other ports. It effects crosstalk and isolation parameters.

$$ \text{DIR} = 10 \log \frac{P_{\text{refl}}}{P_{ic}} \tag{3-74} $$

Uniformity (%).   Uniformity is the measure of the consistency of the coupling ratio as various ports are used as the input port, or in an $N \times N$ coupler, and is the equivalency of the power split between ports. It is either expressed as a percent or as a loss term in dB.

$$ U \text{ (dB)} = 10 \log \left(1 - \frac{U\,(\%)}{100} \right) \tag{3-75} $$

Polarization sensitivity (%).   The structure of some designs, particularly fused biconical, causes birefringence, which can cause a rotation in the plane of polarization of the light. If the design has a coupling ratio that is polarization-dependent, then the polarization of the light upstream of the coupler can effect the coupling ratio. Likewise, vibration of the fiber and coupler can be converted to amplitude modulations [41].

3.7.3 Coupler Insertion Loss (L_t)

Coupler insertion loss (L_t) mechanisms are also shown in Figure 3.47. Insertion loss is defined here as the total optical power loss realized between the input port (P_{ic}) and the coupled output port (P_{oc}):

$$L_t = 10 \log \frac{P_{ic}}{P_{oc}} \tag{3-76}$$

Mean Loss (dB):

$$L_t = L_e - 10 \log \text{CR} - 10 \log \left(1 - \frac{U\ (\%)}{100} \right) + D_{TW} \tag{3-77}$$

Standard Deviation:

$$\sigma_{l_t} = \{\sigma_{cr}^2 + \sigma_u^2 + \sigma_T^2\}^{1/2} \tag{3-78}$$

where D_{TW} is the combined derating for temperature and wavelength in dB, σ_{cr} is the coupling-ratio tolerance, σ_u is the uniformity expressed in statistical terms, and σ_T is the variation with temperature. These loss elements are described in what follows.

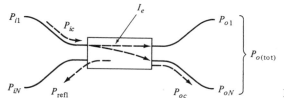

Figure 3.47   Coupler performance model.

   Power-splitting loss is a function of the ratio of the power out of the port of interest (P_{oc}) divided by the sum of the power available from all output ports ($P_{o1} + \cdots + P_{oN}$). This ratio is defined as the coupling ratio (CR). The loss component associated with the power split is, therefore, $-10 \log \text{CR}$, as shown in Equation (3-74). In the case of an $N \times N$ star coupler, where there exist multiple "uniform" ports, then CR $= 1/N$, where N is the total number of output ports:

$$L_{sp} = -10 \log N \qquad \text{for an } N \times N \text{ coupler} \tag{3-79}$$

   Connection loss, or splice loss, is also experienced in the interconnection of the coupler to the fiber link or electrooptical device. For convenience, we treat the connector losses, or splice losses as external to the coupler and combine these losses with those of other connectors and splices in the link-budget analysis.

3.7.4 Wavelength-Dependent Couplers

Although wavelength-independent single-mode couplers can be fabricated [43], there is a sinusoidal relationship between coupling ratio and wavelength in single-mode fused biconical-coupler designs [42]:

$$P_{o2} = P_{ic} \sin^2(CL)$$

$$C = \frac{3\pi\lambda}{32 n_2 a^2 \left(1 + \dfrac{1}{V}\right)^2} \qquad (3\text{-}80)$$

where V is the relationship given in Equation (1-4) in Chapter 1, and $2a$ is the rectangular cross section, L is the length, and n_2 is the refractive index of the tapered mixing region. The results are shown in Figure 3.48 for a single-wavelength design (solid line) and a dual-wavelength design (dashed line). The dual-wavelength effect produces a 2 × 2 wavelength-division multiplexing coupler.

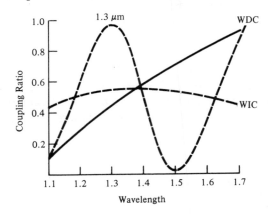

Figure 3.48 Wavelength dependence of fuse tapered single-mode coupler.

3.7.5 Performance Variation

Sensitivity to temperature is somewhat unpredictable and is totally dependent on the structure of the coupler. The parameter most apt to change, however, is the coupling ratio. The range σ_T that might be expected of single-mode fused biconical couplers, for example, is on the order of from 0.001 to 0.01%/°C.

There is no specific aging mechanism with couplers. Constant-temperature cycling, however, can cause stresses that degrade performance. Couplers with connectors suffer degradation effects with mating cycles.

Since many couplers depend on optical-waveguide shaping, mode, coupling, and refractive-index change in the coupling components, they are sensitive to wavelength. Relatively predictable mean coupling-ratio changes, on the order of 0.1 to 0.4% nm can be expected of WDC fused biconical single-mode couplers. Margin should be included for wavelength drift of the source or when couplers are to be used outside of their specified wavelengths.

REFERENCES

1. Institute of Electrical and Electronics Engineers, *Standard Definition of Terms Relating to Fiber Optics*, ANSI/IEEE Standard 812-1984, New York, 1984.

2. J. Gowar, *Optical Communications Systems,* Prentice-Hall International, London, 1984, pp. 79, 245, 376.

3. R. J. McIntire, "Multiplication Noise in a Uniform Avalanche Photodiode," *IEEE Transactions Electronic Devices,* **ED-13** (January 1966): 164–168.

4. Tran Muoi (PCO), contribution from his unpublished work, 1989.

5. Dr. Bernstein, *Fiber and Integrated Optics,* George Washington University, Continuing Engineering Education, Washington, D.C., 1979, chap. 4, p. 35.

6. H. Sunak and M. Zampronio, *Applied Optics* **22,** 1983, p. 2337.

7. W. H. Cheng, *Proceedings of the IEEE* **69** (1981): 396.

8. J. Gowar, *Optical Communications Systems,* Prentice-Hall International, London, 1984, p. 280.

9. C. J. Hwang (President, General Optronics Corp., Edison, N.J.), information provided with that of J. Gowar. The $\sqrt{3}$ factor relates the 1.5-dB relationship from Gowar to the 3-dB bandwidth.

10. C. J. Hwang, President, General Optronics, 201 Sen Ave., Edison, N.J., previously unpublished information, 1987.

11. R. A. Linke, "Direct Gigabit Modulation of Injection Lasers, Structure Dependent Speed Limitations," *Journal of Lightwave Technology* **LT-2,** 1 (February 1984): 40–43.

12. This analysis follows some of the original work performed by R. Baird at Honeywell, Richardson, Texas, 1976.

13. C. J. Hwang, N. Patel, M. Sacilotti, F. Prince, D. Bull, *Journal of Applied Physics* **49** (1978): 29.

14. C. A. Burrus and B. I. Miller, *Optical Fiber Communication* **4** (1971): 307.

15. M. K. Barnoski, *Fundamentals of Optical Fiber Communications,* Academic Press, New York, 1976.

16. H. Kobrinski and C. A. Brackett, "A Survey of Optical Frequency Multiplexing Techniques for Subscriber Local Loop Applications," paper presented at the *Lightwave* Symposium, "Fiber in the Subscriber Loop," 1986.

17. P. Anthony, "Reliability of GaAIAs and InGaAsp Semiconductor Lasers," *Proceedings from the FOC/LAN Conference,* 1983, page 247.

18. J. Ryan, "Lasers and LEDs", *Lightwave* (November 1986): 25–35.

19. D. Charlton (Corning Glass), previously unpublished information, 1988.

20. American National Standards Institute, *Generic Specification for Optical Waveguide Fibers,* ANSI/EIA-4920000-A-1987, New York, 1987.

21. Data derived from laboratory measurements by the author on biconical tapered single-mode couplers.

22. D. Gloge, *Bell System Technological Journal* **54** (1975): 243.

23. W. B. Gardner, *Bell System Technological Journal* **54** (1975): 457.

24. Curves were derived from test observations by the author as well as a general analysis from supplier literature.

25. K. Nassau, "The Diffusion of Water in Optical Fibers," *Mater. Res. Bull.* **13** (1978): 67.

26. M. Buckler and W. Gadner, "Measured Change with Time in Optical Fiber 0.94 μm Loss," paper presented at the Topical Meeting on Optical Fiber Transmission (OSA), 1979.

27. M. K. Barnoski, *Fundamentals of Optical Fiber Communications,* Academic Press, New York, 1976, pp 22–38.

28. Material-dispersion range developed as an average compiled from various fiber suppliers.

29. D. Knecht, Manager of Connector Development, Siecor Corporation, Hickory, N.C., previously unpublished information, 1988.

30. D. Marcuse, "Loss Analysis of Single Mode Fiber Splices", *Bell System Technological Journal* **56,** 5 (May–June 1977): 703–718.

31. W. A. Gambing, H. Matsumua, and C. Ragdale, "Joint Loss in Single Mode Fibers", *Electronics Letter* **14,** 15 (July 1978): 491–499.

32. I. H. Keyama, H. Tsuchiya, "Fusion Splices for Single Mode Optical Fibers", *IEEE Journal of Quantum Electronics* **QE-14,** 8 (August 1987): 614–619.

33. Results obtained from data provided to the author by Sumitomo Electric and GTE Fiber Optic Product.

34. D. Taylor and S. Daikkonen, "Factory Splicing of Optical Waveguide Fiber," *Wire Journal International* (August 1984): 46–52.

35. Measurement of fusion splices of identical fiber made on Sumitomo Electric fusion splicer.

36. Field test data measured by the author on over 300 fusion splices.

37. "A Splice for All Seasons," GTE elastomeric splice test report, extracted from document #QR85-7700-001, 1985.

38. Results derived from PSI Lightlinker Burbank, CA, Fiber Optic Splice System data sheet, 1986.

39. Effect was extrapolated from results reported to the author by Bell Laboratories and measurements taken by the author on fusion-spliced single-mode fiber.

40. J. Lidh, "Splice Losses: How Measurements Can Mislead," *Lightwave* (June 1986): 43–47.

41. Gould Electronics Inc. OSD, *Coupling Ratio Stability,* Tech. Note #2, Glen Burnie, MD, 1986.

42. F. P. Payne, C. D. Hussey, and M. S. Yataki, *Electronics Letter* **21** (1985): 461.

43. Gould Electronics Inc. OSD, *Wavelength Dependence,* Tech. Note #3, Glen Burnie, MD, 1986.

4

Signal-to-Noise Ratios in the Optical System

4.1 INTRODUCTION

Signal quality is generally specified in terms of (a) bandwidth and signal-to-noise ratio for analog transmission, and (b) bit rate, bit-error rate, and jitter for digital transmission. This chapter defines these performance parameters and provides the basis for the relationships between parameters in the electrical realm and those on the optical realm.

The signal-quality terms are interrelated. For instance, SNR cannot be defined without also defining the bandwidth in which the noise is measured, since total noise power (P_n) is the product of noise density (P_n) and bandwidth:

$$P_n = P_n' \, \text{BW} \tag{4-1}$$

The SNR varies in different parts of the system as bandwidth varies. In order to eliminate confusion, we use a subscript on the BW and SNR terms to define the bandwidth in which the noise power or SNR is measured. $(\text{SNR})_b$ and $(\text{BW})_b$ indicate the baseband signal bandwidth, $(\text{SNR})_c$ and $(\text{BW})_c$ indicate parameters measured in the intermediate per-channel signal bandwidth (prior to multiplexing at the line layer), and $(\text{SNR})_t$ and $(\text{BW})_t$ are parameters measured in the transmission channel bandwidth within the section layer. $(\text{SNR})_o$ and $(\text{BW})_o$ represent measurements in the optical realm (optical-link layer) and $(\text{SNR})_e$ and $(\text{BW})_e$ represent measurements in the electrical realm of the higher layers.

Analog and binary signaling is treated individually. Signal-to-noise ratio primarily defines the quality of signal transmission on a signal-amplitude and noise-amplitude basis. This is appropriate for defining the quality of analog signals. For data signals, however, SNR has little meaning except within the detector/regenerator. The quality of a baseband data signal is defined in terms of bit-error rate (BER)

141

and timing jitter (t_j). Bit-error rate is related to the rate of erroneous symbol detections due generally to noise levels that exceed the detection threshold between symbol states. For this reason, BER and SNR can be related. Jitter is a function of noise that causes the signal to cross the detection threshold at the wrong point in time, therefore, causing nonuniform timing between symbol states. Jitter, therefore, can also be related to received SNR.

4.2 SNR RELATIONSHIPS

4.2.1 Signal-to-Noise Ratio

The signal-to-noise ratio (SNR) is the ratio between the information signal component of a composite waveform and the noise component. Signal and noise components can be represented as voltages, currents, or as power, therefore, creating many possible definitions for SNR. SNR is also represented differently depending on where the SNR is measured and the signal format. In Figure 4.1, (SNR)$_e$ is represented as a ratio of voltages at the receiver output, as with most radio or copper communications systems. Within the optical receiver, it is represented in terms of the ratio of signal photocurrent (i_s) to noise current (i_n) at the detector/preamplifier interface. Prior to detection, it is measured in terms of the ratio of optical-signal power (P_s) to noise equivalent power (NEP); NEP is the optical-power equivalent of the noise current at the detector.

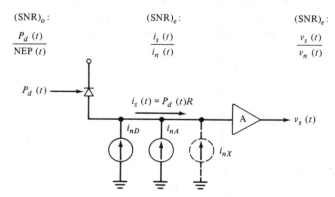

$$(SNR)_o: \quad \frac{P_d(t)}{NEP(t)} \qquad (SNR)_e: \quad \frac{i_s(t)}{i_n(t)} \qquad (SNR)_e: \quad \frac{v_s(t)}{v_n(t)}$$

Figure 4.1 Signal-to-noise relationships at the optical receiver.

SNR relationships differ, depending on whether it is a power ratio or a current or voltage ratio. In order to keep track of the nomenclature, subscript capital letters are used to indicate power ratios, whereas subscript lowercase letters indicate current or voltage ratios. The basic definitions for the SNR are as follows:

$$(SNR)_{\text{level}} = \frac{V_s}{V_n} = \frac{I_s}{I_n} \tag{4-2}$$

$$(SNR)_{\text{POWER}} = \frac{V_s^2/R}{V_n^2/R} = \frac{I_s^2 R}{I_n^2 R} = \frac{\text{signal power}}{\text{noise power}} \tag{4-3}$$

It is common to relate SNR in decibels (dB). A decibel is the relation of powers. SNR (dB) is therefore represented as follows:

$$\text{SNR (dB)} = 10 \log \frac{P_s}{P_n} \tag{4-4}$$

$$\text{SNR (dB)} = 10 \log \frac{V_s^2}{V_n^2} = 20 \log \frac{V_s}{V_n} \tag{4-5}$$

$$\text{SNR (dB)} = 10 \log \frac{I_s^2}{I_n^2} = 20 \log \frac{I_s}{I_n} \tag{4-6}$$

Since signal level and noise level are generally time varying (not constant), they are treated over long time periods as average or statistical relationships. Some of the most common relationships follow, represented as photocurrent ratios:

(a) Instantaneous SNR: This is the ratio of signal-to-noise levels at an instant in time.

$$\text{SNR}(t) = \frac{i_s(t)}{i_n(t)} = \frac{\text{signal level}}{\text{noise level}} \tag{4-7}$$

(b) Peak signal-to-rms noise: Since noise is generally random, a more useful definition is the ratio of signal level to rms noise level. When the peak signal level is constant, such as with digital signaling, the peak-to-rms SNR is sometimes used:

$$(\text{SNR})_{\text{pk/rms}} = \frac{i_{s(\text{pk})}}{\langle i_n \rangle} \tag{4-8}$$

(c) Average signal-to-rms noise current: This is useful when the signal has relatively constant characteristics, such as with digital signaling.

$$(\text{SNR})_{\text{avg/rms}} = \frac{i_{s(\text{avg})}}{\langle i_n \rangle} \tag{4-9}$$

For a series of rectangular pulses of amplitude I_p, pulse width T_p, and period T:

$$(\text{SNR})_{\text{avg/rms}} = \frac{I_p(T_p/T)}{\langle i_n \rangle} \tag{4-10}$$

(d) RMS-signal-to-rms noise level: If the signal is a constant-amplitude sinusoid of peak amplitude $i_{s(\text{pk})}$

$$(\text{SNR})_{\text{rms/rms}} = \frac{i_{\text{pk}}/\sqrt{2}}{\langle i_n \rangle} \tag{4-11}$$

For a rectangular pulse train:

$$(\text{SNR})_{\text{rms/rms}} = \frac{I_p(T_p/T)^{1/2}}{\langle i_n \rangle} \tag{4-12}$$

(e) RMS-signal-to-rms-noise power: This is the most common relationship in communications systems since, in most cases, both the signal and the noise are somewhat random with time.

$$(\text{SNR})_{\text{RMS/RMS}} = \frac{\langle i_s^2 \rangle}{\langle i_n^2 \rangle} \tag{4-13}$$

(f) Carrier-to-noise ratio (CNR): When the signal of interest is transmitted by modulation of a sinusoidal carrier, there exists a direct relationship between the original baseband SNR and the level of the carrier above the noise in the passband of the transmission channel. This relationship is a function of the modulation parameters, as is defined in Chapter 5. The relationship of carrier power to noise power (CNR) is, therefore, a useful performance parameter for analog-modulated transmission. The carrier is a constant-level sinusoid and, therefore, peak-level relationships can be used:

$$(\text{CNR})_{\text{RMS/RMS}} = \frac{i_{c(\text{pk})}^2 / 2}{\langle i_n^2 \rangle} \tag{4-14}$$

4.3 SNR IN THE OPTICAL SYSTEM

In analyzing performance at the photonics layer, minimum required power at the detector (MDP) is calculated by defining the desired SNR at the output of the photodetector for desired systems performance. All noise at the receiver, including the receiver amplifier noise, is therefore referenced as noise current to this point. The signal-to-noise ratio is represented as a ratio of received signal photocurrent (i_s) to equivalent rms noise currents $\langle i_n \rangle$.

4.3.1 Signal to Noise at the Optical Receiver

Figure 4.1 shows the relationships at the optical receiver. The noise sources are all related as current sources at the input to a noiseless gain stage. The received signal is shown on the optical side of the detector as signal power $p_d(t)$. After detection, on the electrical side, it is shown as the signal component of the photocurrent (i_s). The relationships are as follows:

(a) Photocurrent (i_p): From Figure 4.1, the detected photocurrent as a result of detected power P_d is

$$i_p = rGP_d = RP_d \tag{4-15}$$

where $R = rG$ is the photodetector responsivity times gain, P_d is in watts, and R is in amperes/watt.

$$r = \text{responsivity} = \eta \left\{ \frac{e\lambda}{hc} \right\} \text{ A/W}$$

$$= \frac{\eta(1.6 \times 10^{-20})\lambda}{(6.626 \times 10^{-34})(3 \times 10^8)} = \frac{\eta\lambda}{1.24} \ (\lambda \text{ in } \mu\text{m}) \tag{4-16}$$

where η = external quantum efficiency

$\eta = (1 - p)\eta_i$

η_i = internal quantum efficiency

p = surface reflectance = $\dfrac{(n_d - n_o)^2}{(n_d + n_o)^2}$

n_d and n_o = refractive indexes of the detector and the surrounding medium, respectively

(b) Noise equivalent power $[(NEP)_o]$: This is defined as the received optical power equivalent of all noise sources referenced to the detector:

$$(NEP)_o = \frac{i_{nT}}{R} = \frac{i_{nT}}{rG} \qquad (4\text{-}17)$$

where $i_{nT} = i_{nD} + i_{nA} + i_{nX}$, detector, amplifier and other noise referenced to the detector.

(c) Detected optical power: If the optical source is biased such that the detector receives power P_o with no signal present, then detected power in the presence of signal $s(t)$ is

$$P_d(t) = [1 + M_o s(t)]P_o \qquad (4\text{-}18)$$

where M_o is the modulation index of the source power by the signal. When the signal is defined in terms of rms values $\langle s(t) \rangle$ then

$$\langle P_d(t) \rangle = P_o + M_o P_o \langle s(t) \rangle \qquad (4\text{-}19)$$

(d) Detected signal current: Substituting Equation (4-18) into Equation (4-15) we get

$$i_p(t) = rGP_o[1 + M_o s(t)] \qquad (4\text{-}20)$$

Eliminating the nonsignal steady-state component and expressing the signal in rms $\langle s(t) \rangle$, we get the rms detected signal current:

$$\langle i_s(t) \rangle = M_o rG[P_o \langle s(t) \rangle] \qquad (4\text{-}21)$$

(e) Detected noise current: The total rms noise current is the square root of the sum of the squares of the individual noise sources:

$$\langle i_{nT} \rangle = \sqrt{\langle i_{nD}^2 \rangle + \langle i_{nA}^2 \rangle + \langle i_{nX}^2 \rangle} \qquad (4\text{-}22)$$

The signal-to-noise ratio can be related in two ways, depending on which side of the photodetector is referenced. If the relationship is in terms of optical-power ratios on the optical side of the detector, the term is $(SNR)_o$. If the relationship is between current or power ratios on the electrical side of the detector, then the term is $(SNR)_e$. From Figure 4.1, the following can be defined:

(a) Predetected optical signal-to-noise ratio, $(SNR)_o$:

$$(SNR)_o = \frac{p_d(t)}{NEP(t)} = \frac{p_d(t)}{i_{nT}(t)/R} \qquad (4\text{-}23)$$

(b) Post-detected electrical signal-to-noise ratio, $(SNR)_e$:

$$(SNR)_e = \frac{i_s(t)}{i_{nT}(t)} \tag{4-24}$$

When signal-to-noise ratio is related in terms of dB, then $(SNR)_e = 2\,(SNR)_o$ as follows:

$$(SNR)_e = 20 \log \frac{i_s(t)}{i_{nT}(t)} \tag{4-25}$$

$$(SNR)_o = 10 \log \frac{p_d(t)}{NEP(t)} \tag{4-26}$$

If we multiply the numerator and denominator of $(SNR)_e$ by responsivity R, then:

$$
\begin{aligned}
(SNR)_e &= 20 \log \frac{i_s/R}{i_{nT}/R} \\[4pt]
&= 20 \log \left\{ \frac{p_d(t)}{NEP(t)} \right\} \\[4pt]
&= 2 \times 10 \log \frac{p_d(t)}{NEP(t)} \\[4pt]
&= 2\,(SNR)_o
\end{aligned} \tag{4-27}
$$

These relationships have been defined for instantaneous values of signal and noise. $(SNR)_e$ at the optical receiver can be defined by substituting Equations (4-21) and (4-22) into the equations in Section 4.2. Some of the more commonly used relationships follow.

(a) Peak-signal-to-rms-noise level: If we define the peak signal power as $P_o\langle s_{pk}\rangle = P_{d(pk)}$, then from Equation (4-8):

$$(SNR)_{e,\,pk/rms} = \frac{I_{s(pk)}}{\langle i_{nT}\rangle} = \frac{M_o r G P_{d(pk)}}{\sqrt{\langle i_{nD}^2\rangle + \langle i_{nA}^2\rangle + \langle i_{nX}^2\rangle}} \tag{4-28}$$

(b) RMS signal power to rms noise power: By substituting Equations (4-21) and (4-22) into Equation (4-13), we get

$$
\begin{aligned}
(SNR)_{e,\,RMS/RMS} &= \frac{\langle i_s^2(t)\rangle}{\langle i_{nT}^2(t)\rangle} \\[6pt]
&= \frac{(M_o r G P_o)^2 \langle s(t)^2\rangle}{\langle i_{nD}^2\rangle + \langle i_{nA}^2\rangle + \langle i_{nX}^2\rangle}
\end{aligned} \tag{4-29}
$$

4.3.2 Carrier to Noise in the Optical System

When the baseband signal modulates a sinusoidal carrier for transmission, a relationship can be formed between baseband signal-to-noise, $(SNR)_b$; carrier modulation index, m; and the ratio of the transmitted unmodulated carrier power to the noise,

$(CNR)_c$, in the modulated intermediate channel bandwidth, $(BW)_c$. Once the relationship is established, then the required detected power (MDP) can be determined by measuring the ratio of unmodulated carrier power to noise power in $(BW)_c$. These relationships are discussed in detail in Chapter 5. Since this relationship can be established, we can then work with simple sinusoidal carriers in computing SNR within an analog optical transmission system.

Since the carrier can be assumed to be a constant-level sinusoid, then transmitted carrier power C is [1]

$$C(t) = 2P_x[1 + M_o \cos(w_c t)] \cos^2(w_o t) \qquad (4\text{-}30)$$

where w_c = signal carrier
w_o = optical carrier
P_x = rms optical-carrier electric field = $A^2/2$
M_o = modulation index of the optical carrier, and is equal to the ratio of peak-to-peak carrier signal power to total available power

From Equations (4-20) and (4-30), the instantaneous detected photocurrent is, therefore:

$$i_c(t) = rGP_o[1 + M_o \cos(w_c t)]$$

The rms detected carrier power C_{RMS} is, therefore:

$$C_{\text{RMS}} = \langle i_c^2(t) \rangle = \{M_o rGP_o\}^2 \langle c^2(t) \rangle$$

where $\langle c(t)^2 \rangle$ is $\frac{1}{2}$ for a sinusoid. Therefore:

$$C_{\text{RMS}} = \frac{\{M_o rGP_o\}^2}{2} \qquad (4\text{-}31)$$

Substituting Equation (4-31) for signal power into Equation (4-29), we get the rms carrier-to-noise ratio:

$$\text{CNR}_{e,\text{RMS}} = \frac{\langle i_c^2(t) \rangle}{\langle i_{nT}^2(t) \rangle} = \frac{\{M_o rGP_o\}^2/2}{\langle i_{nD}^2 \rangle + \langle i_{nA}^2 \rangle + \langle i_{nX}^2 \rangle} \qquad (4\text{-}32)$$

4.4 BANDWIDTH

Signal-to-noise ratio and bandwidth are generally the key signal-performance parameters associated with analog signal transmission. Because noise magnitude or power is defined within a specific measurement bandwidth, there is a direct relationship between bandwidth and SNR. This section defines bandwidth and its relationships within the optical system.

4.4.1 Definitions

Consistent relationships for bandwidth are required throughout the system design so that the noise within each signal passband can be computed to establish the SNR and the receiver bandwidth and to determine the response times of the various components in the signal path. Depending on what is being described, bandwidth can be defined in terms of the spectral occupancy of the signal, the response of filter networks in the system, or that equivalent bandwidth that defines the amount of noise in the system.

(a) *Ideal Bandwidth:* This is shown as the sharp cutoff (square response shape) in Figures 4.2 and 4.3. It assumes a constant level or gain within the passband and zero level beyond the passband.

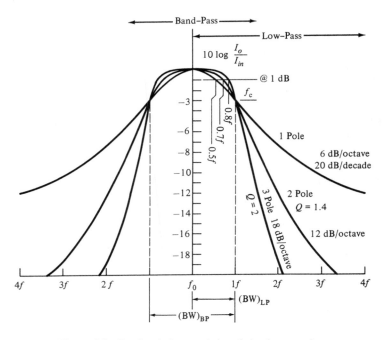

Figure 4.2 Passband characteristics of simple networks.

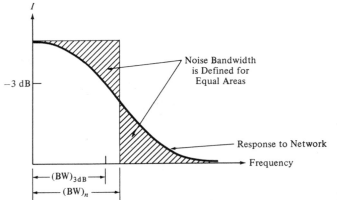

Figure 4.3 Noise bandwidth.

(b) Response Passband: This is the actual response of a network in terms of gain versus frequency. The passband bandwidth is defined as the frequency band between lower and upper frequency points on the response curve, which are at the defined signal, gain, or power level. The most commonly used definition is the 3-dB bandwidth, $(BW)_{3dB}$, which defines the frequency band between the points at which the gain, voltage, or current level is 3 dB below the maximum level. This corresponds to the frequency of the poles of the network. Figure 4.2 shows the 3-dB bandwidth response for simple 1-, 2- and 3-pole networks, with the ideal 3-dB bandpass superimposed (dashed line).

(c) Noise Bandwidth, $(BW)_n$: The relationship between noise bandwidth and signal or network response is shown in Figure 4.3. Where the noise is considered "white" (equal power per unit bandwidth anywhere in the spectrum), the equivalent bandwidth is defined as the ideal bandwidth describing the points where the area under the noise bandwidth curve is equal to that under the response curve. Table 4.1 describes the relationship between noise bandwidth and the 3-dB bandwidth for simple networks [2]. In the case of three or more poles, $(BW)_{3dB}$ can be substituted for noise bandwidth $(BW)_n$ with only about a 10 to 15% error.

TABLE 4.1   NOISE BAND-
WIDTH RELATED TO 3-dB
BANDWIDTH

No. poles	$(BW)_n$
1	1.57 $(BW)_{3dB}$
2	1.22 $(BW)_{3dB}$
3	1.15 $(BW)_{3dB}$
4	1.13 $(BW)_{3dB}$
5	1.11 $(BW)_{3dB}$

(d) Signal Bandwidth: A transmitted signal can be characterized in the frequency domain by a frequency spectrum. Although this spectrum is constantly changing with time as the signal varies, most of the energy in the signal is limited to occupying a finite band of frequencies. The width of this frequency band, between points defined in terms of power or signal levels, is called the bandwidth.

Definitions can vary widely for signal bandwidth since different signal spectra, modulation techniques, and applications use different definitions to determine the power points. For the purposes of optical-systems design, we generally use the 3-dB definition. In this case, it is the band of frequencies between the points at which the power spectral density has dropped to half the power of the peak value. This can also be called the half-power bandwidth.

Bandwidth can also be defined based on the percentage of power contained between two spectral frequencies ($\%P_{BW}$). Since the percentage is often specified as 90% or more, 90% P_{BW} is greater than the 3-dB bandwidth. For signals with well-defined lobes or null frequencies, such as binary signals, the bandwidth of the main lobe between null points can be specified. In some cases, such as for voice or video transmission, the signal bandpass is specified in terms of a "weighting-filter" passband. This is an attempt to recognize only the signal components noticeable to the listener or viewer when defining SNR.

If these alternate definitions for bandwidth are used, then conversions are required to translate between them and the 3-dB bandwidth. This is necessary in order to establish an equivalent noise bandwidth for computation of total noise power.

We use the following additional bandwidth terminology to discriminate between the interfaces at the path, line, section, and photonics layers. These terms are illustrated in Figure 4.4 for the analog transmission case. Also refer to Figure 1.9 in Chapter 1.

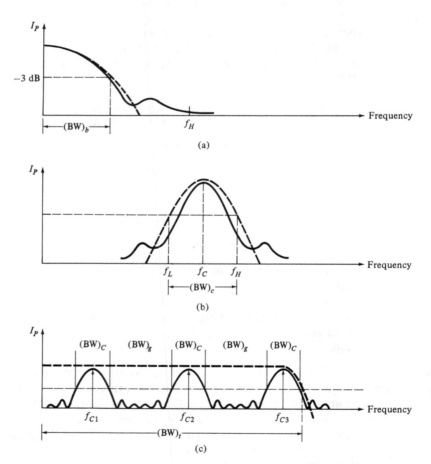

Figure 4.4 Channel-bandwidth terminology. (a) Baseband channel bandwidth, $(BW)_b$; (b) intermediate-channel bandwidth, $(BW)_c$; and (c) transmission-channel bandwidth, $(BW)_t$.

(a) Baseband Bandwidth, $(BW)_b$: This is the band occupied by the original signal at the input port (path- or service-layer interface) of the system.

(b) Intermediate-Channel Bandwidth $(BW)_c$: This term is shortened in some sections to "channel bandwidth." It describes the passband occupancy of the signal at the line-layer interface once it has been processed for transmission by the path layer. For analog signals, processing generally consists of carrier modulation or digital encoding, and sometimes lower-level multiplexing, all of which transform the power and spectral content from that of the original baseband channel. Without modulation or further processing within the path layer, $(BW)_c = (BW)_b$.

(c) Transmission Bandwidth $(BW)_t$: $(BW)_{ts}$ and $(BW)_t$ are the total spectral occupancy of the composite signal or signals within the transmission channel as presented to the section and photonics layers, respectively. If the bandwidth requirements of the section layer are different than that of the photonics layer, the term $(BW)_{ts}$ is used at the section layer. With multiplexing or further signal processing within the line layer, the transmission bandwidth is different from that of the individual intermediate-channel bandwidths. With no intermediate-channel processing or multiplexing, $(BW)_t = (BW)_c$.

Figure 4.4 shows the three definitions for a case where three baseband analog signals modulate a carrier and then are frequency-division multiplexed (FDM). Here three hypothetical signal spectra are illustrated. The baseband signal occupies a spectrum from DC to a high-frequency point f_H with a 3-dB bandwidth of $(BW)_b$. The intermediate channel is indicative of how the baseband signal might appear if it modulated a carrier of frequency f_c. In this case, the signal occupies a wider spectrum extending from f_L to a new f_H, centered around f_c, with a 3-dB channel bandwidth of $(BW)_c$. In the transmission band, the three intermediate-channel signals, all of bandwidth $(BW)_c$, are frequency multiplexed at different carrier frequencies f_{c1}, f_{c2}, and f_{c3}. They cannot be totally adjacent at the 3-dB points or there would be interference of the higher-frequency components. For this reason, there exist guard bands $(BW)_g$ between the three channels. The transmission-channel bandwidth $(BW)_t$, therefore, is much larger than $(BW)_c$, not only by the multiplexing multiple, but by an inefficiency factor due to guard bands as well.

For all cases, a dashed line is superimposed indicating how the corresponding channel passband filter might appear.

4.4.2 Bandwidth Conversion

SNR performance is measured in the receiver noise bandwidth on the electrical side of the detector/receiver. Fiber bandwidth and dispersion limitations, however, do effect the spectrum of the received signal, making it necessary to compensate by increasing receiver bandwidth, as is discussed in Chapter 6. Since fiber performance is often specified in terms of optical bandwidth, conversion between optical and electrical bandwidth parameters is often a necessary part of the analysis. Bandwidth on the optical side is represented by a reduction in optical power, and on the electrical side by a reduction in detected photocurrent. The relationships are as follows:

(a) 3-dB optical bandwidth: The 3-dB optical bandwidth defines the modulation frequency f at which the optical power has reduced to 50% of the power level at zero modulation frequency, $P_o(0)$:

$$10 \log \frac{P_o(f)}{P_o(0)} = -3 \text{ dB}$$

$$P_o(f) = 0.5P_o(0)$$

$$(4\text{-}33)$$

(b) 3-dB electrical bandwidth: The 3-dB electrical bandwidth defines the frequency f at which the detected modulated photocurrent has reduced to 3 dB of its value at zero modulation frequency, $I(0)$:

$$20 \log \frac{I(f)}{I(0)} = -3 \text{ dB}$$

$$I(f) = 0.707I(0)$$

$$(4\text{-}34)$$

(c) Optical-to-electrical conversion: From the previous relationships, it can be seen that the optical-power reduction to electrical-current reduction at the corresponding 3-dB points are related by $\sqrt{2}$. This is shown in Figure 4.5.

$$(\text{BW})_{e,\,3\,\text{dB}} = 0.707 \, (\text{BW})_{o,\,3\,\text{dB}} \qquad (4\text{-}35)$$

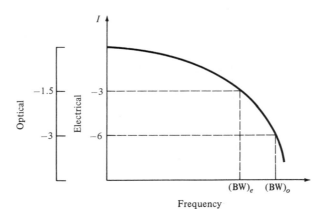

Figure 4.5 Relationship between optical and electrical bandwidth.

(d) Bit rate versus bandwidth: Gowar [3] gives the following relationships for bandwidth versus bit rate:

$$\text{BR} = 1.92(\text{BW})_e = 1.01(\text{BW})_o \qquad \text{for a rectangular pulse} \qquad (4\text{-}36)$$

$$\text{BR} = 1.89(\text{BW})_e = 1.34(\text{BW})_o \qquad \text{for a Gaussian pulse} \qquad (4\text{-}37)$$

These are based on the assumption that the impulse response $h(t)$ and the fiber frequency response $H(f)$ is for a rectangular pulse:

$$h(t) = 1/T \qquad \text{for } t < T/2$$

$$H(f) = \frac{\sin(\pi fT)}{\pi fT}$$

and for a Gaussian pulse:

$$h(t) = \frac{1}{\sqrt{2\pi}\sigma} \exp(-t^2/2\sigma^2)$$

$$H(f) = \exp(-2\pi^2 f^2 \sigma^2)$$

(e) Optical-pulse dispersion, t_d: In systems transmitting binary signals, the response of the fiber medium is measured in terms of dispersion instead of bandwidth. Dispersion has the effect of (1) spreading the received pulse in time while reducing its amplitude and (2) reducing the rise time of the received pulse from that of its transmitted value.

Figure 4.6 shows the optical power (or detected photocurrent) from a transmitted impulse emerging at the receiving end of a fiber. For the purpose of analysis, the

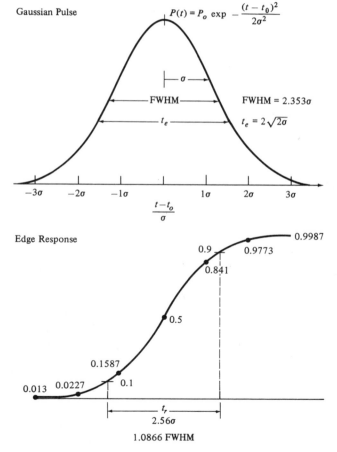

Figure 4.6   Gaussian pulse relationships.

resultant pulse shape due to the dispersed power is assumed to be Gaussian (which is fairly accurate in practice). Dispersion is generally defined and measured in either of two ways:

1. *Full-Width Half Maximum* (FWHM): This is the pulse width at the points where power is 50% of its maximum value.

2. *RMS Pulse Width* (σ): This is the rms value of the pulse width.

$$\text{FWHM} = 2.353\sigma \qquad (4\text{-}38)$$

(f) Dispersion versus bandwidth: If we assume that the Gaussian pulse shape holds, the 3 dB electrical bandwidth of the Gaussian pulse is related to its rms width by $(\text{BW})_{e,\,3dB} = \sqrt{\ln 2}/2\pi\sigma$, therefore we get

$$(\text{BW})_{e,\,3dB} = \frac{0.132}{\sigma} = \frac{0.312}{\text{FWHM}} \text{ Hz/s} \qquad (4\text{-}39)$$

$$(\text{BW})_{f,\,\text{optical}} = \frac{0.187}{\sigma} = \frac{0.441}{\text{FWHM}} \text{ Hz/s} \qquad (4\text{-}40)$$

(g) Dispersion versus rise time: As is further discussed in Chapter 6, the response of a digital or analog transmission system can be analyzed in terms of the rise times of its components. Since pulse dispersion is the spreading of a pulse in time, it has the direct effect of reducing the pulse rise time. Figure 4.6 also geometrically plots the edge response of the Gaussian pulse (power under the curve). Rise time t_r, defined as the time between the 10 and 90% amplitude points, is compared to FWHM dispersion t_d, yielding the following relationship [4]:

$$t_r = 1.0866t_{d,\,\text{FWHM}} = 2.56t_{d,\,\text{rms}} \qquad (4\text{-}41)$$

4.5 BER AND BIT RATE FOR DIGITAL TRANSMISSION

When the transmitted signal is in digital form, the performance parameters are generally bit rate (BR), bit-error rate (BER), and jitter (t_j). Within an optical receiver, however, the signal is treated as analog prior to threshold detection and regeneration. The regenerator performs the function of determining what logical "state" is represented by the received signal waveform at a specific sample period. It "decodes" the received signal by comparing signal amplitude with a set threshold. The presence of noise at the threshold can cause false crossings and, therefore, "bit errors." Because proper decoding depends on signal level relative to noise, there is a direct relationship between BER and SNR at the receiver. The amount of noise present is a function of receiver bandwidth, which in turn is a function of signal BR.

4.5.1 SNR and BER

BER is defined as the ratio of the number of information bits detected erroneously to the total number of bits transmitted. Transmitted bits are detected in error because noise can cause the signal to appear on the wrong side of the threshold detector during the sampling instant, as shown in Figure 4.7. BER is, therefore, a function of the received SNR.

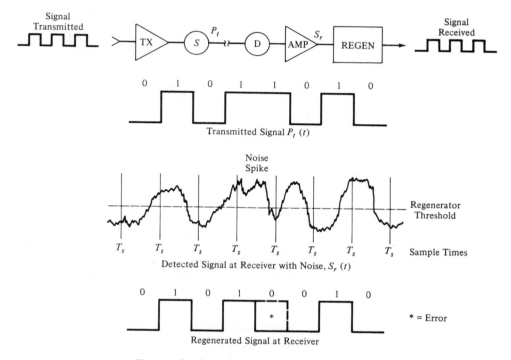

Figure 4.7   Error mechanism in binary signaling.

Binary encoding.   The relationship is statistical, that is, the probability of an error increases as the probability of the noise level exceeding threshold increases. The relationship for binary encoding (on–off) is shown in Figure 4.8, where noise current at the detector is shown with Gaussian distributions for both the binary "1" state and the binary "0" state. Note that the noise distribution is larger for the "1" state than for the "0". This is because the added optical power for a "1" state increases quantum noise. The threshold, therefore, must be offset for equal probability of error for the two states. The threshold voltage is related to its equivalent photocurrent I_{TH} in this diagram.

A rigorous calculation of error probability can be found in various sources [1, 5, 6]. A simplification is given here. BER or error probability P_e is defined by the complementary error function following [1]:

$$P_e = \text{erfc}\,(Q) = \frac{1}{\sqrt{2\pi}} \int_{Q}^{\infty} \exp\left(-\frac{Z^2}{2}\right) dZ \qquad (4\text{-}42)$$

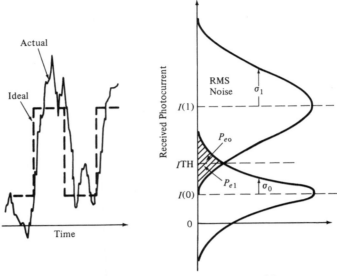

Figure 4.8 Error probability for binary states. $I(0)$ = mean photocurrent for a binary "0," $I(1)$ = mean photocurrent for a binary "1," I_{TH} = decision threshold level, σ_0 = rms signal variance due to noise for a binary "0," σ_1 = rms signal variance due to noise for a binary "1," P_{eo} = probability that a "1" was detected when a "0" was sent, and P_{e1} = probability that a "0" was detected when a "1" was sent.

where noise is assumed gaussian with deviation σ

Q = mean signal/rms noise = $\langle s \rangle / \sigma$

Z = actual noise level/rms noise

$\quad = (i_1 - I_1)/\sigma(1)$ or $(i_0 - I_0)/\sigma(0)$

Error probability is defined in terms of the probability of a signal i_s failing to cross the threshold level I_{TH}. The error probability for each state is, therefore [following 1 and 6]:

$$P_{e,0/1} = \text{erfc} \frac{I_1 - I_{TH}}{\langle i_{n1} \rangle} \qquad (4\text{-}43)$$

$$P_{e,1/0} = \text{erfc} \frac{I_{TH} - I_0}{\langle i_{n0} \rangle} \qquad (4\text{-}44)$$

where I_1 = mean photocurrent for a binary 1

$\quad I_0$ = mean photocurrent for a binary 0

$\quad \langle i_{n1} \rangle$ = rms noise current for a binary 1 state

$\quad \langle i_{n0} \rangle$ = rms noise current for a binary 0 state

If I_{TH} is chosen so that $P_{1/0} = P_{0/1} = P_e$, then $P_e = \frac{1}{2}(P_{1/0} + P_{0/1})$, and, therefore:

$$P_e = \text{erfc} \frac{I_1 - I_0}{\langle i_{n0} \rangle + \langle i_{n1} \rangle} = \text{erfc} \frac{I_s}{\langle i_{n0} \rangle + \langle i_{n1} \rangle} \qquad (4\text{-}45)$$

where I_s is the peak signal current. If we define $P_e = \text{erfc}\ (Q)$, then:

$$Q = \frac{I_s}{\langle i_{n0} \rangle + \langle i_{n1} \rangle} \qquad (4\text{-}46)$$

The noise in either state $\langle i_n \rangle$ is composed of quantum, leakage, and amplifier noise from Table 3.3 and Equation (4-22):

$$\langle i_n \rangle = \sqrt{\langle i_{nQ}^2 \rangle + \langle i_{nL}^2 \rangle + \langle i_{nA}^2 \rangle}$$

Note that for $P_d(0) = 0$, the quantum noise term in $\langle i_{n0} \rangle$ disappears. Therefore, $\langle i_{n0} \rangle$ becomes smaller than $\langle i_{n1} \rangle$ by that amount. For this case,

$$Q = \frac{I_s}{\langle i_{n0} \rangle + \sqrt{\langle i_{nQ}^2 \rangle + \langle i_{n0}^2 \rangle}} \qquad (4\text{-}47)$$

where:

$$\langle i_{n0}^2 \rangle = \langle i_{nL}^2 \rangle + \langle i_{nA}^2 \rangle$$

If quantum noise is not dominant, such as is the case with PIN devices, then $\langle i_{n0} \rangle = \langle i_{n1} \rangle = \langle i_n \rangle$, and the numerator becomes $2\langle i_n \rangle$.

$$P_e = \text{erfc}\ (Q) = \text{erfc} \frac{I_s}{2\langle i_n \rangle} \qquad (4\text{-}48)$$

Therefore:

$$2Q = \frac{I_s}{\langle i_n \rangle} \qquad (4\text{-}49)$$

and

$$Q^2 = \frac{(I_s/2)^2}{\langle i_n \rangle^2} = \frac{\text{average signal power}}{\text{rms noise power}} \qquad (4\text{-}50)$$

The power relationships are, therefore:

$$(\text{SNR})_{\text{PK/RMS}} = 20 \log 2Q \qquad (4\text{-}51)$$

$$(\text{SNR})_{\text{AVG/RMS}} = 20 \log Q \qquad (4\text{-}52)$$

If we plot the error function (erfc) and $(\text{SNR})_{\text{P/RMS}}$ versus Q, we get the curve shown in Figure 4.9. It can be seen, for example, that for a BER of 10^{-9}, a Q of 6 is required and

$$(\text{SNR})_{\text{PK/RMS}} = 20 \log 12 = 21.6\ \text{dB}$$

$$(\text{SNR})_{\text{AVG/RMS}} = 20 \log 6 = 15.6\ \text{dB}$$

Bipolar encoding (three states).   The relationship for error probability is dependent on the signal encoding at the threshold detector. We have been addressing the simple case of a unipolar binary signal (on–off). There are cases even in optical

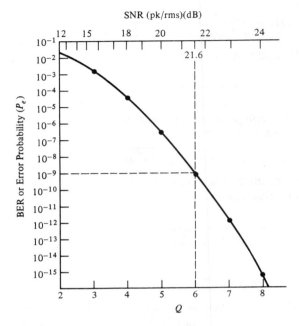

Figure 4.9 Relationship of Q to BER and received $(SNR)_{pk/rms}$.

systems where the signal may be transmitted in a bipolar fashion by biasing the source at midpoint and modulating around it (see Figure 4.10). A "zero" state (or "off" state in some codes) is a level midpoint between the peak signaling states. This corresponds to zero current in a twisted-pair system. Applications include some forms of manchester encoding and certain databus approaches. The relationship for bipolar encoding following Smith [7] is

$$\text{BER} = \frac{2}{3} \text{ erfc } \frac{I_s}{4\langle i_n \rangle} \tag{4-53}$$

If we define Q in the same fashion as before, $Q = I_s/2\langle i_n \rangle$, then:

$$\text{BER} = \frac{2}{3} \text{ erfc } \frac{Q}{2} \tag{4-54}$$

The relationship between Q and SNR remains the same by definition, however, the value of Q for the same BER is higher than that for a unipolar binary signal.

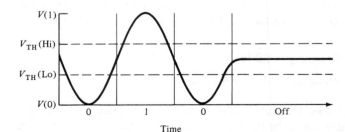

Figure 4.10 Bipolar signaling.

4.5.2 Bit Rate versus Bandwidth for Rectangular Pulses

When the information to be transmitted is binary or in a pulsed format, pulse rate or bit rate (BR) is the specified frequency-domain parameter. Knowing the relationship between bit rate and bandwidth is necessary in order to define receiver noise bandwidth and required optical-link bandwidth. To define SNR at the receiver, bit rate must be translated into equivalent noise bandwidth $(BW)_n$. The relationship between bit rate and bandwidth (or spectral occupancy) differs depending on signal encoding. The translation depends not only on bit period T, which equals $1/BR$, but on pulse shape as well.

Nyquist bandwidth.   The ideal relationship between noise bandwidth and bit rate is called the Nyquist bandwidth. This relationship assumes that in a train of pulses, the pulses are shaped at the output of the filter such that there is no intersymbol interference. This requirement is met by the impulse response of an ideal filter, as shown in Figure 4.11:

$$h(t) = \frac{\sin\,(\pi t/T)}{\pi t/T} \qquad (4\text{-}55)$$

With this output pulse shape, pulses can be transmitted, and sampled (decoded), with a period T and passed by an ideal filter of bandwidth $BW = 1/2T$. Therefore, ideally, a bit rate of 2BW bits per second can be passed by a filter of bandwidth BW hertz. This is the case for simple codes such as NRZ (Non Return to Zero), where $T_p = T$. For RZ (Return to Zero) codes or phase encoding, where $T_p = T/2$, then the bandwidth requirement would double. Ideally, we have the following relationship:

NRZ:

$$BW = \frac{BR}{2} \qquad (4\text{-}56)$$

RZ:

$$BW = BR \qquad (4\text{-}57)$$

In reality, however, this ideal filter characteristic is not realizable. A more realizable filter characteristic, which meets the Nyquist criteria, is the raised cosine characteristic [8, 23], which is shown as the dashed lines in Figure 4.11.

$$h(t) = \frac{\sin\,(2\pi t/T)}{(2\pi t/T)[1 - (2t/T)^2]} \qquad (4\text{-}58)$$

Using this filter increases the band occupancy over that of the ideal. The bandwidth multiples of 1.5 and 2, which allow the pulse shape to meet intersymbol interference criteria, are shown in Figure 4.11. Note that since the characteristic rolloff is sinusoidal and symmetric, the noise bandwidth remains the same as the ideal. This filter design is assumed for the receiver noise relationships in Section 4.9. Parameters J_2 and J_3 are used there to relate noise bandwidth to bit rate.

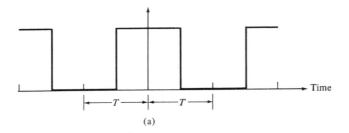

(a)

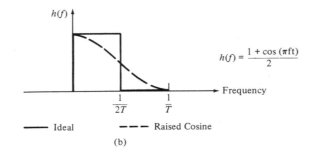

$$h(f) = \frac{1 + \cos(\pi ft)}{2}$$

——— Ideal   — — — Raised Cosine

(b)

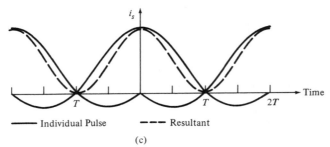

——— Individual Pulse   — — — Resultant

(c)

Figure 4.11 Nyquist pulse response and raised cosine filtering. (a) Input pulse train: Representative of an NRZ signal of pulse width $T_p = T$ and pulse period $= T$. (b) Filter characteristic. (c) Response.

Response time (rise time).   When transmitting pulses, the responses of the various components in the system are often specified in terms of rise and fall times. Rise time t_r is defined as the time it takes for the voltage or current to rise from 10 to 90% of its final value. The relationship between rise time and bandwidth for an ideal rectangular pulse stream is defined in terms of the response of a low-pass filter to a step-voltage input [9]:

$$t_r = \frac{0.35}{(\text{BW})_{e,\,3\text{dB}}} \tag{4-59}$$

The ideal relationship between rise time and bit rate can be derived by assuming the Nyquist criteria. This relationship is shown in Figure 4.12.

For NRZ signaling:

$$(\text{BW})_{e,\,3\text{dB}} = \frac{\text{BR}}{2} = \frac{1}{2T} \tag{4-60}$$

$$t_r = 0.7/\text{BR} \tag{4-61}$$

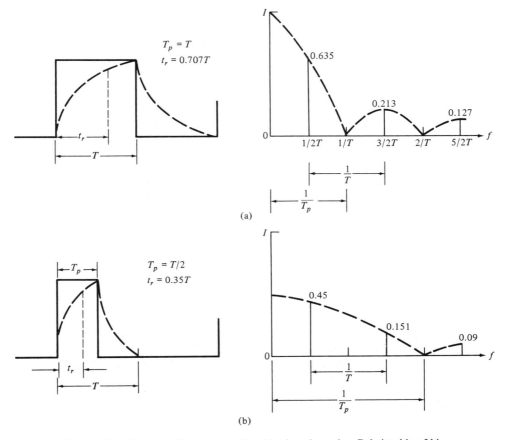

Figure 4.12   Time and frequency relationships in pulse trains. Relationship of bit rate $(1/T)$ to raise time (t_r). (a) NRZ coding and (b) RZ coding.

For RZ signaling:

$$(\text{BW})_{e,\,3\text{dB}} = \text{BR} = \frac{1}{T} \tag{4-62}$$

$$t_r = \frac{0.35}{\text{BR}} \tag{4-63}$$

Pulse series.   Figures 4.12 and 4.13 show the frequency spectrum of a series of rectangular pulses following transform theory [10]. Note that the amplitude envelope of the spectrum is a function of the pulse width (T_p) given by the following relationship:

$$H(f) = \frac{\sin (\pi f T_p)}{\pi f T_p} \tag{4-64}$$

and that the spectral lines are spaced at the pulse-repetition rate (bit rate) of $1/T$.

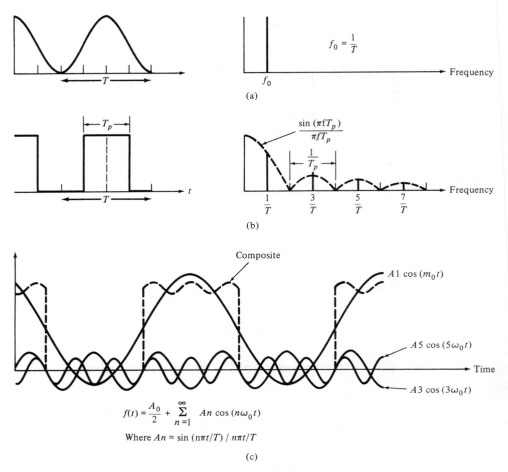

$$f(t) = \frac{A_0}{2} + \sum_{n=1}^{\infty} An \cos(n\omega_0 t)$$

Where $An = \sin(n\pi t/T) / n\pi t/T$

(c)

Figure 4.13   Spectrum of a binary pulse train. (a) Spectrum of a fundamental frequency equal to the bit rate ($f_0 = 1/T = $ BR). (b) Spectrum of a periodic square-wave pulse train of bit rate f_0. (c) Harmonic composition of the periodic square-wave pulse train.

4.6 SNR AND JITTER

Jitter is defined as short-term timing or phase variations in a received regenerated signal pulse from that of the ideal. Jitter is generated in (a) repeaters and receivers, wherever threshold detection and signal regeneration takes place, and in (b) multiplexers, wherever bits are buffered, stuffed, and retimed. Jitter can be classified into three categories [7, 11]:

  1. Systematic jitter (SJ): This is pattern-dependent (data-dependent) jitter accumulated in a systematic and predictable manner, such as might be related to varying data patterns or duty cycle, creating predominant spectral components or DC baseline drift in the transmitted signal.

2. Nonsystematic jitter: This comes from uncorrelated jitter sources, such as timing variations between repeaters or waiting-time jitter introduced by multiplexing.

3. Random jitter (RJ): This is uncorrelated Gaussian-noise-related jitter, such as might be caused by amplitude noise converted to phase noise at the sampling instant.

The elements of systematic pattern-dependent jitter include (a) data-dependent jitter (DDJ), caused by bandwidth limitations and intersymbol interference, which results in baseline drift and pulse distortions at the sampling threshold as data patterns vary, and (b) duty-cycle distortion (DCD), which is the deviation of pulse duration from the nominal width. DDJ is present at both the thresholding circuitry as well as at the sampling clock. These are generally defined in terms of peak-to-peak timing errors.

Random jitter (RJ) is generally related to thermal noise, or other random-noise sources, referenced to the receiver, and, therefore, is a function of received SNR. A component of RJ is also present in the local oscillator or clock. It is generally defined in terms of rms Gaussian timing error.

A mechanism in regenerators that contributes to pattern-dependent jitter and random jitter is static alignment error (SAE), which is a fixed or slowly varying clock frequency or threshold-setting error from the optimal sampling position. SAE-related jitter contains components of both DDJ and RJ since it is the nonoptimum sampling threshold or timing that contributes to DDJ and RJ. It is generally related in terms of peak-to-peak values.

The regenerator (at the repeater or receiver) is a point at which jitter is introduced into the optical transmission portion of the system. This is where the degenerated signal is sampled and a binary pulsed waveform is reconstructed based on the level of the signal at the sampling instant. Figure 4.14 is a simplified functional

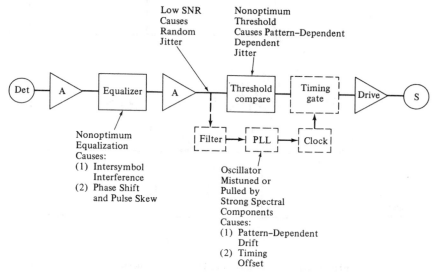

Figure 4.14   Sources of timing jitter in a regenerative repeater.

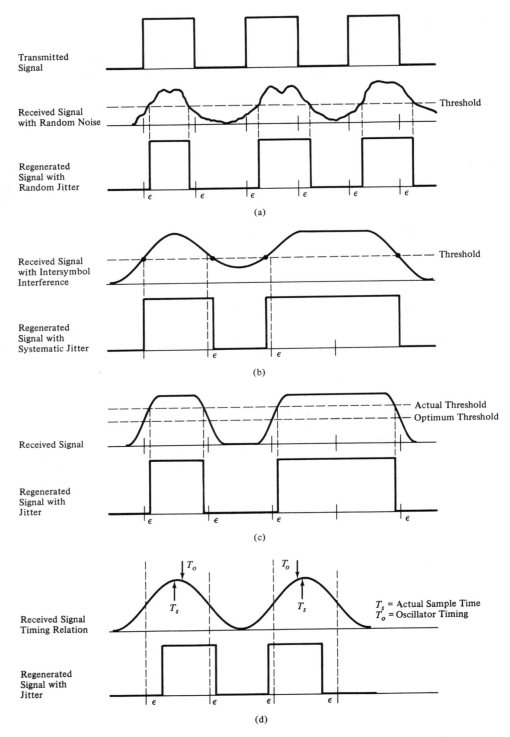

Figure 4.15 Jitter mechanisms. (a) SNR degradation, (b) intersymbol interference, (c) nonoptimum threshold, and (d) mistuned oscillator.

block diagram of a repeater. The various jitter mechanisms discussed as they apply to the repeater are shown in Figure 4.15. Figure 4.16 shows how these sources create the jitter at the threshold comparator circuit.

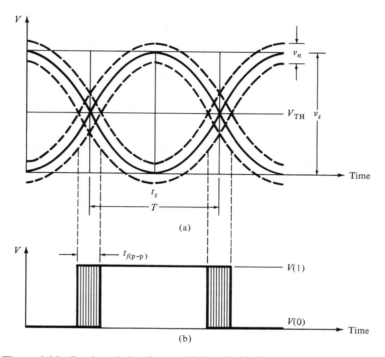

Figure 4.16 Random timing jitter as displayed with the eye pattern. (a) Eye pattern and (b) regenerated signal.

In the simple, but somewhat unusual, case that the repeater does not regenerate timing, the binary signal is approximately reconstructed by simply switching the output between binary states as the signal voltage crosses a reference threshold. The output is a pulsed waveform with pulse-width and timing relationships strongly effected by the input signal-to-noise ratio and waveform distortion.

When the repeater is fully regenerative, timing recovery and regeneration components (dashed-line blocks in Figure 4.14) are included. In this case, a clock circuit (with a phase-locked-loop or high-Q oscillator) synchronizes to the average input data rate and "samples" the signal at midpulse. The output is a pulsed waveform with proper timing and pulse-width relationships, and, therefore, jitter is reduced by the averaging effects of the clock circuit and sampling function.

Total jitter at a repeater station, following the ANSI X3T9 standard [11], can be related as:

$$t_j = \text{DCD} + \text{DDJ} + \text{SAE} + \text{DDJ}_{\text{clk}} + \sqrt{\text{RJ}_{\text{in}}^2 + \text{RJ}_{\text{clk}}^2} \tag{4-65}$$

The ANSI X3T9.5 standard referenced in [11] also provides suggested test methods.

4.6.1 Jitter at the Threshold Crossing

If equalization and threshold setting are optimum, jitter at the threshold crossing can be related roughly to the ratio of signal to noise at the input of a binary threshold detector. A signal with a finite rise time crossing the threshold results in conversion of amplitude noise directly to timing noise, i.e., jitter.

Figure 4.16 shows an eye-pattern measurement as might be observed on an oscilloscope. The eye pattern displays the relative amplitudes of signal and noise in successive received binary pulses. It also relates this signal-to-noise ratio to the decoder threshold level, the opening at the center of the eye being related to error probability. The smaller the opening, the greater the chance of a decoding error as noise pulses cross the threshold.

The eye pattern also displays timing jitter. The pattern is marked with respect to the ideal threshold crossing points in time, which correspond to bit period T. The amplitude-noise peak-to-peak envelope is marked with a dashed line. Because the signal has a finite rise-time slope (di/dt referenced to the detector) as it crosses the threshold, the noise envelope is transformed into a corresponding jitter envelope. Noise current (i_n), peak signal (i_s), and rise time (t_r) can be geometrically related to timing jitter (t_j), from Figure 4.17, as

$$t_{j,\text{p-p}} = \frac{i_{n,\text{p-p}}}{di/dt} \qquad \text{at the threshold}$$

Since di/dt is related to signal level $i_{s,\text{p-p}}$ and rise time t_r, we can relate jitter to SNR as follows:

$$t_{j,\text{p-p}} = \frac{t_r}{i_{s,\text{p-p}}/i_{n,\text{pp}}} \qquad (4\text{-}66)$$

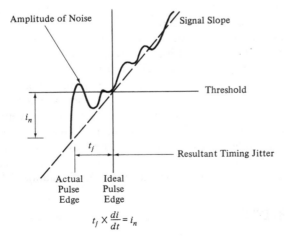

Figure 4.17 shows the diagram labeled with: Amplitude of Noise, Signal Slope, Threshold, i_n, t_j, Resultant Timing Jitter, Actual Pulse Edge, Ideal Pulse Edge, and $t_j \times \dfrac{di}{dt} = i_n$.

Figure 4.17 Conversion of random amplitude noise to timing jitter at a threshold detector.

$$t_{j,\text{rms}} = \frac{t_r}{\sqrt{(\text{SNR})_{\text{rms/rms}}}} \tag{4-67}$$

If we assume that the noise is Gaussian, then the conversion from rms to peak-peak jitter is approximately [11]:

$$t_{j,\text{p-p}} = 12.6t_{j,\text{rms}} \tag{4-68}$$

When viewing an oscilloscope, rms noise can roughly be determined as 1/8 the peak-to-peak visual envelope of the noise.

Note that when the jitter equation is applied to the receiver SNR relationships in Sections 4.3 and 4.10, the rise time and bandwidth factors divide in such a way that an optimum bandwidth for minimum jitter can be obtained.* For example, if we take the case of a PIN/FET receiver dominated by amplifier noise, from Equation (4-93), assuming f $\gg$ fc:

$$\text{SNR} = \frac{(rP_o)^2}{\dfrac{4kT}{R_i}(\text{BW})_r + \dfrac{16\pi^2 kT t C_t^2}{gm}\text{BW3}^3}$$

By substituting this into Equation (4-67), and assuming $t_r = 0.35/(\text{BW})_r$, and $\text{BW3}^3 = (\text{BW})_r^3/3$ for a low-pass network, we get

$$t_{j,\text{rms}} = \frac{\left(\dfrac{4kT}{(\text{BW})_r R_i} + \dfrac{5.3\pi^2 kT \tau C_t^2}{gm}(\text{BW})_r\right)^{1/2}}{(1/0.35)rP_o}$$

Finding the minimum with respect to bandwidth, we get minimum jitter at the optimum bandwidth $(\text{BW})_{r,\text{opt}}$:

$$(\text{BW})_{r,\text{opt}} = \left(\frac{0.75\,gm}{R_i \tau (\pi C_t)^2}\right)^{1/2} \tag{4-69}$$

4.6.2 Regenerative Retiming

In a regenerative repeater, the clock oscillator is a slowly responsive high Q circuit that stabilizes to the average rate of the incoming signal. Q_{osc} is the quality factor of the oscillator. Rather than the threshold detector reacting to the incoming pulse edge, the clock oscillator reacts to these pulse transitions, but with the averaging effect that in essence filters the timing jitter. The filtering or attenuation of jitter is in

---

*The author wishes to acknowledge Dr. Tran Muoi for the general information upon which this section was developed.

direct proportion to the loop bandwidth $(BW)_q$ or Q of the clock recovery oscillator circuitry. The narrower the loop bandwidth, the more jitter is attenuated. The practical limits or desired response time of the loop sets the lower limits on bandwidth. For this reason, the regenerator is susceptible to slowly varying jitter or patterns.

Jitter can be shown to be a function of noise within the timing loop (N_q), i.e., loop SNR. Since this is attenuated by loop bandwidth, the relationship to received $(SNR)_r$ is

$$(SNR)_q = (SNR)_r \frac{(BW)_q}{(BW)_r} \qquad (4\text{-}70)$$

where $(BW)_r$ is the receiver bandwidth, and

$$(BW)_q = \frac{f}{Q_{osc}}$$

4.7 NOISE SOURCES IN THE OPTICAL LINK

In order to evaluate a system's signal-to-noise performance, it is first necessary to determine the contribution of the various noise sources in the system. Noise sources generated in a fiber-optic link generally consist of

(a) source noise,
(b) modal noise,
(c) detector noise,
(d) preamplifier noise,
(e) unwanted distortion products due to nonlinearity, and
(f) crosstalk and reflections in optical couplers.

All of these various noise sources contribute to the total noise experienced at the optical receiver. In Chapter 6 we account for these in the link-budget analysis by adding signal power to compensate, either by increasing the SNR requirements of the receiver or by adding a power penalty.

External noise such as electromagnetic interference, ambient light (on open detectors), and vibration or environmentally induced distortions are also of concern, but are generally negligible with proper design, and are beyond the scope of this chapter. Crosstalk and reflections are dealt with in Chapters 5 and 6; the other sources are addressed here.

4.7.1 Source Noise

The origin of optical-source noise in coherent lasers is treated in detail in references [12] through [17]. LED noise is caused by incoherent intensity fluctuations and beat frequencies between modes. Laser-source noise is the result of quantum noise in the light-generation process; nonlinearities; mode "hopping" within the laser cavity; and interaction with coupled fiber, causing reflections back into the cavity, which in turn reduces coherence. Since it is not possible to totally isolate source noise from the fiber link to which it is coupled, it is a difficult parameter to measure. It is also difficult to quantify because there are so many variables. Usually, source noise is of most concern with multimode lasers operating into multimode fibers. Single-mode systems, when properly coupled, appear more stable and have less opportunity for interaction between multiple modes. Table 4.2 lists some of the sources of laser noise.

TABLE 4.2   NOISE GENERATED AT THE OPTICAL SOURCE

Noise mechanism	Cause	Solution
Relaxation oscillation ($\geq$100 MHz)	Nonlinearities, Lateral carrier diffusion, Spontaneous emissions in lasing mode	Generally only in conventional stripe DH lasers, Narrow the stripe, Mode stabilize (BH, CSP, etc.)
Self-pulsations (0.2 to 2 GHz)	Aged or poor-quality lasers	External feedback or driver reactance
Quantum noise ($\leq$100 MHz $\approx$ −60 dB per mode)	Quantum formation of carrier	Multimode quantum noise cancels for short-distance transmission (modal dispersion separates over distance)
Kinks	Poor-quality laser, Poor confinement	Mode stabilize (BH, CSP, etc.)
Reflections back into cavity	Reflection of fiber end facet	Termination with lens, film, isolator to reduce reflections
Signal distortion and intermodulation	Source nonlinearity, Improper matching	Linear source, Lower modulation depth to driver, Adjust bias point

SNR limitations of coherent sources.   A coherent light wave, such as might be emitted by a single-mode laser, contains random intensity fluctuations called "particle fluctuations." These are proportional to the square root of the optical power [18] and as such the SNR capability of the source is proportional to optical power. The relationship is illustrated in Figure 4.18 [following 18, 19].

Incoherent source noise.   With incoherent sources, each lightwave emission is independent and, therefore, the intensity fluctuation with time is random.

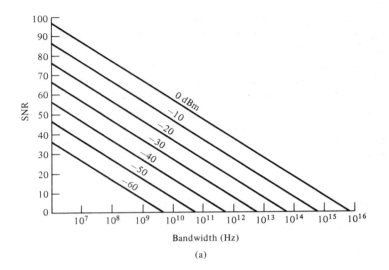

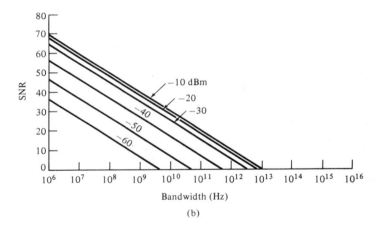

Figure 4.18 Modulation bandwidth versus SNR limitations imposed by coherence time (t_c) of optical sources for various optical-source output powers. $t_c = 1/(BW)_w$, where $(BW)_w$ is the bandwidth of the incoherent light wave. (a) Coherent source and (b) incoherent source.

Within the optical spectrum, incoherent light is analogous to radio noise in the RF spectrum. It is only useful for transmission if the signal bandwidth is orders of magnitude smaller than the bandwidth of the lightwave, i.e., $(BW)_t \ll (BW)_\lambda$. In this way, the signal "envelope modulates" the incoherent lightwaves, and the fluctuations are averaged by the narrower signal-transmission bandwidth.

Since incoherent light is essentially noise, there is a direct relationship between signal bandwidth and source SNR limitation. In other words, noise is a function of $(BW)_t/(BW)_\lambda$, where $(BW)_t$ is the signal bandwidth, and $(BW)_\lambda$ is the optical wave bandwidth. Figure 4.18 also shows the SNR limitations for incoherent light sources, such as LEDs. Note that SNR is limited by the incoherent noise fluctuations regardless of source power, much beyond -30 dBm output power.

Beat noise in incoherent sources.   LED systems operate with such a large number of modes that averaging occurs and the resultant noise is negligible. A simple relationship can be given for beat noise between modes in incoherent sources [1]:

$$\langle i_{ns}\rangle^2 = \frac{2(rGP)^2}{M\,\Delta\lambda}(BW)_t\left(1 - \frac{(BW)_t}{2\,\Delta\lambda}\right)$$ (4-71)

where M = number of spatial modes propagated in the fiber
  $\Delta\lambda$ = source spectral width
  $(BW)_t$ = transmission bandwidth

Because $(BW)_t$ is generally very small in comparison with $\Delta\lambda$, this noise component is usually negligible in comparison to others at the receiver.

4.7.2 Modal Noise in Multimode Fibers

Modal noise within fibers is generally caused by an interaction of multiple coherent modes, traveling with different time delays in a fiber, at an interface such as a connector or splice. The different time delays for each mode are caused by intramodal (material) dispersion and intermodal (waveguide) dispersion in multimode fibers. The interaction, in the form of constructive and destructive interference of wavefronts at the interface, causes amplitude-modulated noise power when only a portion of the signal at the interface is coupled. This noise condition is often called "speckle noise."

A detailed discussion of these noise mechanisms can be found in references [12] through [17]. A simplified diagram showing the coherence time effect is given in Figure 4.19. Since a laser is an oscillating cavity, it produces modes that are coherent and in phase over a time period called "coherence time," which is a function of the source spectral width $(\Delta\lambda)$ [26]:

$$t_c = \frac{\lambda^2}{c\,\Delta\lambda} = \frac{1}{(BW)_\lambda}$$ (4-72)

where c = speed of light
  λ = mean wavelength of the source
  $\Delta\lambda$ = spectral width
  $(BW)_\lambda$ = bandwidth of the incoherent lightwave

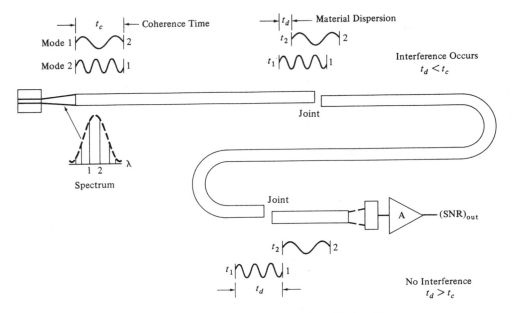

Figure 4.19 Modal noise mechanisms in multimode fiber.

Although coherent, the modes change in wavelength and intensity with time as the laser is modulated. As these modes propagate down the fiber, they are delayed differently in time either by different path lengths (waveguide dispersion) or by the different refractive index experienced by each wavelength (material dispersion). If the two modes shown in Figure 4.19 are shifted in phase as they reach a joint, there is constructive and destructive interference at that joint that varies with time as the relative phases of the modes change. If all of the modes are coupled by the receiving fiber, there is no resultant noise. However, if partial (mode-selective) coupling occurs, then amplitude-modulated noise occurs as a function of the number of modes interacting and the contrast between the resultant patterns.

The effect is length-dependent. As the modes propagate and are separated in time, the effect reduces. There is no interference between modes with a difference in delays greater than the coherence time (t_c). The length at which this occurs is called the "coherence length" (l_c), which is the coherence time (t_c) multiplied by the propagation velocity at the specific wavelength:

$$l_c = \frac{c}{n} t_c \tag{4-73}$$

The resultant noise is a function of contrast between patterns. The more modes propagating, the more contrast C decreases and likewise noise [17]:

$$C = \frac{1}{\sqrt{N}} \tag{4-74}$$

where N is the number of independent patterns due to an independent spectral source.

Modes can be increased by chosing a multiline laser. In binary systems, biasing the laser just below threshold, where multiple modes are stimulated, in effect cancels out much of the noise effect [12, 13]. In order to reduce speckle noise in multimode fibers, the following are important:

(a) laser-to-fiber coupling should be done in a fashion to reduce reflections;

(b) pigtail fiber should be long, with much care taken in the selection of low-loss nonreflecting connectors at the source;

(c) lasers should be selected with multiple (at least four) stable modes to reduce contrast between speckle patterns and thus noise;

(d) care should be taken in making low-loss splices at the source end;

(e) proper choice of bias conditions should be made for binary transmission.

4.7.3 Partition Noise in Single-Mode Systems

Partition noise is caused by the effects of intramodal (material) dispersion on a single propagating mode with a finite spectral width. The result is the spread of the coherent wave (containing various wavelengths) into coherent waves traveling at different speeds down the fiber. The effect is proportional to distance, that is, the further down the fiber, the more phase separation is evidenced between the different wavelengths. The phenomenon at the receiving end is similar to beat noise between coherent modes.

In a binary system, the partition-noise limitation to the signal-to-noise ratio and bit-error rate (as characterized by the factor Q_r) is [20]

$$Q_r = \frac{\sqrt{2}}{[\pi (\text{BR}) D \sigma_\lambda \, dt/d\lambda]^2} \qquad (4\text{-}75)$$

where BR = bit rate (bits/sec)
D = length of fiber (km)
σ_λ = rms-source spectral width (nm)
$dt/d\lambda$ = material dispersion (ps/nm/km) = t_{dm}

4.7.4 Source Nonlinearity Effects

Distortions at the source can convert some of the intended signal power to unwanted signal components (noise power), thus causing an effective power penalty when equating performance to average received power. The effect is negligible with binary transmission, but with analog signaling, the distortion power can cause significant degradation.

Signal distortion mechanisms include:

(a) saturation: signal currents drive the source beyond its operating region;
(b) nonlinear operating region: results in harmonic distortion and intermodulation products present in the transmitted signal; and
(c) operation within the region of the lasing threshold in ILDs.

Harmonic products become of particular concern when multiple-signal frequency bands are present, causing harmonics and harmonic mixing products to fall within the passbands of some or all of the signals. This unwanted signal (and, there-

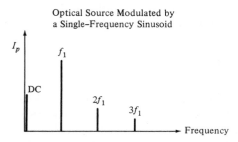

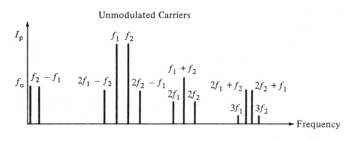

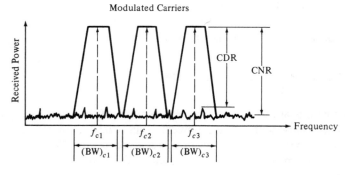

Figure 4.20 Relationship of harmonic distortion to SNR.

fore, noise), in effect, creates a noise floor at the level of the in-band harmonic product.

Nonlinearities contribute to power penalties in two ways: (1) the power in the harmonics represents lost out-of-band power; and (2) unwanted harmonics falling in band represent noise power, and thus decreased SNR, as shown in Figure 4.20. Generally, the power lost to harmonics is negligible. The in-band harmonic products, however, can have a significant interference effect due to their power levels at specific frequencies. A detailed analysis is given in Chapter 5 in the discussion on frequency-division multiplexing. The analysis determines the level of harmonic power (P_h) within a band of interest, referenced to the peak signal or carrier power (P_c). The resultant transmitted $(CNR)_t$ is, therefore:

$$(CNR)_t = 10 \log \frac{P_c}{P_h} \text{ dB} \qquad (4\text{-}76)$$

Refer to Chapter 5 for an analytical treatment of harmonic distortion products.

4.8 DETECTOR NOISE

Detector noise is related to two mechanisms: (1) current flow through the detector not associated with incident optical energy (leakage or "dark" current), and (2) quantum noise associated with the creation of electron-hole pairs through the absorption of photons. Leakage current is present through and around the surface layer of the structure, since the semiconductor is not a perfect insulator. Quantum noise is caused by the uncertainty in time that an electron-hole pair (current) is generated after a photon enters the detector. Both the number of electron-hole pairs as well as the time of creation are random quantities. This random uncertainty in time shows up as random noise in the current signal created from the absorption of the optical-signal energy incident on the device.

Following Personick [21], in the absence of gain, the rms noise power in noise bandwidth $(BW)_n$ attributed to quantum noise is

$$\langle i_{nQ}^2 \rangle = 2\frac{\eta^2 e}{h\upsilon} P_d(BW)_n \qquad (4\text{-}77)$$

where η = quantum efficiency (internal or external)
$\quad e$ = electron charge = 1.6×10^{-19} coulombs
$\quad h$ = planck's constant = 6.63×10^{-34} J-s
$\quad \upsilon$ = c/λ = optical frequency in Hertz
$\quad P_d$ = detected optical power in watts

Since responsivity can be defined as

$$r = \frac{\eta e}{h\upsilon} \text{ amperes/watt} \qquad (4\text{-}78)$$

then

$$\langle i_{nQ}^2 \rangle = 2erP_d(\text{BW})_n \tag{4-79}$$

The rms noise attributed to nonmultiplied leakage (dark) current I_d is simply

$$\langle i_{nL}^2 \rangle = 2eI_d(\text{BW})_n \tag{4-80}$$

Noise current differs depending on whether the detector is a PIN device without gain or an avalanche device with gain. In a PIN detector, rms-noise power is composed of a quantum-noise component, Equation (4-79), and a dark-current component, Equation (4-80).

In a PIN detector, the total noise power in noise bandwidth $(\text{BW})_n$ is, therefore:

$$\langle i_{nD}^2 \rangle = 2e(rP_d + I_d)(\text{BW})_n \tag{4-81}$$

In an avalanche detector, the gain mechanism effects all current flowing within the active depletion region of the device. The gain mechanism not only multiplies the signal (electron-hole pair creation), but the noise currents that flow through and are generated within the depletion region as well. Unfortunately, since the holes and electrons drift through the high-gain region, the probabilities per unit length that they produce new electron-hole pairs (ionization probability k) is unequal. This produces an excess gain factor F that multiplies noise at a greater rate than signal. From Chapter 3, Section 3.2.4, it should be recalled that APDs have an optimum gain because multiplication of quantum and leakage noise current is at a greater rate than signal-current multiplication. Following McIntyre [22] and Personick [21], it can be shown that the excess noise factor is approximately

$$F = kG + (1 - k)\left(2 - \frac{1}{G}\right) \tag{4-82}$$

This is often expressed as the approximation $F = G^x$. The plot of F versus gain, varying k and x, is shown in Figure 4.21 [27].* Note that the linear approximation is not accurate for all but a certain small range of values.

APD detector noise power is composed, therefore, of a quantum-noise component $(2erP_dG^2F)$ and a dark current component $(2eI_{dm}G^2F)$, both of which are multiplied by gain mechanism G, just as is the signal current, and an excess-noise multiplication factor F. A third unmultiplied current component (I_{du}) is associated with currents outside the gain mechanism. The total detector-noise power in bandwidth $(\text{BW})_n$ is, therefore:

$$\langle i_{nD}^2 \rangle = 2e[(rP_d + I_{dm})G^2F + I_{du}](\text{BW})_n \tag{4-83}$$

where I_{dm} is the leakage current subject to multiplication and I_{du} is the surface leakage not subject to multiplication.

Detector-noise performance is often specified in terms of noise equivalent power (NEP). This is the amount of received optical power it would take to produce

---

*The author wishes to acknowledge Dr. Tran Muoi of Plesscor for submitting this previously unpublished figure to this publication.

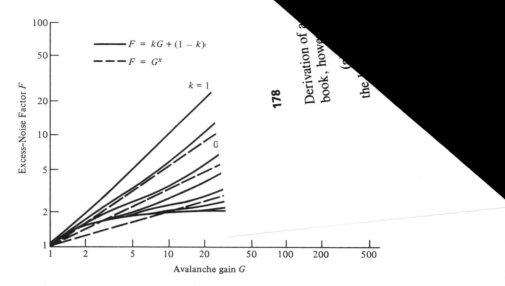

Figure 4.21 Excess-noise factor of avalanche photodiodes. (Courtesy of Tran Muoi, unpublished [27].)

a signal-photocurrent level equivalent to the noise-current level. Specifically, it is the optical-power level that produces a signal-to-noise ratio of unity in a 1-Hz bandwidth. It is expressed in watts/$\sqrt{\text{Hz}}$ and is usually specified for the case where dark current dominates (P_r is small). External η is assumed in r.

As is discussed in the next section, detected rms-signal current is

$$\langle i_s \rangle = rGP_r \tag{4-84}$$

The signal-to-noise ratio at the detector is, therefore:

$$(\text{SNR})_{\text{RMS/RMS}} = \frac{(rGP_r)^2}{2e[(rP_r + I_{dm})G^2F + I_{du}](\text{BW})_n} \tag{4-85}$$

Setting SNR = 1, solving for $P_r/\sqrt{(\text{BW})_n}$, and assuming quantum noise is small:

$$\text{NEP} = \frac{1}{rG}\sqrt{2eI_{du} + 2eI_{dm}G^2F} \tag{4-86}$$

4.9 AMPLIFIER NOISE

In an optical receiver, the two elements that most contribute to the noise are the detector and the input stage of the amplifier that follows it. Although the amplifier transforms both internal and external noise sources to noise voltages at its output, we can relate these sources to an equivalent noise current at the input of the amplifier (i_{nA}). Noise-current contribution differs depending on amplifier design.

amplifier noise for various amplifier designs is beyond the scope of this
...ver a summary is provided in what follows.*

...) Resistive-load-dominant preamplifiers: In simple voltage amplifiers, where
...oad resistor following the detector dominates, the noise current is

$$\langle i_{nA}^2 \rangle = \frac{4KT\,(\mathrm{BW})_n}{R_i} \tag{4-87}$$

where k = Boltzmann's constant = $1.38\ 10^{-23}$ J/K
$\quad T$ = temperature in degrees Kelvin = °C + 273
$\quad R_i$ = total input resistance in ohms

(b) Bipolar amplifiers: When bipolar transistors are used, noise is dominated
by thermal noise of the bias resistors and "shot" noise within the transistor. If shot
noise is minimized by optimizing the bias current, then the minimum total noise becomes:

$$\langle i_{nA}^2 \rangle = \frac{4kT}{R_i}(\mathrm{BW})_c + \frac{8\pi kTC_o}{\sqrt{\beta}}(\mathrm{BW2})^2 + 16\pi^2 kT\,(t_f C_o + r_{bb}\,C_{dsf}^2)(\mathrm{BW3})^3 \tag{4-88}$$

When the signal is binary, this becomes [23]

$$\langle i_{nA}^2 \rangle = \frac{4kT}{R_i}(\mathrm{BR})J_2 + 8\pi kTC_o(\mathrm{BR})^2\sqrt{\frac{J_2 J_3}{\beta}} + 16\pi^2 kT\,(t_f C_o + r_{bb}\,C_{dsf}^2)(\mathrm{BR})^3 J_3 \tag{4-89}$$

The optimum bias current is [23]

$$I_{c,\,\mathrm{opt}} = \frac{2\pi kT}{e}C_o(\mathrm{BR})\sqrt{\beta\frac{J_3}{J_2}} \times \frac{1}{\sqrt{1 + [2\pi t_f(\mathrm{BR})]^2\beta J_3/J_2}} \tag{4-90}$$

When R_i is made large through use of an integrating or transimpedance design,
then the dominant terms are

$$\langle i_{nA}^2 \rangle = \frac{8\pi kTC_o}{\sqrt{\beta}}(\mathrm{BW2})^2 \tag{4-91}$$

$$\langle i_{nA}^2 \rangle = \frac{8\pi kTC_o}{\sqrt{\beta}}(\mathrm{BR})^2\sqrt{J_2 J_3} \tag{4-92}$$

where R_i = amplifier input or feedback resistance
$\quad C_o = C_{dsf} + (C_{je} + C_{jc})$
$\quad C_{dsf} = C_d + C_s + C_f$
$\quad C_t = C_o + \alpha I_c$
$\quad C_d$ = detector capacitance (0.2- to 2-pF range)
$\quad C_s$ = stray capacitance

* The author wishes to acknowledge Dr. Tran Muoi of PCO for providing information from
which these formulas were derived [27] based on his original work [23] and that of Smith and
Personick [24] and Ogawa [25].

C_f = feedback capacitance, if any, across R_i

C_{je}, C_{jc} = base-to-emitter and collector capacitances, 0.2 to 1 pF and 0.1 to 0.5 pF respectively

α = t_f/V_T where $V_T = kT/e$

β = transistor current gain (h_{fe}) (100 to 300)

r_{bb} = base spreading resistance

t_f = forward transit time of transistor = $1 \times 2\pi f_t$

f_t = cut-off frequency, typically 5 to 10 GHz

$(BW)_c$ = 3-dB channel bandwidth = $f_h - f_l$

BR = bit rate

$BW2, BW3$ = f and f_2 channel noise power bandwidth, where f_h and f_l are the high and low band edges, respectively, and noise power = $\int_{f_l}^{f_h} P_n'(f)\, df$.

$$(BW2)^2 = \frac{f_h^2 - f_l^2}{2}$$

$$(BW3)^3 = \frac{f_h^3 - f_l^3}{3}$$

J_2, J_3 = noise BW factors for binary signaling [23] assuming $H_{f/T}$ of raised cosine filter:

$$
\begin{array}{ccc}
 & \text{NRZ} & \text{RZ} \\
J_2 = \int_0^\infty |H_{f/T}|^2\, df = & 0.563 & 0.403 \\[2mm]
J_3 = \int_0^\infty |H_{f/T}|^2 f^2\, df = 0.0868 & & 0.0361
\end{array}
$$

(c) FET amplifiers: For amplifiers with FET front ends, the dominant noise contribution is from the input-gate resistor R_i and the FET-channel noise. For analog signaling, this becomes

$$\langle i_{nA}^2 \rangle = \left(\frac{4kT}{R_i} + 2eI_g\right)(BW)_c + \frac{16\pi^2 kT\tau C_t^2}{g_m}(BW3)^3\left(1 + \frac{f_c}{f}\right) \tag{4-93}$$

For binary signaling [23]:

$$\langle i_{nA}^2 \rangle = \left(\frac{4kT}{R_i} + 2eI_g\right)(BR)J_2 + \frac{16\pi^2 kT\tau C_t^2}{g_m}(BR)^3 J_x J_3 \tag{4-94}$$

where $C_t = C_{dsf} + (C_{gs} + C_{gd})$

C_{gs}, C_{gd} = FET-gate-source and gate-drain capacitances, 0.2 to 0.8 pF and 0.05 to 0.1 pF, respectively

J_x = $1 + [J_f f_c/(BR)J_3]$

f_c = $1/f$ corner frequency of the FET 20 to 50 MHz for GaAs

τ = channel-noise factor = 1.03 for silicon MOSFETs, 1.75 for GaAs MESFETs

J_f = $\int_0^\infty |H_{f/T}|^2 f\, df$ = 0.184 for NRZ, 0.0984 for RZ [23]

I_g = FET-gate leakage current, less than 0.01 nA for silicon MOSFETs 1 to 3 nA for GaAs MESFETs

g_m = transconductance in mhos = .03 to 0.05 for silicon MOSFETs .015 to .05 for GaAs MESFETs

Generally, an integrating design is used, making R_i very large (in the 1-megohm range), and, therefore, the second term becomes dominant at high frequencies.

Figure 4.22 [27] shows the amplifier-noise contribution of various designs.

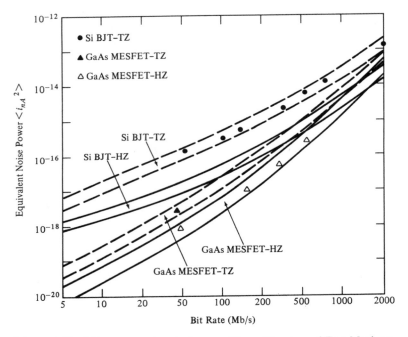

Figure 4.22 Noise power of typical preamplifiers. (Courtesy of Tran Muoi, unpublished, [27].)

4.10 RECEIVED SNR AND REQUIRED OPTICAL POWER

The previous sections of this chapter defined the signal and noise elements present within a fiber optic system. When allocating signal-to-noise performance to the elements of the photonics layer, many of the noise and distortion mechanisms can be conveniently handled as power penalties (added signal power required to compensate for noise) and, therefore, are not involved in the receiver-sensitivity analysis. When transmitted noise and distortion elements are referenced to signal power (as SDR), an option exists to reference these elements to the receiver as added noise current $\langle i_{nx} \rangle$. The elements of detector and amplifier noise are always referenced to the receiver as the defining parameters for receiver sensitivity. In this section, we define the relationships for SNR at the receiver as a function of detected optical power. We can then define receiver sensitivity in terms of the minimum required detected optical power (MDP) for a desired SNR performance.

4.10.1 Analog-Signal Modulation

From Equation (4-29), the rms signal-to-noise power for direct intensity modulation by one of N signals $s(t)$, each occupying bandwidth $(\mathrm{BW})_c$, is

$$(\mathrm{SNR})_{e,\mathrm{RMS/RMS}} = \frac{(M_o r G P_o/N)^2 \langle s^2 \rangle}{\langle i_{nD}^2 \rangle + \langle i_{nA}^2 \rangle + \langle i_{nX}^2 \rangle} \qquad (4\text{-}95)$$

In this case, $\langle i_{nX} \rangle$ is related to the electrical rms signal-to-distortion ratio, $(\mathrm{SDR})_{e,\mathrm{RMS/RMS}}$, as

$$\langle i_{nX}^2 \rangle = \frac{I_s}{SDR} = (rGM_o\langle s \rangle P_o/N)^2/SDR \qquad (4\text{-}96)$$

If we substitute Equation (4-83) for the detector-noise terms for $\langle i_{nD} \rangle$, we obtain a relationship for the general case of the APD, which can be solved for the mean detected power (P_o).

$$(\mathrm{SNR})_{e,\mathrm{RMS}} = \frac{(M_o r G P_o^2)\langle s^2 \rangle}{2e[G^2 F(rP_o + I_{dm}) + I_d](\mathrm{BW}) + \langle i_{nA}^2 \rangle + \langle i_{nX}^2 \rangle} \qquad (4\text{-}97)$$

Grouping terms, we get

$$(\mathrm{SNR})_{e,\mathrm{RMS}} = \frac{SP_o^2}{qP_o + A + L + SP_o^2/\mathrm{SDR}} \qquad (4\text{-}98)$$

where S = signal power = $(rGM_o\langle s \rangle/N)^2$
  q = quantum noise = $2erG^2F(\mathrm{BW})_c$
  L = leakage noise = $2e(G^2FI_{dm} + I_{du})(\mathrm{BW})_c$
  A = amplifier noise = $\langle i_{nA}^2 \rangle$

Using SNR to represent $(\mathrm{SNR})_{e,\mathrm{RMS/RMS}}$ and solving for P_o, we get

$$\left[S\left(1 - \frac{\mathrm{SNR}}{\mathrm{SDR}}\right) \right] P_o^2 + [-q(\mathrm{SNR})]P_o - (A + L)(\mathrm{SNR}) = 0 \qquad (4\text{-}99)$$

Substituting MDP for P_o and solving the quadratic:

$$MDP = \frac{q(\mathrm{SNR}) + \sqrt{q^2(\mathrm{SNR})^2 + 4S(A + L)(1 - \mathrm{SNR}/\mathrm{SDR})\mathrm{SNR}}}{2S(1 - \mathrm{SNR}/\mathrm{SDR})} \qquad (4\text{-}100)$$

Substituting $(\mathrm{SNR})_r = \mathrm{SNR}/(1 - \mathrm{SNR}/\mathrm{SDR})$ as the new receiver SNR required to compensate for SDR, and simplifying, we get

$$\mathrm{MDP} = \frac{(\mathrm{SNR})_r}{2S}\left(q + \sqrt{\frac{4S(A + L)}{(\mathrm{SNR})_r} + q^2} \right) \qquad (4\text{-}101)$$

Substituting $\langle i_{nL}^2 \rangle$ for L and the other relationships from Equation (4-98), we get

$$\text{MDP} = \frac{N^2 eF(\text{BW})_c(\text{SNR})_r}{M_o^2 \langle s^2 \rangle r}\left(1 + \sqrt{1 + \frac{M_o^2 \langle s^2 \rangle(\langle i_{nA}^2 \rangle + \langle i_{nL}^2 \rangle)}{[NeGF(\text{BW})_c]^2(\text{SNR})_r}}\right) \quad (4\text{-}102)$$

If the signal is a single channel occupying total transmission bandwidth $(\text{BW})_t$, then make $N = 1$ and substitute $(\text{BW})_t$ for $(\text{BW})_c$.

4.10.2 Receiver MDP with Carrier Modulation

The value of $\langle s^2 \rangle$ for an unmodulated sinusoidal carrier is $\frac{1}{2}$. If we assume N channels of bandwidth $(\text{BW})_c$ are frequency multiplexed, then the unmodulated rms carrier-to-noise ratio is

$$(\text{CNR})_{e,\text{RMS}} = \frac{\frac{1}{2}(M_o rGP_o/N)^2}{\langle i_{nD}^2 \rangle + \langle i_{nA}^2 \rangle + \langle i_{nX}^2 \rangle} \quad (4\text{-}103)$$

Substituting the noise terms of Equations (4-81) and (4-83) in Equation (4-103), and assuming $\langle i_{nX} \rangle = 0$, for a PIN detector, we get

$$(\text{CNR})_{e,\text{RMS}} = \frac{\frac{1}{2}(M_o rP_o/N)^2}{2e(rP_o + I_d)(\text{BW})_c + \langle i_{nA}^2 \rangle} \quad (4\text{-}104)$$

or for an APD, we get

$$(\text{CNR})_{e,\text{RMS}} = \frac{\frac{1}{2}(M_o rGP_o/N)^2}{2e[G^2 F(rP_o + I_{dm}) + I_{du}](\text{BW})_c + \langle i_{nA}^2 \rangle} \quad (4\text{-}105)$$

Substituting CNR for SNR and $\langle s^2 \rangle = \frac{1}{2}$ in Equation (4-102), for the general case of an APD, we get

$$\text{MDP} = \frac{2eN^2}{rM_o^2}F(\text{CNR})_r(\text{BW})_c\left[1 + \sqrt{1 + \frac{M_o^2(\langle i_{nA}^2 \rangle + \langle i_{nL}^2 \rangle)}{2(\text{CNR})_r[NeGF(\text{BW})_c]^2}}\right] \quad (4\text{-}106)$$

4.10.3 MDP for Digital Receivers

In Equation (4-46), we derived a parameter Q to represent signal-to-noise ratio in a binary transmission system assuming the process is Gaussian [1, 6]:

$$Q = \frac{I_s}{\langle i_{n0} \rangle + \langle i_{n1} \rangle} \quad (4\text{-}46)$$

where

$$\begin{aligned}
\langle i_{n0}^2 \rangle &= \langle i_{nL}^2 \rangle + \langle i_{nA}^2 \rangle \\
\langle i_{n1}^2 \rangle &= \langle i_{nQ}^2 \rangle + \langle i_{nL}^2 \rangle + \langle i_{nA}^2 \rangle \\
I_{s(\text{pk})} &= \text{peak signal current} = I_{s(1)} - I_{s(0)} \\
&= I_{s(1)} - I_{s(1)}\text{ER} = I_{s(1)}(1 - \text{ER})
\end{aligned} \quad (4\text{-}107)$$

Substituting the signal and noise parameters from Equation (4-83) for the general APD case, and assuming $I_{s(0)} = 0$ and P_o is the average detected power in the symbol period, we get

$$\langle i_{nQ}^2 \rangle = [2erG^2F\,(\text{BW})P_o = qP_o \tag{4-108}$$

$$\langle i_{nL}^2 \rangle = 2e\,(G^2FI_{dm} + I_{du})\text{BW} = 1 \tag{4-109}$$

$$I_s = (rG)P_{p(1)} = sP_{p(1)} = 2sP_o \tag{4-110}$$

$$\langle i_{nA}^2 \rangle = a \tag{4-111}$$

Substituting Equations (4-108) through (4-111) into Equation (4-46), we get

$$Q = \frac{2sP_o}{\sqrt{1 + a} + \sqrt{qP_o + 1 + a}} \tag{4-112}$$

This can be simplified with the approximation:

$$Q = \frac{sP_o}{2\sqrt{qP_o/2 + 1 + a}} \tag{4-113}$$

We can solve for the minimum required average detected power by substituting MDP for P_o and solving for MDP:

$$s^2(\text{MDP})^2 - \frac{Q^2q}{2}\text{MDP} - Q^2(1 + a) = 0 \tag{4-114}$$

Solving the quadratic for MDP, we get

$$\text{MDP} = \frac{Q}{s}\left(\frac{Qq}{4s} + \sqrt{\frac{q'}{16s'} + (1 + a)}\right) \tag{4-115}$$

Substituting parameters in Equations (4-108) through (4-111) into Equation (4-115), and simplifying we get [23, 24]:*

$$\text{MDP} = \frac{Q}{r}\left(QeF\,(\text{BR})J_1 + \sqrt{\frac{\langle i_{nA}^2 \rangle}{G^2} + 2eI_{dm}F\,(\text{BR})J_2}\right) \tag{4-116}$$

where the noise bandwidth factor is per Personick [28]:

$$J_1 = \int_0^\infty H_{i,f/T}(H_{f/T} * H_{f/T})\,df \tag{4-117}$$

$H_{i,f/T}$ is the normalized input optical pulse shape at the detector, and $H_{f/T}$ is the normalized transfer function of the filter $= H_{\text{out}}/H_{\text{in}}$.

	NRZ	RZ
$J_1 =$	0.5	0.5
$J_2 =$	0.563	0.403

*The author wishes to acknowledge Dr. Tran Muoi for the general information upon which this relationship was developed.

This can be reduced at optimum gain [27] to

$$\text{MDP} = \frac{2e}{r} Q^2(\text{BR}) J_1 \left(\sqrt{\frac{k \langle i_{nA}^2 \rangle^{1/2}}{eQ(\text{BR}) J_1}} + 1 - k \right) \tag{4-118}$$

where k is the ratio of ionization rates, 0.02 to 0.035 for silicon, about 0.4 for GaAs, and 1.0 for germanium.

$$G_{\text{opt}} = \sqrt{\frac{\langle i_{nA}^2 \rangle^{1/2}}{keQ(\text{BR}) J_1}} \tag{4-119}$$

For PIN detectors, where amplifier noise and dark current (I_{du}) dominate, the expression can be reduced to

$$\text{MDP} = \frac{Q}{r} \sqrt{\langle i_{nA}^2 \rangle + 2eI_{du}(\text{BR}) J_2} \tag{4-120}$$

REFERENCES

1. F. R. McDevitt, I. B. Slayton, *Optical Cable Communications*, RADC Contract No. F30602-74-C-0193, Melbourne, FL, Final Report RADC-TR-187 (A016846), Harris Corp., July 1978. Also available as *Fiber Optics Design Aid Package*, Vol. IV, Part II, Information Gatekeepers, Boston.

2. H. W. Ott, *Noise Reduction Techniques in Electronic Systems*, Wiley, New York, 1976.

3. J. Gower, *Optical Communications Systems*, Prentice-Hall International, London, 1984, pp 61–66.

4. C. D. Miller and V. E. Heeren, *Mathematical Ideas: An Introduction*, Scott, Foresman, Glenview, IL, 1969, Table C in the Appendix (values derived from area under a normal curve).

5. J. E. Midwinter, *Optical Fibers for Transmission*, Wiley, New York, 1979, pp. 347–357.

6. W. M. Hubbard, "Efficient Utilization of Optical-Frequency Carriers for Low and Moderate Bandwidth Channels," *Bell System Technical Journal* **52** (May–June 1973).

7. D. R. Smith, *Digital Transmission Systems*, Van Nostrand Reinhold, New York, 1985, p. 207.

8. J. Gower, *Optical Communication Systems*, Prentice-Hall International, London, 1984, pp. 434–437.

9. J. Millman and H. Taub, *Pulse Digital and Switching Waveforms*, McGraw-Hill, New York, 1965, p. 44.

10. S. C. Gupta, *Transform and State Variable Methods in Linear Systems*, Wiley, New York, 1966, pp. 76–78.

11. American National Standards Institute, ANSI Draft Standard X3T9.5/84-88 REV 8, New York, July 1, 1988, Appendix A.

12. R. E. Epworth, "Modal Noise, Causes and Cures," *Laser Focus* (September 1981): 109–115.

13. R. E. Epworth, "The Phenomenon of Modal Noise in Analogue and Digital Optical Fibre Systems", paper presented at the Fourth European Conference on Optical Fibre Optical Communications, Genoa, 1978.

14. R. E. Epworth, "The Phenomenon of Modal Noise in Fiber Systems," paper presented at the Topical Meeting on Optical Fiber Communications, Washington, D.C., March 1979.

15. R. E. Epworth, "Polarization Modal Noise and Fiber Birefringence in Single Mode Fiber Systems," paper presented at the IOOC '81, San Francisco, 1981.

16. E. G. Rawson, J. W. Goodman, and R. E. Norton, *Dependence of Fiber Modal Noise on Source, Fiber and Splice Parameters,* FOC 80 Proceedings, San Francisco, September 16–18, 1980: 118–122.

17. E. Rawson, J. Goodman, and R. Norton, "Analysis and Measurement of Modal Noise Probability Distribution for a Step Index Optical Fiber," *Optical Letters.* **5**, 8 (August, 1980): 357–358.

18. Howes and Morgan, "Optical Fibre Communications," in D. Russer (ed.), Wiley, New York, 1980, pp. 1–20.

19. G. Grau, "Temperatur-und Laserstrahlung als Informationstrager," *Arch Elektron Ubertragungstech,* **18** (1964).

20. K. Ogawa, "Analysis of Mode Partition Noise in Laser Transmission Systems," *IEEE Journal of Quantum Electronics.* **QE-18** (May 1982): 849–855.

21. S. D. Personick, "Photodetectors for Fiber Systems," *Fundamentals of Optical Fiber Systems* in M. K. Barnoski (ed.), Academic Press, New York, 1976, pp. 155–201.

22. R. J. McIntyre, "Multiplication Noise in Uniform Avalanche Diodes," *IEEE Transactions, Electronic Devices* **ED-13** (1966): 164–168.

23. T. V. Muoi, "Receiver Design for High Speed Optical Fiber Systems," *IEEE/OSA Journal Lightwave Technology* **LT-2** (1984): 243–267.

24. R. G. Smith and S. D. Personick, "Receiver Design for Optical Fiber Communication Systems," *Semiconductor Devices for Optical Communication,* Springer-Verlag, New York, 1980.

25. K. Ogawa, "Considerations for Optical Receiver Design," *IEEE J. Selective. Area Communication* **SAC-1** (April, 1983): 524–532.

26. American National Standards Institute, *IEEE Standard Definition of Terms Relating to Fiber Optics,* ANSI/IEEE Standard 812–1984, New York, 1984.

27. Tran Muoi (PCO), Previously unpublished information, 1989.

28. S. D. Personick, "Receiver Design for Digital Fiber Optic Communications Systems," *Bell Systems Technical Journal* **52** (July–August, 1973): 843–886.

5

Modulation Encoding and Multiplexing

5.1 DEFINITIONS

5.1.1 Modulation

Modulation is a process whereby a carrier waveform is varied in amplitude, frequency, or phase by either a digital or analog information signal. Modulation of a carrier by the signal source conserves the information content while changing the frequency and power relationships of the transmitted signal. In a fiber-optics system, modulation prior to transmission is performed for a number of reasons including:

(a) conserving optical power while maintaining desired baseband $(SNR)_b$ by trading transmission bandwidth for power;

(b) creating a convenient path or line-layer channel format for subsequent multiplexing; and

(c) formatting the signal for compatability with line-, section- or photonics-layer interfaces.

5.1.2 Encoding

Encoding is a process used with binary information signals whereby a unique group of output transmission symbols and a format are produced for each group of binary input signals. In the encoding process at the section layer, the format of the output is made suitable for fiber-optic transmission. Encoding is generally performed prior to transmission at:

(a) the section layer in order to make the spectral characteristics of the transmitted signal compatible with the fiber-optic transmission system (achieving either a fixed or negligible zero-frequency component); and

(b) the higher layers in order to provide a format compatible with subsequent multiplexing electronics.

In all cases of modulation and encoding, there is a transformation between the SNR of the channel before and after the process. The transform relationship depends on the modulation or encoding approach. Encoding and modulation directly relate to the optical-receiver requirements, as seen in the receiver-sensitivity relationships in Chapters 4 and 6. Figure 5.1 provides a graphic illustration of those relationships.

5.1.3 Multiplexing

Channel multiplexing is performed in order to combine multiple- lower-capacity channels, and any overhead information, into one high-capacity transmission channel for interfacing at the next lower layer. Electronically, signals can be multiplexed in the time (TDM) or frequency (FDM) domain. Within the photonics layer, by using multiple optical sources, each at a separate wavelength, signals can be further multiplexed in wavelength (WDM) onto a single fiber. In essence, WDM is a higher level of frequency multiplexing, where the carriers are in the infrared region of the electromagnetic spectrum.

Certain multiplexing approaches lend themselves best to certain modulation approaches. Typical combinations are given in Table 5.1. The multiplexing approach utilized has an effect on the received signal-to-noise and required optical power. Multiplexing is used in fiber-optic systems, taking advantage of the broad bandwidth of the fiber, by combining more than one channel onto a fiber. In doing so, the available optical power is shared among the number of channels multiplexed, either directly as a power division or indirectly as a result of transmission-bandwidth sharing. The effects of this sharing, as well as the effects of multiplexer efficiency and added noise, are discussed in Sections 5.6 through 5.8.

This chapter addresses the most common modulation, encoding, and multiplexing approaches and their SNR- and bandwidth-transformation relationships. Although the application to fiber transmission is addressed here, it should be noted that these relationships are not peculiar to fiber-optic transmission, but are the same relationships that are found in radio, satellite, and coaxial transmission and in communications engineering texts. The reader should, therefore, refer to such texts e.g., Carlson [1], Smith [2], Freeman [3], McDevitt and Slayton [4], Shanmugam [5], and Haykin [6], for more detailed treatments.

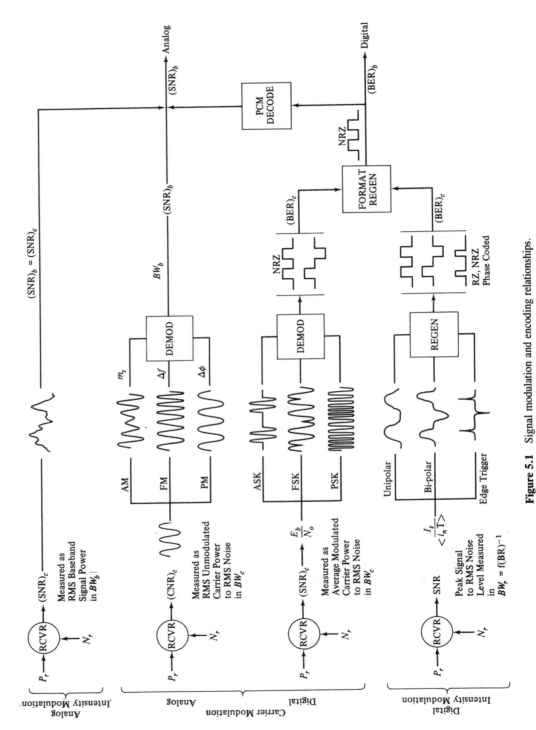

Figure 5.1 Signal modulation and encoding relationships.

TABLE 5.1   TYPICAL MULTIPLEXING AND MODULATION DESIGN COMBINATIONS

Multiplexing	Modulation
Frequency-division (FDM)	Amplitude (AM)
	Vestigial sideband (VSB) AM
	Frequency (FM)
	Frequency, Amplitude, or Phase
	Shift keying (FSK, ASK, PSK)
Time-division (TDM)	Pulse-code (PCM)
	Differential (and other forms) of PCM
Wavelength-division (WDM)	Direct baseband transmission
	All forms of modulation

5.2 CARRIER MODULATION

Analog carrier modulation is an approach whereby a sinusoidal carrier is amplitude, frequency, or phase modulated by the information signal. The form of the transmitted signal is a complex spectrum of signal and carrier components that requires a linear transmission system. Analog carrier modulation is generally used with fiber optics in order to conserve optical power while taking advantage of the larger transmission bandwidth that fiber offers. A SNR-versus-bandwidth trade-off occurs, as shown in Figure 5.2 for a frequency-modulation example. Carrier modulation is generally chosen over digital approaches when cost or complexity of digital encoding is prohibitive.

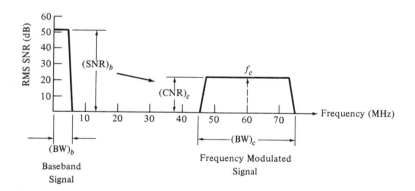

Figure 5.2   SNR-versus-bandwidth trade-off: FM example.

5.2.1 Carrier-Modulated Analog Transmission

The information signal that modulates the carrier can be of any form, analog or digital. With analog modulation, the signal voltage to amplitude, frequency, or phase conversion is a continuous, generally linear, function. Three approaches are discussed here, with time and frequency relationships shown in Figure 5.3:

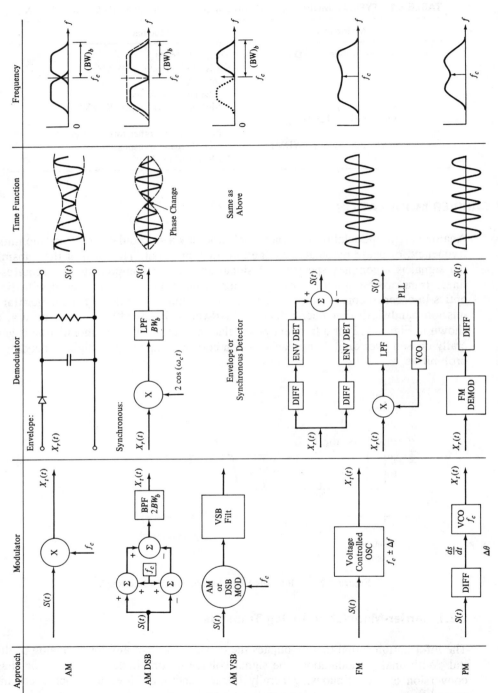

Figure 5.3 Analog carrier modulation.

190

1. Amplitude modulation: This is performed by mixing the information signal with a sinusoidal carrier at a higher frequency and filtering out all but the desired sideband(s) for transmission.

2. Frequency modulation: This is performed by deviating the frequency of a fixed amplitude sinusoidal carrier as a function of the information signal voltage.

3. Phase modulation: This is performed by deviating in phase the phase of a fixed-frequency fixed-amplitude sinusoidal carrier as a function of information-signal voltage.

The choice of carrier frequency depends on the modulated-signal bandwidth and on the method of subsequent multiplexing, if any. The choice of modulation approach depends on the SNR improvement desired versus bandwidth occupancy, and on practical factors such as cost and complexity.

Basically, with fiber-optic systems, there exists a large amount of bandwidth but little signal power (limited systems gain). The designer, therefore, generally employs signal modulation in order to maintain the required baseband SNR while minimizing transmission-channel signal power. The trade-off is generally to trade bandwidth for power, i.e., a reduction in transmission-channel SNR or CNR can be traded for larger transmission-channel bandwidth, $(BW)_c$ and still maintain a constant baseband SNR, as shown in Figure 5.2. The resultant difference between transmission-channel $(CNR)_c$ and the baseband $(SNR)_b$ is called the modulation "improvement factor." Often, modulation approaches are chosen for other reasons than minimization of power. These may, therefore, have no improvement factor (as with VSB) or a negative one (as with subcarrier AM). Generally, the methods with negative factors are not the choice for fiber transmission.

Relationships between transmission-channel $(CNR)_c$ and baseband signal $(SNR)_b$ are summarized in Tables 5.2 and 5.3. Table 5.2, based on Shanmugam [5] combined with the original work performed by McDevitt and Slayton [4], provides the signaling relationships at the source, detector, and demodulator output. Table 5.3 summarizes the channel parameters and improvement factors. All relationships assume single-channel transmission.

Application.

(a) Baseband AM intensity modulation: This approach, in general, is used for low-cost single-channel systems or broadband high-SNR systems where only a single channel exists or where the degradation due to multiplexing cannot be tolerated. Applications include single-channel instrumentation and control links, and high-performance video links.

(b) Subcarrier AM DSB and VSB: This approach is not generally used with fiber unless the signal is already in that format or unless a subsequent frequency-division-multiplexing format requires it. Applications include the transmission of broadcast or CATV video signals.

(c) Subcarrier FM: This is generally used when FM modulation is required to preserve a high SNR over long distances or when multiple frequency-division-multi-

TABLE 5.2 SNR PARAMETERS AT STAGES WITHIN THE FIBER LINK FOR VARIOUS MODULATION APPROACHES*

Modulation	Modulating signal	Detector photocurrent	SNR at demodulation output
Baseband AM/IM	$1 + m_s \cos(\omega_s t)$ for $m_s = m_c = m_o \leq 1$ and $(BW)_c = (BW)_b$	$rGP_o[1 + m_o \cos(\omega_s t)]$	$\dfrac{(rGP_o)^2 m_o^2}{2N_b}$ for $N_b = N_c$
Subcarrier AM	$1 + m_c[1 + m_s \cos(\omega_s t)] + \cos(\omega_c t)$ $m_s < 1,\ m_c = A_c/i_b$ $m_o = m_c(1 + m_s) \leq 1$ $(BW)_c = 2(BW)_b$	$rGP_o\{1 + m_c[1 + m_s \cos(\omega_s t)]\cos(\omega_c t)\}$	$\dfrac{(rGP_o)^2 m_s^2 m_c^2}{4N_b}$ $= \dfrac{(rGP_o)^2 m_o^2}{8N_c}$ for $m_s = 1,\ m_c = m_o/2$
Subcarrier DSB	$1 + m_s \cos(\omega_s t)\cos(\omega_c t)$ for $m_c = m_o \leq 1$ and $(BW)_c = 2(BW)_b$	$rGP_o[1 + m_o \cos(\omega_s t)\cos(\omega_c t)]$	$\dfrac{(rGP_o)^2 m_o^2}{4N_b}$ $= \dfrac{(rGP_o)^2 m_o^2}{2N_c}$ since $(BW)_c = (BW)_b$
Subcarrier SSB	$1 + m_s \cos(\omega_s + \omega_c)t$ for $m_c = m_o \leq 1$ and $(BW)_c = (BW)_b$	$rGP_o[1 + m_o \cos(\omega_s + \omega_c)t]$	$\dfrac{(rGP_o)^2 m_o^2}{2N_b}$ and $N_b = N_c$ since $(BW)_b = (BW)_c$
Subcarrier FM	$1 + m_c \cos[\omega_c t + B \sin(\omega_s t)]$ for $m_c = m_o \leq 1$ and $B = \Delta f/(BW)_b$ and $(BW)_c = 2(B + 1)(BW)_b$	$rGP_o\{1 + m_o \cos[\omega_c t + B \sin(\omega_s t)]\}$	$\dfrac{3B^2(B + 1)(rGP_o)^2 m_o^2}{2N_c}$ $= \dfrac{3B^2(rGP_o)^2 m_o^2}{4N_b}$

Subcarrier PM

$1 + m_c \cos[\omega_c t + \phi \sin(\omega_s t)]$
for $m_c = m_o \leq 1$
and ϕ = peak angular deviation between 0 and π

$rGP_o\{1 + m_o \cos[\omega_c t + \phi \sin(\omega_s t)]\}$

$$\frac{\phi^2(\phi+1)(rGP_o)^2 m_o^2}{2N_c}$$
$$= \frac{\phi^2(rGP_o)^2 m_o^2}{4N_b}$$

B = peak frequency deviation/baseband bandwidth
N_b = rms noise power in baseband bandwidth = $N_o(BW)_b$
N_c = rms noise power in modulated channel bandwidth = $N_o(BW)_c$
N_o = rms noise power density in W/Hz
P_o = optical power at zero modulation
m_o = modulation index of optical carrier
m_s = modulation index of information signal on electrical carrier
m_c = modulation index of unmodulated subcarrier on source drive current
ω_s, ω_c = ω of information signal and subcarrier

TABLE 5.3 MODULATION PARAMETERS FOR ANALOG TRANSMISSION

Modulation approach	Channel bandwidth	$(CNR)_{c, RMS/RMS}$ to $(SNR)_{c, RMS/RMS}$ relationship	SNR improvement factor in $(BW)_b$ $[(SNR)_b - (CNR)_b]$
Baseband AM/IM	$(BW)_c = (BW)_b$	$(SNR)_c = (SNR)_b$	0 dB
Subcarrier AM	$(BW)_c = 2\,(BW)_b$	$(CNR)_c = (SNR)_b + 6$ dB	-9 dB
Subcarrier DSB	$(BW)_c = 2\,(BW)_b$	$(CNR)_c = (SNR)_b$	-3 dB
Subcarrier SSB or VSB	$(BW)_c = (BW)_b$	$(CNR)_c = (SNR)_b$	0 dB
Subcarrier FM	$(BW)_c = 2[\Delta f + (BW)_b]$ where Δf = peak frequency deviation	$(CNR)_c = -20 \log [\Delta f/(BW)_b] -10 \log [(BW)_c/(BW)_b] -1.76$	$20 \log [\Delta f/(BW)_b] + 1.76$
Subcarrier PM	$(BW)_c = 2[\Delta\phi + (BW)_b]$ where $\Delta\phi$ = peak phase deviation in radians	$(CNR)_c = (SNR)_b -20 \log \Delta\phi -10 \log [\Delta\phi + 1]$	$20 \log \Delta\phi - 3$

plexed channels on a single fiber are required. Applications include single-channel and multichannel FM/FDM video trunking.

(**d**) Subcarrier phase modulation: This is generally only used when the signal to be transmitted is already phase modulated. Applications include the direct replacement of microwave or coaxial cable with fiber.

5.2.2 Carrier-Modulated Digital Transmission

Carrier-modulated digital transmission includes such signaling schemes as amplitude shift keying (ASK), phase shift keying (PSK), and frequency shift keying (FSK). The modulation waveforms and associated frequency spectra are shown in Figure 5.4. In general, carrier-modulated binary transmission is not used with fiber-optics systems; it is intended for broadband, frequency-multiplexed, coaxial, or microwave systems. Some fiber transmission architectures, however, do require a mixture of analog and digital signals, and, for economical reasons, frequency multiplexing is chosen as the transmission approach. The separate digital channels are, therefore, transmitted on separate carriers, and these sinusoidal carriers are phase, frequency, or amplitude modulated by the binary signal. An example is the use of coaxial-cable data modems directly with fiber. The modem signals are generally PSK or a combination of ASK and PSK.

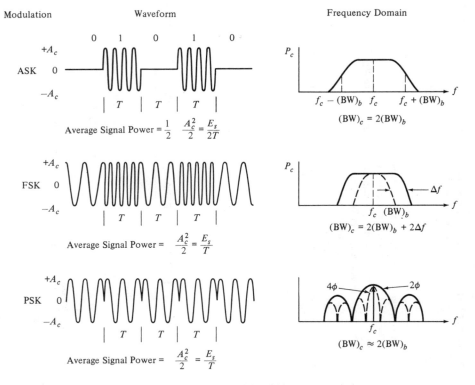

Figure 5.4   Subcarrier-modulated binary transmission.

For carrier-modulated signaling, it is more common to equate error probability to the signal energy (E_s) in the bit period (T_p) after detection. The relationships are as follows:

$$S = \frac{E_s}{T_p} \qquad (5\text{-}1)$$

where S = average signal power in a bit of duration T_p
E_s = total signal energy in a signal state (value for each state)

$$E_b = \frac{S}{\text{BR}} \qquad \text{or} \qquad S = E_b(\text{BR}) \qquad (5\text{-}2)$$

where E_b = average signal energy per bit.
    Combining Equations (5-1) and (5-2), we get

$$E_b(\text{BR}) = \frac{E_s}{T_p} \qquad \text{or} \qquad E_b = \frac{E_s}{T_p(\text{BR})} = \frac{E_s T}{T_p} \qquad (5\text{-}3)$$

where BR is the transmission bit rate = $1/T$.

$$N = \text{noise power} = N_o \times (\text{BW})_n \qquad (5\text{-}4)$$

where N_o = one-sided (positive frequency) noise-power spectral density of Gaussian noise, W/Hz
BWn = noise bandwidth = $(\text{BW})_c$ in this case

$$N_o W = \text{noise power in the binary signal Nyquist BW}$$

where W = Nyquist bandwidth:

$$W = \frac{1}{2T} \qquad \text{if NRZ} \qquad (5\text{-}5)$$

E_b/N_o = energy per bit per noise power density

We can relate E_b/N_o to SNR power by equating E_b to the signal component of the modulated carrier and by measuring received noise in the postdetection noise bandwidth $(\text{BW})_n$, which is usually wider than the Nyquist bandwidth W. Combining Equations (5-1), (5-2), and (5-4), we get:

$$\frac{S}{N} = \frac{E_s/T_p}{N_o(\text{BW})_n} = \frac{E_b}{N_o} \frac{(\text{BR})}{(\text{BW})_n} \qquad (5\text{-}6)$$

The signal-to-noise power defined in the Nyquist bandwidth is

$$\frac{S}{N} = \frac{E_s}{T_p} \frac{1}{N_o(W)} \qquad (5\text{-}7)$$

Substituting Equations (5-3) and (5-5) into Equation (5-7), we get for NRZ signaling

$$\text{SNR} = \frac{2E_b}{N_o} \qquad (5\text{-}8)$$

The ratio of energy per bit to noise density (E_b/N_o) is generally the parameter used to compare modulation approaches with regard to relative power efficiency. In order to equate E_b/N_o to received optical power, we must determine the average received signal-to-noise ratio. See Figure 5.4 for the signal waveforms.

The average signal-to-noise ratio for a constant-level carrier (FSK, PSK), in channel bandwidth $(BW)_c$, is found by combining Equations (4-103) and (5-6):

$$\frac{C}{N} = \frac{\frac{1}{2}(M_o r G P_o)^2}{\langle i_{nT}^2 \rangle} = \frac{E_s}{T_p} \frac{1}{N_o(BW)_c} \tag{5-9}$$

In order to use this relationship, we must convert to E_b/N_o. For NRZ, where $T_p = T$, by substituting Equation (5-3) into (5-9), we get

$$CNR = \frac{E_b(BR)}{N_o(BW)_c}$$

or

$$\frac{E_b}{N_o} = \frac{C}{N} \frac{(BW)_c}{BR} \tag{5-10}$$

For ASK, where the carrier is not constant, but rather zero energy for the "0" state, then $E_b = E_s/2$ and

$$\frac{E_b}{N_o} = \frac{1}{2} \frac{C}{N} \frac{(BW)_n}{BR} \tag{5-11}$$

Table 5.4 gives bit-error rate versus E_b/N_o for the various carrier-modulation and detection approaches; these comparisons can be found in most any transmission text [1, 2, 3, 4].

TABLE 5.4 PARAMETERS FOR SUBCARRIER DIGITAL MODULATION

Modulation approach	Channel bandwidth, $(BW)_c$	BER to E_b/N_o relationship
ASK		
Coherent	$(BW)_c = 2(BW)_b \approx BR$	$BER = erfc \sqrt{\dfrac{E_b}{N_o}}$
Noncoherent	$(BW)_c = 2(BW)_b \approx BR$	$BER = 0.5 \exp \dfrac{-E_b}{2N_o}$
FSK		
Coherent	$(BW)_c = 2[\Delta f + (BW)_b]$	$BER = erfc \sqrt{\dfrac{E_b}{N_o}}$
Noncoherent	$(BW)_c = 2[\Delta f + (BW)_b]$	$BER = 0.5 \exp \dfrac{-E_b}{2N_o}$
PSK		
Coherent	$(BW)_c = 2(BW)_b \approx BR$	$BER = erfc \sqrt{\dfrac{2E_b}{N_o}}$
DPSK	$(BW)_c = 2(BW)_b \approx BR$	$BER = 0.5 \exp \dfrac{-E_b}{N_o}$

5.3 PULSE-CODE MODULATION

Fiber-optic transmission systems are characterized by limited power gain and only moderate linearity. For this reason, binary pulse transmission is generally the chosen format for transmission. As was discussed in Section 4.5, a binary signal can be transmitted with a very low received SNR requirement (15.6 dB AVG/RMS) and the nonlinearity of the source is of little concern.

Other pulse-coding formats (pulse-position, pulse-frequency, and pulse-duration modulation) can be used to gain a similar SNR advantage. These approaches, however, cannot be easily multiplexed and, therefore, are usually limited to single-channel transmission whereby the pulse encoding takes place at the section layer, as is described in Section 5.6.

Pulse-code modulation (PCM) is by far the most common technique used for analog transmission over digital medium and over fiber. The quality of the received and decoded baseband signal is almost totally a function of the encoding parameters and only a weak function of received SNR. If the received signal remains above a certain threshold where bit errors are low, then decoded $(SNR)_b$ is fixed as a function of the number of encoding bits.

Figure 5.5 shows the PCM encoding process. An analog signal at the input is first filtered to $(BW)_b$ to eliminate any unwanted noise components. The signal amplitude is then sampled at a frequency f_s equal to or greater than the Nyquist frequency:

$$f_s \geq 2(BW)_b \qquad (5\text{-}12)$$

The amplitude levels stored by the sampler are gated into a quantizer. Here each sample is compared in amplitude to a reference with 2^n discrete levels. The quantizer outputs a binary coded word that represents the amplitude level of the sample measured in each sample period $(1/f_s)$. Each coded word has n bits of encoding representing the 2^n levels of the quantizer. The stream of binary words are then formatted for transmission into a serial data stream with framing and synchronization bits. The transmission rate is slightly greater than nf_s.

At the receiving end, the timing is recovered and the sample words are separated for decoding. Each word enters a digital-to-analog (D/A) converter. The D/A outputs a discrete voltage level corresponding to the value of the binary code word received. The output is a serial pulse-amplitude-modulated (PAM) waveform that is then filtered by a baseband filter $(BW)_b$ in order to reconstruct the original analog waveform.

For standard PCM, the SNR of a PCM-encoded signal is primarily a function of (a) the number of quantizing levels (the number of bits per code word), (b) the quantizing characteristic (compression), or (c) the amplitude or amplitude statistics of the input signal as compared to the range of the quantizer. For delta modulation (DM) or differential PCM (DPCM), the SNR is also a function of the ratio of sampling frequency to the bandwidth of the signal. These relationships are listed in Table 5.5. For a more detailed analysis, refer to Smith [2], Shanmugam [5], and Haykin [6] from which the data in the table were derived.

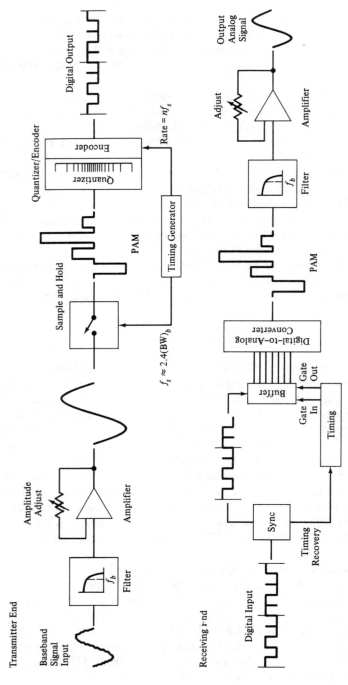

Figure 5.5 PCM encoding process.

198

TABLE 5.5 PULSE-CODE MODULATION PERFORMANCE RELATIONSHIPS BASED ON SINUSOIDAL INPUTS[a]

Encoding method	Bit rate per channel $(BR)_c$ (bps)	Decoded received $(SNR)_b$ $(SNR)_{b, (RMS/RMS)}$ (dB)
Linear PCM[b]	$(BR)_c = nf_s$ $f_s > 2\,(BW)_b$ Typically $(BR)_c = 2.35n\,(BW)_b$	$(SNR)_b = 6n + 1.8 - QF$ $QF = 10\log \dfrac{\text{quantize range}}{\text{signal level}}$
Compressed PCM	$(BR)_c = 2.35n\,(BW)_b$	$(SNR)_b = 6n + 4.8$ $\qquad - 20\log\,[\ln\,(1 + \mu)]$
Delta Modulation	$(BR)_c = f_s$	$(SNR)_b = 20\log\,(f_s/f_o)$ $\qquad + 10\log\,[\,f_s/(BW)_b]$ $\qquad - 14.2\text{ dB}$
Differential PCM	$(BR)_c = nf_s$ $(BR)_c = 2.35n\,(BW)_b$	$(SNR)_b = 20\log\,(f_s/f_o)$ $\qquad + 10\log\,[\,f_s/(BW)_b]$ $\qquad + 6n - 14.2\text{ dB}$

[a] All relationships assume negligible effects from bit-error distortion, i.e., BER $> 10^{-4}$. Typically, BER is specified at 10^{-9}, which corresponds to a received $SNR_{c, AVG/RMS}$ of 15.6 dB, or a $Q = 6$.
[b] f_s = sampling frequency, where $f_s > 2(BW)_b$, typically 2.35 $(BW)_b$ to 2.4 $(BW)_b$ for practical filtering
  n = number of bits per sample
  f_o = frequency of received signal (sinusoid assumed)
  $(BW)_b$ = baseband signal bandwidth, also assumed to be the bandwidth of the filter at the decoder output
  μ = quantizer compression shape factor; $\mu > 100$ for logarithmic compression

In general, most of these encoding techniques have been designed to reduce bandwidth utilization at the expense of increased coder complexity and cost. With the greater bandwidth offered by fiber optics, the simpler linear encoding techniques may gain greater utilization, particularly for high-quality video. For this reason, linear encoding is addressed in more detail here.

5.3.1 Linear PCM

The signal-to-noise ratio in a PCM system is determined by two distortion sources within the encoder, quantizer distortion and decoding-error distortion. Above a certain received digital signal-to-noise threshold, distortion due to decoding errors becomes negligible. Most optical systems specify a received $SNR_{AVG/RMS}$ of 15.6 dB, which corresponds to a 10^{-9} BER. At this level, only quantizing noise is significant.

For linear encoding, the quantizer has 2^n signal-voltage-measurement steps of uniform step size q. Because the steps are discrete, there is measurement error present with a maximum value of $+q/2$ to $-q/2$. Haykin [6] gives the average power of the resultant quantizing noise N_q:

$$N_{q, RMS} = \frac{q^2}{12} \qquad (5\text{-}13)$$

The maximum peak-to-peak signal-measurement range of a linear quantizer (A_q) is defined as

$$A_q = q \times 2^n \qquad (5\text{-}14)$$

Then:

$$(\text{SNR})_{\text{RMS/RMS}} = \frac{\text{RMS signal power}}{\text{quantizer noise power}} = \frac{(\text{rms signal level})^2}{q/12} \qquad (5\text{-}15)$$

If the rms signal level is designated $V_{s,\text{rms}}$ and since $q = A_q/2^n$, then from Equation (5-14)

$$\text{SNR}_{\text{RMS}} = \frac{V_{s,\text{rms}}^2}{(A_q/2^n)^2/12} = 12(2^{2n})\left(\frac{V_{s,\text{rms}}}{A_q}\right)^2 \qquad (5\text{-}16)$$

Expressed in decibels, we get

$$\text{SNR}_{\text{RMS/RMS}} = 6n + 10.8 + 20 \log \frac{V_{s,\text{rms}}}{A_q} \qquad (5\text{-}17)$$

For a full-load sinusoid of peak level V:

$$V_{s,\text{rms}} = \frac{V}{\sqrt{2}}$$

$$A_p = 2V$$

Then:

$$(\text{SNR})_{\text{RMS/RMS}} = 6n + 10.8 + 20 \log \frac{1}{\sqrt{2}} \qquad (5\text{-}18)$$

$$= 6n + 1.8 \text{ dB}$$

Often, the signal does not fully load the quantizer or is not expressed in peak-to-peak terms. For example, when the signal is a complex waveform, it is generally expressed in terms of rms voltage or some statistical terms that can be related to an average or rms level. When this is the case, a quantizer conversion factor (QF) is added:

$$(\text{SNR})_{\text{RMS/RMS}} = 6n + 1.8 \text{ dB} - \text{QF} \qquad (5\text{-}19)$$

where

$$\text{QF} = 20 \log \frac{\text{quantizer range}}{\text{signal level}}$$

$$= 20 \log \frac{A_{q,\text{p-p}}}{V_{s,\text{p-p}}} \qquad (5\text{-}20)$$

For example, with voice encoding, the quantizer is often designed to match the statistically determined rms value of the voice waveform. It is common for the pattern to be considered Gaussian with zero volts as the mean, and an overload at 4 sigma (four times the rms signal level). For this design, the quantizer range is, therefore

$$A_q = 2(4\sigma) = 8\sigma$$

SNR is, therefore,

$$(\text{SNR})_{\text{RMS/RMS}} = 6n + 10.8 - 20 \log \frac{8\sigma}{\sigma}$$

$$= 6n - 7.2 \text{ dB}$$

5.3.2 Compression or Companding

At times it is desireable to have a quantizer with a nonuniform step size, i.e., smaller steps near zero voltage to encode the lower-level signals, and larger steps for the higher-level signals. In this way, the average SNR is improved in a system with varying signal level. The advantage can be a reduction in the number of encoding bits required and thus the bit rate.

As an example, a commonly used characteristic is the logarithmic μ-law curve [1, 2, 5]. A value of $\mu = 0$ represents linear encoding. A value of $\mu > 100$ is generally used to gain a relatively flat SNR over a wide dynamic signal range. Following Smith [2], the approximate SNR for compressed PCM with $\mu > 100$ is given in Table 5.5.

5.3.3 Differential PCM (DPCM) and Delta Modulation (DM)

With DPCM and DM, only the difference in voltage from one sample to the next is transmitted. Since this is small, in most cases, the number of bits required per sample is smaller than standard PCM, which contains the number of bits required for the entire signal range in each sample.

With DPCM, the voltage difference between samples is measured and encoded. In this way, SNR becomes a function of both the number of encoding bits and the sampling frequency. With DM, a single pulse is sent if the sample magnitude increases, and no bit is sent if the magnitude decreases. SNR, therefore, becomes almost totally a function of sampling frequency. Errors, and thus distortion noise, are created when the change in signal level (slope) exceeds the ability of the encoder to follow it or properly represent the change with a code word.

The advantage of DPCM or DM is the fewer required encoding bits and thus the greatly reduced bit rate over standard PCM. Following reference [2], the approximate relationships for SNR and bit rate are given in Table 5.5.

5.4 FREQUENCY-DIVISION MULTIPLEXING

Frequency-division multiplexing (FDM) is shown in Figure 5.6. Basically, channel signals of bandwidth $(\text{BW})_1$ through $(\text{BW})_N$ are upconverted in a mixer/modulator by carrier frequencies f_1 through f_N to form individual transmission channels $(\text{BW})_{c1}$ through $(\text{BW})_{cN}$. The modulator also filters out unwanted sidebands and modulation products. If the function is only upconversion, then the transmission channels $(\text{BW})_{cN}$ have roughly the same bandwidth as the original signal $(\text{BW})_N$. The mixer in some systems may form the dual function of AM, FM, or PM modulator. In this case, $(\text{BW})_{cN}$ is greater than $(\text{BW})_N$ as a function of the modulation parameters.

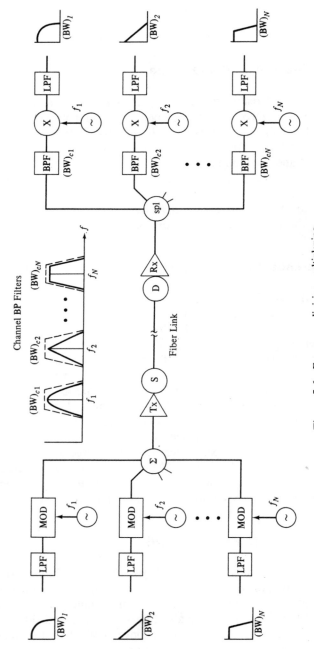

Figure 5.6 Frequency-division multiplexing.

The individual transmission channels are then combined (usually passively) and the resultant composite waveform transmitted on the fiber link. At the receiver, the signals are amplified and split, then filtered in band-pass filters, matched to the transmitted channel bands, of bandwidths $(BW)_{c1}$ through $(BW)_{cN}$. The signals are each then mixed back down to their original frequency bands and filtered to their original bandwidths, $(BW)_1$, $(BW)_2$, . . . , $(BW)_N$.

The effects that FDM has on the transmission analysis are as follows:

(a) the transmission bandwidth required is a function of the number of channels, N; the channel bandwidth of each channel, $(BW)_{c1}$ through $(BW)_{cN}$; and the guard band, $(BW)_g$, required between each channel to eliminate crosstalk;

(b) the optical receiver bandwidth may have to be larger than the transmission bandwidth so that the characteristics of the receiver filter do not distort the channels near the band edge;

(c) the available optical power P_t is divided among the N channels such that each is allocated P_t/N plus or minus power differences required for compensation of one channel over another; and

(d) harmonic interference and intermodulation between channels add another noise component that must be accounted for.

5.4.1 Bandwidth Requirements

The bandwidth requirements of an FDM transmission system depend on the channel frequency plan, which in turn is dictated by the type of signal to be transmitted and the transmission medium. There are many standard frequency plans depending on type of signal and transmission medium. For example, for FDM voice, CCITT Recommendation G.233 gives frequency groupings in terms of Groups (12 channels), Supergroups (60 channels), Mastergroups (300 channels), and Supermastergroups (3600 channels). Line-frequency allocations for coax are given in CCITT Rec. G.333. MIL-STD-188-100 addresses FDM for tactical military voice communications. For video, the standard L3 channelization plan for coax can be found in reference [7], and the plan for radio transmission of video in CCIR Rec. 402. The list becomes extensive for voice, video, and data transmission over balanced pairs, coax, microwave radio, and satellite. There are no standards specifically described for FDM on fiber, however, since fiber transmission is generally intended to accommodate only digitized signals. To some degree, the standard plans used for coax can be adapted to fiber directly, with care as to intermodulation and SNR constraints. Generally, when fiber is to be used with FDM, it is a custom application and the frequency plan is tailored to that application and the specific modulation parameters used.

Figure 5.7 illustrates the bandwidth relationships for an FDM transmission. Channels 1 through N occupy bandwidths $(BW)_{c1}$ through $(BW)_{cN}$, respectively. These signals must be spaced with some guard band $(BW)_g$ between them to allow for tolerances and drift and so that practical receiver band-pass filters can select each channel and eliminate crosstalk from adjacent channels. The amount of guard band required depends on the modulation approach selected, its band-pass requirements,

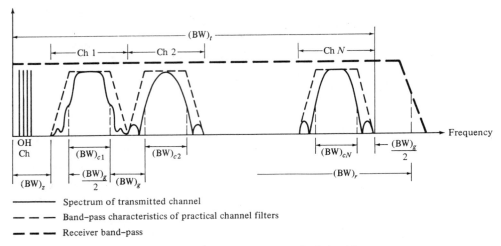

Spectrum of transmitted channel

— — — Band–pass characteristics of practical channel filters

▬ ▬ ▬ Receiver band–pass

Figure 5.7 FDM Frequency-band relationships.

its spectral band occupancy, and the crosstalk susceptibility. The total transmission bandwidth requirement (for adjacent channel spacing) is, therefore, the sum of the individual channel passbands and the guard bands. For practical systems, depending on the circumstances, receiver bandwidth can be greater than $(BW)_t$ by some extension factor (F_{BW}) so as not to degrade the channels near the band edge.

$$(BW)_r = F_{BW}(BW)_t \qquad (5\text{-}21)$$

$$(BW)_t = (BW)_z + [(BW)_1 + \cdots + (BW)_N] + N(BW)_g \qquad (5\text{-}22)$$

For example, a F_{BW} of $\sqrt{2}$, where $(BW)_r = (BW)_t\sqrt{2}$, causes only a 1.5-dB degradation in the highest-frequency signal, whereas $F_{BW} = 1$, that is, $(BW)_r = (BW)_t$, results in a 3-dB degradation in the highest channel.

5.4.2 Power Penalties for FDM

Power penalties unique to FDM transmission arise from the following:

(a) the division of available power between channels;

(b) the reduction of carrier power in the higher-frequency channels due to transmission-link band limiting; and

(c) the increase in noise power density with frequency for bipolar and FET receivers where transistor noise dominates.

Figure 5.8 shows these effects along with the effects of intermodulation products, which are discussed in the next section.

Channel modulation index.   In calculating the link-power budget, the effects of FDM is accommodated in the calculation of MDP and power penalties. The MDP calculation is performed from the basis of the carrier-to-noise ratio of the indi-

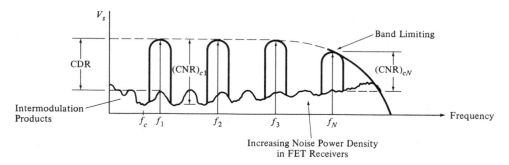

Figure 5.8   Practical limits on CNR in an FDM system.

vidual worst-case channel. The MDP-versus-CNR relationship for analog carrier-modulated signals is given in Chapter 4, Section 4.10. The most pronounced effect is that the signal power is divided by N, assuming all channels are given equal power. The source modulation index for a particular channel, $M_{o,\text{ch}}$, is, therefore,

$$M_{o,\text{ch}} = \frac{M_o}{N} \tag{5-23}$$

For nonequivalent carrier levels, M_o is factored accordingly.

Effect on MDP.   In determining the link-power budget, generally, the link-bandwidth budget is performed in order to determine the band-limiting effect on signal level within the transmission channel, and, therefore, to assign a power penalty. With FDM, however, each received channel is filtered by band-pass filters $(\text{BW})_{c1}$ through $(\text{BW})_{cN}$, and $(\text{SNR})_c$ is therefore treated separately for each channel. Minimum required detected optical power (MDP) is therefore calculated individually based on channel bandwidths. The transmission bandwidth does not directly factor into the SNR power penalty as would be the case with baseband or digital transmission. The CNR (or SNR) that is used to compute the required received optical power is therefore considered the $(\text{CNR})_{c,m}$ in channel m:

$$(\text{CNR})_{c,m} = \frac{\text{channel } m \text{ received carrier power}}{\text{rms noise power in } (\text{BW})_{c,m}} \tag{5-24}$$

By taking this and $M_{o,\text{ch}}$ into account, the MDP calculation for FDM multiplexed transmission, based on the CNR requirements for channel m, is given by Equation (4-106) if we substitute $(\text{BW})_{c,m} = (\text{BW})_c$ and $(\text{CNR})_{c,m} = (\text{CNR})_c$. From Equation (4-101), if quantum noise is small relative to amplifier and multiplied leakage noise, then Equation (4-106) simplifies to

$$\text{MDP} = \frac{2N}{M_o r}\left[\frac{(\text{CNR})_c eFN\,(\text{BW})_c}{M_o} + \sqrt{\frac{(\text{CNR})_c}{2}\left(\frac{\langle i_{nA}^2 \rangle}{G^2} + 2eI_{dm}F\,(\text{BW})_c \right)} \right] \tag{5-25}$$

The effect is that MDP increases as a direct function of N if amplifier or leakage noise dominates (such as with a PIN/FET receiver), and as a function of N^2 if quantum noise dominates (as with an APD).

Transmission band limiting.   Band limiting in the transmission channel affects the levels of those channels that are near the band edge. If transmission bandwidth is inadequate, then the result is a channel power penalty due to the overall decrease in amplitude of the channel:

$$p_{m(-)} = \text{decreased carrier power in channel } m$$

If equal channel levels are desired, band limiting can be compensated for by increasing the transmitted power of those channels affected. Figure 5.9 shows two cases and uses some simple conservative assumptions to analyze the power penalty imposed by equalizing the channel power. For example, assume only one of N channels, channel m, is reduced by an amount $p_{m(-)}$ due to band limiting, Figure 5.9(a). If power $p_{m(+)}$ is added to channel m in order to compensate, power in each of the other carriers automatically reduces by $p_{c(-)}$ since total power transmitted remains constant:

$$p_{m(+)} = (N - 1)p_{c(-)} \tag{5-26}$$

and for equal carrier levels:

$$p_{m(-)} = p_{m(+)} + p_{c(-)} \tag{5-27}$$

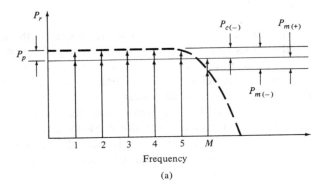

(a)

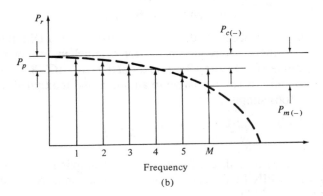

(b)

Figure 5.9   Compensation for unequal channel power in frequency-division multiplexed systems. (A) Highest-frequency channel affected and (b) all channels affected.

By solving Equation (5-27) for $p_{c(-)}$, the power penalty, we get the channel-m decrease $p_{m(-)}$ equally divided among the carriers:

$$P_p = \frac{\text{channel } m \text{ decrease}}{\text{number of channels}} = \frac{p_{m(-)}}{N} \tag{5-28}$$

Generally, in a transmission passband, the power decreases steadily from DC to the corner frequency (frequencies), and then more rapidly (10 to 20 dB/decade) beyond. This case is shown in Figure 5.9(b). Here the approximate relationship for power penalty due to equalization is

$$P_p = \frac{p_{m(-)}}{2} \tag{5-29}$$

In some cases, the solution may not be as easy as simple power-level compensation. A channel placed near or beyond the corner frequency of the link band-pass experiences a difference in gain and phase across its channel frequency band. With some modulation approaches, this differential gain and phase affects signal performance. The designer must consider this problem for those applications where small gain and phase differences across the band are critical. In these cases, it would be best to increase receiver and or fiber bandwidth, if possible.

Effect of frequency-dependent receiver noise.   When using avalanche photodiodes with receiver systems where thermal noise dominates, the noise power density across the transmission band is nearly constant. This means that the noise power densities of each of N FDM channels are the same. When PIN receivers are used however, amplifier noise dominates and the noise power density increases with frequency, as indicated in Chapter 4, Equation (4-91) for bipolar transistors and Equation (4-93) for FETs. This means that the noise power density in the higher-frequency channels is greater than that of the lower-frequency channels (see Figure 5.8). If bandwidth and carrier power are equivalent, then the $(CNR)_c$ decreases as channel carrier frequency increases. As in the band-limiting case, this effect can be compensated for by increasing carrier power to the affected channels, at the expense of a reduction in power to the remaining channels. The overall effect is a power penalty when compared with the case of a constant-noise power density.

This condition can be observed by examining the MDP equations for both the APD thermal-noise dominated case and the amplifier-noise dominated PIN/bipolar and PIN/FET cases. For a given $(CNR)_c$ and $(BW)_c$, MDP becomes a square-root function of receiver and photodetector noise. Since detector noise density is essentially constant with frequency, the only variable is amplifier noise. For a receiver that is thermal-noise dominated, the noise-current relationship is, from Equation (4-87),

$$\langle i_{nA}^2 \rangle = \frac{4kT}{R_i}(BW)_c$$

For a bipolar transistor, where transistor noise dominates, the noise-current relationship is, from Equation (4-91),

$$\langle i_{nA}^2 \rangle = \frac{8\pi kTC_o}{\sqrt{\beta}}(BW2)^2$$

where

$$(BW2)^2 = \frac{f_h^2 - f_l^2}{2}$$

For an FET receiver, where FET noise dominates, the noise-current relationship is, from Equation (4-93),

$$\langle i_{nA}^2 \rangle = \frac{16\pi^2 kTtC_i^2}{g_m}(BW3)^3(1 - fc/f)$$

where

$$(BW3)^3 = \frac{f_h^3 - f_l^3}{3}$$

With the APD thermal-noise case, the amplifier noise is simply a function of channel bandwidth $(BW)_c$, and, as such, the noise in any channel (1 through N) is equal across the FDM band. With the bipolar- and FET-noise-dominant case, noise is not only a function of bandwidth, but of the center frequency of the channel as well. The BW multiplier is a function of the frequency of the band edges (f_h and f_l) in which the channel noise is measured. The effect is shown in Figure 5.8 as an increasing noise level below fc and with increasing frequency, and is quantified in the FDM example in Chapter 7.

If we want to equalize the $(CNR)_c$ of all carriers for a given received power, we must add power to the higher-frequency channels and reduce power to the channels in the lower half of the grouping. If we use a rough assumption that the noise power increases linearly over the band, then the resultant average MDP after compensation is, approximately,

$$(MDP)_{comp} = \frac{MDP_{chN} + MDP_{ch1}}{2} \tag{5-30}$$

5.4.3 Intermodulation Products

Whenever multiple carriers are transmitted through a common channel, any nonlinearities in the transmitter or receiver create the following:

(a) spurious signals that are harmonics of the signal or carrier frequencies ($2f_n$, $3f_n$, etc.); and

(b) a mixing between carriers and their harmonics to form intermodulation products ($f_1 + f_2$, $2f_1 + f_2$, etc.).

The mechanism of harmonic generation is shown in Figure 5.10. The result of intermodulation when two signals (or unmodulated carriers) are introduced is shown in Figure 5.11.

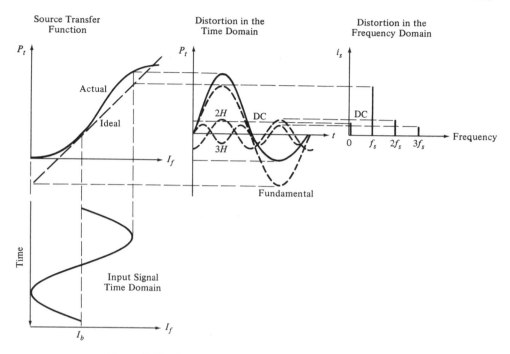

Figure 5.10   Generation of harmonics by a nonlinear source.

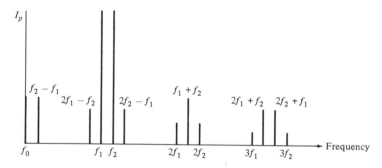

Figure 5.11   Intermodulation products resulting from the transmission of two unmodulated carriers with second- and third-order distortion.

In a fiber transmission system, the detector and receiver are generally very linear and, therefore, the level of the distortion products depends on the linearity of the laser or LED. Typical values are given in Table 5.6. It should be noted that in order to achieve linear performance, the LED or laser diode must be driven from a linear current source.

With a nonlinear device, as shown in Figure 5.10, the output power (and thus the detected photocurrent) can be represented by the following polynomial, which represents the transfer characteristic of the device [8]:

$$P = K_0 + K_1(I + K_2I^2 + K_3I^3 + \cdots) \tag{5-31}$$

TABLE 5.6 HARMONIC CONTENT OF TYPICAL SOURCES AT 50% MODULATION

Device	Second harmonic	Third harmonic
Surface LED	−30 to −40 dBc	−50 dBc or better
Lower-quality multimode laser	−25 to −40 dBc	−35 to −50 dBc
High-quality laser	−40 to −50 dBc	−60 to −70 dBc

where I is the drive current, and K_1 through K_n are constants that are dictated by the shape of the source transfer function. In order to obtain the ratios of the second and third harmonics that are produced to the power of the fundamental, we can substitute a normalized drive current, $I = \cos(\omega t)$, into the polynomial:

$$P = K_0 + K_1[\cos(\omega t) + K_2 \cos^2(\omega t) + K_3 \cos^3(\omega t) + \cdots\} \qquad (5\text{-}32)$$

Simplifying the second- and third-order terms, we get

$$K_2 \cos^2(\omega t) = \frac{K_2}{2}[1 + \cos(2\omega t)] \qquad (5\text{-}33)$$

$$K_3 \cos^3(\omega t) = \frac{K_3 \cos(\omega t)}{3}[1 + \cos(2\omega t)]$$

$$= \frac{3K_3}{4}\cos(\omega t) + \frac{K_3}{4}\cos(3\omega t) \qquad (5\text{-}34)$$

such that:

$$P = K_0 + K_1\left[\frac{K_2}{2} + \left(1 + \frac{3K_3}{4}\right)\cos(\omega t) + \frac{K_2}{2}\cos(2\omega t)\right.$$

$$\left. + \frac{K_3}{4}\cos(3\omega t) + \cdots\right] \qquad (5\text{-}35)$$

The ratios of the harmonic power levels to that of the fundamental are, therefore, as follows:

Fundamental:

$$\cos(\omega t) = 1 + (3K_3)/4 \qquad (5\text{-}36)$$

Second harmonic $\cos(2\omega t)$ to fundamental:

$$\frac{K_2/2}{1 + 3K_3/4} = \frac{K_2}{2} \qquad \text{for } K_3 \ll 1 \qquad (5\text{-}37)$$

Third harmonic $\cos(3\omega t)$ to fundamental:

$$\frac{K_3/4}{1 + 3K_3/4} = \frac{K_3}{4} \qquad \text{for } K_3 \ll 1 \qquad (5\text{-}38)$$

These relationships assume that the transfer function and the signal waveform are symmetrical around some reference. It may be desired, however, to analyze the effects of bias-point selection and signal amplitude on harmonic distortion. The effects of bias point on distortion can be rather significant for any one device. Figure 5.12 shows the nature of the effect that might be observed. This figure uses an extremely nonlinear laser (the second harmonic is approximately -20 dBc at I_o), so the results are somewhat worse than would be observed with typical devices.

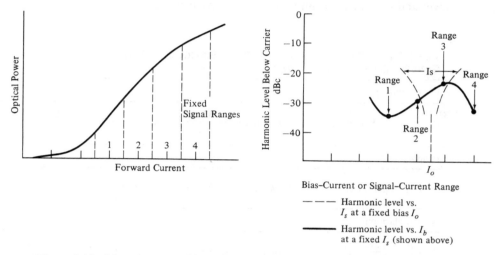

Figure 5.12 Distortion versus bias point and signal magnitude for a hypothetical nonlinear laser.

In order to accomplish an analytical treatment of the effects of bias point on harmonic content, the device transfer function and the signal current are represented with bias current as a factor. If we restrict operation to some region between I_1 and I_2 centered at bias point I_o, then we can represent the drive current for one carrier as

$$I = I_o + A \cos (\omega t) \tag{5-39}$$

If there exist multiple carriers of amplitude A, then this becomes

$$I = I_o + A[\cos (\omega_1 t) + \cos (\omega_2 t) + \cdots] \tag{5-40}$$

If there are N carriers of equal amplitude, equally spaced starting at frequency ω_0, then this becomes

$$I = I_o + A \sum_{n=1}^{N} \cos (n\omega_0 t) \tag{5-41}$$

If these carriers are FM modulated, for example, then the relationship becomes [8]

$$I = I_o + A \sum_{n=1}^{N} \cos [n\omega_0 t + \theta_n(t)] \tag{5-42}$$

Substituting the relationship for the single carrier into the polynomial, for example, we get

$$P = K_0 + K_1[I_o + A \cos (\omega t)] + K_2[I_o + A \cos (\omega t)]^2$$
$$+ K_3[I_o + A \cos (\omega t)]^3 + \cdots \tag{5-43}$$

We get the results shown in Table 5.7 for the total products produced by a source driven with a single unmodulated carrier. If we substitute the multiple-carrier relationships for I into the polynomial, the analysis can be extended to determine the level and placement of intermodulation products for n carriers.

TABLE 5.7 HARMONIC PRODUCTS FOR A SINGLE UNMODULATED CARRIER GENERATED BY A NONLINEAR SOURCE

DC component

$$P_{DC} = K_0 + K_1 I_o + K_2 \left(I_o^2 + \frac{A^2}{2} \right) + K_3 \left[I_o^3 + \frac{(3I_0 A)^2}{2} \right] + K_4 \left[I_o^4 + 3I_o^2 A^2 + \frac{(3A)^4}{8} \right]$$

Fundamental

$$P_0 = A \sin(\omega t) \left[(K_1 + 2 K_2 I_o + 3K_3 I_o^2 + 4K_4 I_o^3) + 3A^2 \left(\frac{K_3}{4} + K_4 \right) \right]$$

Second Harmonic

$$P_2 = \frac{A^2 \cos(2\omega t)}{2} (K_2 + 3K_3 I_o + 6K_4 I_o^2 + K_4 A^2)$$

Third Harmonic

$$P_3 = [A^3 \sin(3\omega t)] \left(\frac{K_3}{4} + K_4 I_o \right)$$

If we can determine the polynomial factors K_n that represent a device's transfer function, then the harmonic products can be derived analytically. This can be accomplished by measuring the static transfer of an operating device and using a polynomial curve-fitting program to derive the K_n factors. Michaels [9] discusses such a method for detailed analysis and prediction of intermodulation products.

Generally, when performing an FDM analysis, we are not so much interested in finding the magnitude of all of the harmonics, but rather are trying to determine the amount (usually, the worst-case amount) of harmonic power in any one signal band. Knowing the worst-case total harmonic distortion sets the carrier-to-noise limit for the link. Since this harmonic power is a function of the frequency plan as well as source nonlinearity, we must make some simplifying assumptions. Daly [10] treats the case for N FM modulated carriers, with equal carrier spacing starting at ω_o. Since the intermodulation terms are uncorrelated, they can be summed as a power distribution. The second- and third-order products in any band can be related (in decibels below the carrier), therefore, as shown in the following.

Second-order distortion.

$$P_2 \text{ (dBc)} = 10 \log (T_1) + L_1 + 20 \log \frac{K_2}{4} - (BF)_2 \qquad (5\text{-}44)$$

where: T_1 = number of $f_a \pm f_b$ terms in the band
$\quad L_1$ = level of each T_1 term above the second harmonic = 6 dB
$\quad K_2/4$ = magnitude of the second harmonic below the carrier
$\quad (BF)_2$ = bandwidth correction factor for the second-order distortion term power
$\qquad$ spread due to carrier modulation

Third-order distortion.

$$P_3 \text{ (dBc)} = 10 \log \left(T_2 + \frac{T_3}{4} \right) + L_2 + 20 \log \frac{K_3}{4} - (BF)_3 \qquad (5\text{-}45)$$

where T_2 = number of $f_a \pm f_b \pm f_c$ terms in the band
$\quad T_3$ = number of $2f_a \pm f_b$ and $f_a - 2f_b$ terms in the band
$\quad L_2$ = level of T_2 term above the third harmonic = 15.6 dB
$\quad K_3/4$ = magnitude of the third harmonic below the unmodulated carrier
$\quad (BF)_3$ = bandwidth correction factor for the third-order distortion term power
$\qquad$ spread due to carrier modulation

Bandwidth correction factors (BF) are factors that account for distortion terms that are actually spread beyond the channel band edges with modulated carriers. For FM signals, assuming that the FM bandwidth is proportional to the rms signal amplitude, second-order terms are spread over $\sqrt{2}$ times the bandwidth of the original signal and third-order terms are over $\sqrt{3}$ times the channel bandwidth [10]. Bandwidth factors are

$$BF = 10 \log \frac{BW \text{ spectrum}}{(BW)_c} \qquad (5\text{-}46)$$

$$(BF)_2 = 10 \log (\sqrt{2}) = 1.5 \text{ dB}$$

$$(BF)_3 = 10 \log (\sqrt{3}) = 2.5 \text{ dB}$$

Table 5.8 summarizes the number of intermodulation terms, following Daly [10], with some modifications in third-order terms. An analysis of terms across a band of equally spaced adjacent FDM channels shows that the center channel (odd number) or center two channels (even number) contain the greatest number of distortion products. For analysis purposes, therefore, the distortion analysis can be performed on one center channel to find worst-case results.

Following the relationships given in this section and Table 5.8, the results shown in Figure 5.13 have been calculated by Muoi [8] for two laser diodes with different linearity characteristics.

TABLE 5.8 NUMBER OF TERMS FALLING AT CARRIER FREQUENCY nf_o IN A SET OF N EQUALLY SPACED CARRIERS WITH FREQUENCIES Xf_o, $(X + 1)f_o$, . . . , Yf_o

Term	Number of Terms	Where
T_1: Second-Order $f_a \pm f_b$ Terms		
$f_a + f_b$	$^a\text{INT}\dfrac{n + 1}{2} - X$	$2X - 1 < n < X + Y$
$f_a - f_b$	$Y - n - X + 1$	$0 < n < Y - X + 1$
T_2: Third-Order $f_a \pm f_b \pm f_c$ Terms		
$f_a + f_b + f_c$	$\text{INT}\left\{n - 3(X + 1) + \dfrac{3(X + 1)}{n}\right\}$ $+ \text{INT}\left\{n - 6(X + 1) + \dfrac{6(X + 1)}{n}\right\}$	$n > 3(X + 1)$
$f_a - f_b - f_c$	$\text{INT}\dfrac{Y - n - 2X + 1}{2}$ $\times \text{INT}\dfrac{Y - n - 2X}{2}$	$0 < n < Y - 2X - 1$; $2X + 1 < Y$
$f_a - f_b - f_c$	$\text{INT}\dfrac{n - X}{2}\,\text{INT}\dfrac{n - X - 1}{2}$ $+ (n - X)(Y - n)$ $+ \text{INT}\dfrac{Y - n}{2}\,\text{INT}\dfrac{Y - n - 1}{2}$	$X < n < Y$
T_3: Third-Order $2f_a \pm f_b$ and $f_a - 2f_b$ Terms		
$2f_a + f_b$	$\text{INT}\dfrac{n - X}{2} - \text{INT}\dfrac{n}{3}$ $+ \text{INT}\dfrac{n - 1}{3} - X + 1$	$3X < n < 2(X + Y)$
$2f_a - f_b$	$\text{INT}\dfrac{Y - n}{2} + \text{INT}\dfrac{n - X}{2}$	$X < n < Y$
$f_a - 2f_b$	$\text{INT}\dfrac{Y - n}{2} - X + 1$	$0 < n < (Y - 2X)$; $Y > 2X$

^{a} INT(x) is the largest whole integer in x, for example, INT$(3.6) = 3$.

5.4.4 Carrier Spacing

In band distortion, products may be reduced to some degree by proper frequency planning. If the application calls for a lot of wide-band modulated channels filling most of the available transmission bandwidth, then it is almost impossible to plan carrier spacings to eliminate the effects of distortion products. Even if we can space adjacent carriers such that second-order carrier frequencies fall between bands, some

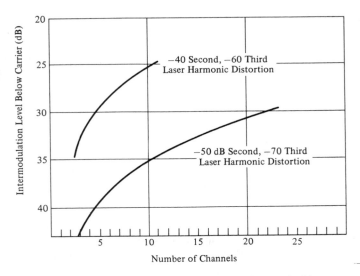

Figure 5.13   FDM Intermodulation distortion, for practical lasers.

of the higher-order products of intermodulation fall in band. If we have a situation where excess transmission bandwidth exists, then frequency planning can be an effective means of reducing the effect.

Adjacent carriers.   For the case of adjacent carriers, some simple rules become obvious by inspecting the equations in Table 5.8. First, if we remove the lower carriers from the group, the number of third-order terms is reduced. Moving the whole group up in frequency moves the second-order ($f_a + f_b$) terms out of band. Groups of terms dissappear entirely under the following conditions.

For third-order terms:

$$(f_a + f_b + f_c) \text{ and } (2f_a + f_b) \text{ terms disappear if } n < 3(X + 1)$$

$$(f_a - f_b - f_c) \text{ and } (f_a - 2f_b) \text{ terms disappear if } X > (Y - n)/3$$

For second-order terms:

$$(f_a + f_b) \text{ and } (f_a - f_b) \text{ terms disappear for the entire group of}$$
$$N \text{ channels if } X > N$$

This method for elimination of second-order terms, shown in Figure 5.14, also eliminates those third-order terms listed before. It requires twice the transmission bandwidth than that occupied by the carriers.

Staggered channel spacing.   When there are only a few channels (2 to 5) to transmit, then the frequency planning becomes analytically more manageable and channels can be staggered to avoid certain products. For example, Michaels [9] suggests an approach that eliminates third-order difference terms. Carriers (f_n) are spaced as follows:

$$f_n = 2f_{N-1} + B_c - f_1 \tag{5-47}$$

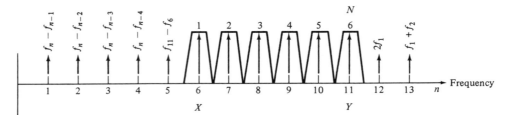

Figure 5.14 Channel-frequency plan for eliminating second-order and reducing third-order harmonics in band.

where B_c = transmission band of each channel = $(BW)_c + (BW)_g$
     f_1 = carrier frequency of the first channel

This approach is useful when bandwidth is not a problem. It requires a transmission bandwidth that increases approximately as $(N - 1)B_c$, where N is the number of channels. For a five-channel group, for example, the transmission bandwidth is

$$(BW)_t = \left[(f_1 + 15B_c) + \frac{B_c}{2} \right] - \left(f_1 - \frac{B_c}{2} \right) = 16B_c$$

5.4.5 Application of FDM

FDM is generally preferred over time-division multiplexing when the following conditions exist:

(a) the channel information is analog and carrier modulated (FM, PM, VSB AM);
(b) the information is wide-band analog (video, for example) and the expense of PCM encoding is not justified;
(c) the previous conditions exist and transmission distances are limited such that multiple repeaters are not necessary.

5.5 TIME-DIVISION MULTIPLEXING

Time-division multiplexing (TDM) interleaves multiple lower-speed digital data channels for transmission over one higher-speed channel. It is the predominant means of transmitting multiple channels of PCM analog information such as voice or video.

5.5.1 Operating Principle

Figure 5.15 shows the concept of time-division multiplexing. Functionally, it is analogous to a rotary commutator. Synchronous or asynchronous digital bit streams are, first, simultaneously received and stored in input buffers that perform the function of synchronizing (SYNC) each data stream to the sampling rate and timing of the multiplexer. A second larger framing buffer sequentially clocks out and stores

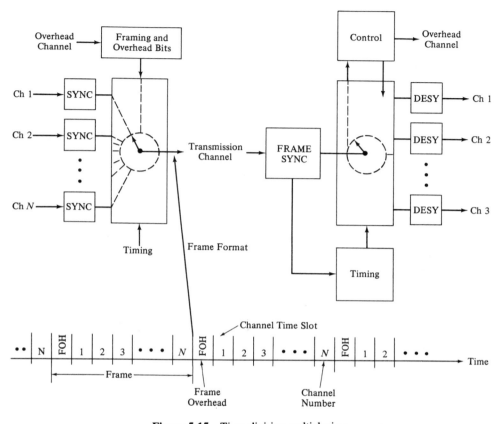

Figure 5.15   Time-division multiplexing.

data words from each channel input buffer, forming a transmission frame of data. Such a frame of N channels is shown at the bottom of Figure 5.15.

As each channel is clocked into the frame, "stuffing" bits can be added to account for differences in timing between channel rate and multiplexer sampling rate. These are extracted on the receiving end. At the frame level, timing and overhead bits are also added, primarily to synchronize the demultiplexer to the multiplexer. Bits can also be added to detect errors and to control and monitor the system. These stuffing, synchronizing, error-encoding, and overhead bits are shown (for simplicity) occupying a separate time slot, FOH, in Figure 5.15.

The resultant serial bit stream is then encoded for transmission. Encoding is performed to scramble and format the signal waveform so that it can be transmitted over the fiber and so that timing can be adequately recovered at the receiving end. At the receiving end, the function is analogous to a decommutator, frame synchronized to the commutator on the transmitting end. Timing is recovered and the signal is decoded and converted back into a bipolar NRZ waveform. The signal is loaded into a large receiving buffer and control and error signals are extracted. The stored frame is then clocked out of the buffer and each word distributed to output buffers and desynchronizers (DESY). The output buffers store the channel words and clock them out serially at the proper channel rate.

5.5.2 TDM Transfer Function

The effects that time-division multiplexing has on an optical link design are

(a) to increase data rate (bandwidth) as a function of the number of channels multiplexed (N);

(b) to alter the jitter requirement due to buffering, retiming, and bit stuffing; and

(c) to alter transmission parameters due as a result of signal encoding at the multiplexer.

Although the number of errors per unit time can increase at the output of the multiplexer, due to summing of individual channel errors, the bit-error rate is theoretically not altered since it is a function of the number of errors to the number of bits transmitted. Ignoring the small effects of frame encoding,

$$(BER)_t = (BER)_c \qquad (5\text{-}48)$$

There is no optical transmitter power sharing as with FDM, and, since the signal remains binary, the required received $(SNR)_{p/rms}$, or Q, is the same for N channels as for one channel.

Transmission bandwidth is a function of data rate per channel, $(BR)_c$; the number of channels multiplexed, N; and the number of overhead bits to data bits in each frame. The ratio of channel bits to total bits transmitted in a frame is called the multiplexing frame efficiency (FEF):

$$(BR)_t = \frac{N(BR)_c}{FEF} \qquad (5\text{-}49)$$

$$FEF = \frac{N(BPC)}{N(BPC) + FOH} \qquad (5\text{-}50)$$

$$FEF = \frac{N(BR)_c}{N(BR)_c + (BR)_{LOH} + MOH/T_f} \qquad (5\text{-}51)$$

where $(BR)_c$ = bit rate per channel
  BPC = bits per channel
  FOH = frame overhead = LOH + MOH
  LOH = line overhead, bits per frame
 $(BR)_{LOH}$ = bit rate of the line overhead channel
  MOH = multiplexer overhead, bits per frame

As an example, we examine the multiplexing efficiency of a DS-1 to DS-3 rate multiplexer used in T-carrier telephony. A DS-1 rate is 1.544 Mb/s. Twenty-eight channels are multiplexed and transmitted at a DS-3 rate of 44.736 Mb/s. The efficiency of such a multiplexer is

$$(FEF)_m = \frac{N\,(BR)_c}{(BR)_t}$$

$$= \frac{28 \times 1.544}{44.736} = 0.96638$$

From Chapter 4, Equation (4-116), it can be seen that the minimum received power is a function of 10 log (BR) where quantum or bipolar amplifier noise dominates, and a function of 15 log (BR) where FET amplifier noise dominates. As a function of the number of channels, the ratio, therefore, becomes a function of 10 log (N) where quantum or bipolar transistor noise dominates and 15 log (N) where FET noise dominates.

5.5.3 TDM Application

TDM is almost always preferred over FDM when the channel information is digital. TDM of PCM-encoded analog channels is generally preferred over FDM of analog modulated carriers when the following conditions exist:

(a) source linearity is of concern and a high SNR is required;

(b) very long transmission distances require multiple repeaters;

(c) fiber or receiver bandwidth is limited;

(d) receiver dynamic range or link-power margin is limited; and

(e) equipment for encoding and multiplexing is standard, available, and reasonably priced.

5.6 SECTION-LAYER MODULATION AND ENCODING

Certain signal modulation and encoding is performed at the section layer. The most common modulation approach is direct intensity modulation of the optical carrier whereby the section layer simply passes the signal through to the optical-source drive electronics such that drive current (and, therefore, optical power) is a direct function of signal voltage.

Other modulation approaches employed at the section layer include pulse-position and pulse-duration modulation whereby the optical carrier is pulsed and the width or time position of the pulse has a direct relationship with input-signal amplitude. This form of modulation is generally used for low-cost single-channel short-distance analog telemetry and control systems.

Digital signal encoding is also performed at the section layer in order to format the signal for compatibility with the AC-coupled unipolar characteristics of the photonics.

5.6.1 Analog Intensity Modulation

With analog intensity modulation, the signal linearly modulates source power. For simplicity, we illustrate the process by assuming that the optical carrier is a coherent sinusoid and the modulating signal is a sinusoid as well. Under these conditions, the modulated optical carrier power at the input to the detector can be characterized as [4]:

$$C(t) = 2P_o[1 + M_o \cos (\omega_s t)] \cos^2 (\omega_o t) \qquad (5\text{-}52)$$

where $2P_o \cos^2 (\omega_o t)$ = intensity of the unmodulated optical carrier, assuming coherence

$\qquad P_o = A^2/2$ = average source power (bias point)

$\qquad M_o = P_s/2P_o$ = the modulation index or fraction of total power available that is modulated

$\qquad \omega_o$ and ω_s = angular frequencies of the optical carrier and the input signal, respectively

At the photodetector, the optical carrier is removed, leaving the photocurrent signal:

$$i_d(t) = rGP_o[1 + M_o \cos (\omega_s t)] \qquad (5\text{-}53)$$

For more complex (nonsinusoidal) modulating signals, we can characterize this as

$$i_d(t) = rGP_o[1 + M_o\langle s(t)\rangle] \qquad (5\text{-}54)$$

At the receiver, the DC component is stripped off so that the signal component is

$$i_d(t) = M_orGP_o\langle s(t)\rangle \qquad (5\text{-}55)$$

The signal-to-noise ratio, from Equation (4-95) is measured and calculated at the optical receiver as the ratio of the rms signal power to the rms noise power in the baseband-signal equivalent noise bandwidth:

$$(\text{SNR})_{\text{RMS/RMS}} = \frac{\langle i_d\rangle^2}{\langle i_n\rangle^2} = \frac{(M_orGP_o)^2\langle s^2\rangle}{\langle i_{nD}^2\rangle + \langle i_{nA}^2\rangle} \qquad (5\text{-}56)$$

5.6.2 Digital Intensity Modulation

When the modulating signal is digital in nature, the process is essentially the same as with analog except that the modulating signal is a pulse waveform instead of sinusoidal. With each pulse representing a binary state, the detected photocurrent is

$$i_d = rGP(1) \qquad \text{for a binary "1"} \qquad (5\text{-}57)$$

$$i_d = rGP(0) \qquad \text{for a binary "0"} \qquad (5\text{-}58)$$

For example, with NRZ encoding, where a binary "1" is represented by pulse presence and "0" by pulse absence, then the detected photocurrent is

$$i_d = rGP_{pk} \qquad \text{for a "1"}$$

$$i_d = 0 \qquad \text{for a "0"}$$

Because of the discrete signal states, the signal-to-noise ratio at the receiver is measured, in terms of peak signal current to rms noise current:

$$(\text{SNR})_{pk/rms} = \frac{I_s}{\langle i_n \rangle} = \frac{rGP_{pk}}{\sqrt{\langle i_{nD}^2 \rangle + \langle i_{nA}^2 \rangle}} \qquad (5\text{-}59)$$

With digital signals, the received signal quality is determined after pulse regeneration, not as a measure of SNR, but rather as a measure of the number of transmitted bits that were received and regenerated correctly. The quality parameter is the bit-error rate (BER) or, in other terms, the error probability (P_e). The relationship of BER or P_e to received SNR for unipolar signaling was derived in Section 4.5, where SNR and BER were related by a quality factor Q from Equation (4-52):

$$(\text{SNR})_{\text{Avg/RMS}} = 20 \log Q$$

For unipolar signaling, from Equation (4-48):

$$\text{BER} = \text{erfc } Q = \text{erfc } \frac{I_s}{2\langle i_{nT} \rangle}$$

For bipolar signaling, from Equation (4-54):

$$\text{BER} = \frac{2}{3} \text{ erfc } \frac{Q}{2} = \frac{2}{3} \text{ erfc } \frac{I_s}{4\langle i_{nT} \rangle}$$

For example, with unipolar signaling, for a BER of 10^{-9}, where the value for Q is 6, SNR is

$$(\text{SNR})_{\text{avg/rms}} = 20 \log (6) = 15.6 \text{ dB}$$

For bipolar signaling, assuming a 10^{-9} BER:

$$\text{BER} = \frac{2}{3} \text{ erfc } \frac{Q}{2} = 10^{-9}$$

$$Q = 2(5.325) = 10.65$$

5.6.3 Pulse-Position Modulation (PPM)

Pulse-position modulation (PPM) is another approach to transmitting analog information by pulse modulating the optical source. As with PCM, the signal is sampled at a frequency f_s greater than or equal to the Nyquist rate, $2(\text{BW})_b$. Unlike PCM, however, PPM does not represent the sample voltage with a code. The sample value is represented by the position (in time) of the transmitted pulse within a time slot T, where $T = 1/f_s$. Figure 5.16 represents the process. The parameters are as follows:

T = width of a time slot = inverse of sampling frequency f_s;

nT = the center of the time slot of the nth sample; this position represents a sample value of zero volts;

T_p = width of a transmitted pulse assuming the pulse is ideal (rectangular); in practice, we assume a cosine pulse shape, therefore, T_p equals the width at the half-power points or the half width at the base of the pulse;

t_n = difference in time between the actual pulse position and the center of the time slot (nT); this is a measure of the sample value and thus a representation of signal value where $t_n = k_1 \text{ (s/V)} \times V_n$;

θ_n = position in time of the nth signal pulse translated to a phase shift where
$\theta_n = k_2(\text{rad/V}) \times V_n$; and

P_{pk} = peak pulse power; i_{pk} is the resultant detected photocurrent.

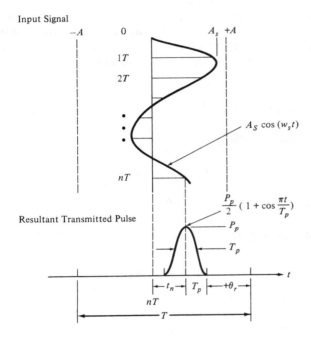

Input Signal

$A_S \cos (w_s t)$

$\frac{P_p}{2}(1 + \cos\frac{\pi t}{T_p})$

Resultant Transmitted Pulse

P_p

T_p

Figure 5.16 Pulse-position modulation.

The signal-to-noise ratio of a PPM modulated signal is related in terms of peak or average received pulse power to rms noise. Since PPM converts voltage amplitude to timing or phase relationships, the SNR analysis is essentially an amplitude-to-time or phase-conversion process. Input signal power is related in terms of the resultant rms phase shift for a fully loaded sinusoidal input. RMS noise is related to the effects of random noise on the time or phase relationships of the reconstructed pulsed waveform as it crosses a reference threshold. Noise in the PPM system is attributed to two mechanisms:

(a) random amplitude noise on the pulse edge that is converted to phase noise (timing jitter) as it crosses the reference threshold; and

(b) pulse decoding errors caused by noise bursts crossing the detection threshold at times other than when a signal pulse is present.

Signal power.   The analysis approach in determining the signal-to-noise ratio follows the approaches of Hubbard [11], Patisaul [12], and McDevitt [4]. The assumption made in the analysis is that the transmitted pulse is a cosine shape. This simplification follows practical pulse shapes, particularly for LEDs. The power in the pulse can be represented, therefore, as

$$P(t) = \frac{P_p}{2}\left(1 + \cos \frac{\pi t}{T_p}\right) \qquad \text{for } -T_p < t < T_p \tag{5-60}$$

The signal amplitude (sinusoidal) can be represented as $S(t) = A_s \cos (\omega_s t)$, and, therefore, the signal amplitude for sample nT as

$$S_{nT} = A_s \cos (\omega_s nT) \tag{5-61}$$

Since the position or value of t_n is a function of S_{nT}, where $t_n = k_1 V_n$, then:

$$t_n = k_1 A_s \cos (\omega_s nT) \tag{5-62}$$

In order to represent SNR as a mean or rms power ratio, the mean signal power must be derived as a function of t_n or θ_n referred to a fully loaded sinusoidal input signal, i.e., one of peak amplitude A. For full signal loading, the maximum phase shift (θ_p) possible (from Figure 5.16) is

$$\theta_p = \frac{T}{2} - T_p \tag{5-63}$$

Since θ_p corresponds directly to A, the maximum signal level at full load, the rms power of a full-load sinusoidal signal can be represented, therefore, as

$$S_{\text{RMS}} = \frac{A^2}{2} = \frac{\theta_p^2}{2} \tag{5-64}$$

By substituting Equation (5-63) for θ_p:

$$S_{\text{RMS}} = \frac{(T/2 - T_p)^2}{2} = \frac{(T - 2T_p)^2}{8} \tag{5-65}$$

Noise power due to timing jitter.   The conversion of noise current (i_n) to timing jitter (t_j) was approximated in Chapter 4, from Figure 4.17, by

$$t_j = \frac{i_n}{di/dt} \tag{5-66}$$

Since di/dt is the slope of the signal pulse at the threshold, the value for PPM can be found by differentiating the relationship given for the transmitted pulse at time $T_{pk}/2$, the time at which the pulse reaches half power (threshold level):

$$\frac{di}{dt} = \frac{\pi}{2T_p} i_{pk} \sin \frac{\pi t}{T_p} = \frac{\pi i_{pk}}{2T_p} \qquad \text{at } T_p/2 \tag{5-67}$$

Timing jitter, therefore, is related as

$$t_j(t) = \frac{2T_p}{\pi} \frac{i_n(t)}{i_{pk}} \tag{5-68}$$

By substituting Equation (5-59), the relationship for rms noise power due to timing jitter can be written in terms of peak received optical power:

$$\langle N_j^2 \rangle = \left(\frac{2T_p}{\pi}\right)^2 \frac{\langle i_{n(1)} \rangle^2}{(rGP_p)^2} \tag{5-69}$$

With the cosine waveform assumed, we can relate average received power to maximum peak power:

$$P_{avg} = \frac{T_p}{T} P_p \tag{5-70}$$

Therefore, in terms of average received power (P_{avg}):

$$\langle N_j^2 \rangle = \left(\frac{2T_p^2}{\pi T}\right)^2 \frac{\langle i_{n(1)} \rangle^2}{(rGP_{avg})^2} \tag{5-71}$$

where $i_{n(1)}$ is the total noise current when a pulse is present.

Noise power due to decoding errors. Noise in the detected signal can occur when noise bursts cross the threshold detector and are decoded as received pulses. By following McDevitt and Slayton [4], the rms noise power for this condition can be derived as follows.

Worst-case rms timing error ($t_{max} = T - 2T_p$):

$$\langle t_e^2 \rangle = \frac{t^2}{3} = \frac{(T - 2T_p)^2}{3} \tag{5-72}$$

Resultant rms noise due to timing errors:

$$\langle N_e^2 \rangle = \frac{(T - 2T_p)^3}{3\sqrt{3}T_p} \exp\left(-\frac{1}{8}\frac{i_{pk}^2}{\langle i_{n(0)}^2 \rangle}\right) \tag{5-73}$$

where $i_{n(0)}$ is the total noise current with no signal pulse present.

Substituting Equation (5-70) into Equation (5-57) and the result into Equation (5-73), we get

$$i_{pk}^2 = (rGP_p)^2 = \left(rG\frac{T}{T_p}P_{avg}^2\right) \tag{5-74}$$

$$\langle N_e^2 \rangle = \frac{(T - 2T_p)^3}{3\sqrt{3}T_p} \exp\left[-\frac{1}{8}\left(\frac{T}{T_p}\right)^2 \frac{(rGP_{avg})^2}{\langle i_{n(0)}^2 \rangle}\right] \tag{5-75}$$

Signal-to-noise ratio for PPM. Combining Equations (5-65), (5-71), and (5-75), we get the following SNR relationship:

$$(SNR)_{RMS} = \frac{(T - 2T_p)^2}{8(\langle N_j^2 \rangle + \langle N_e^2 \rangle)} \tag{5-76}$$

where:

$$\langle N_j^2 \rangle = \left(\frac{2T_p}{\pi}\right)^2 \frac{1}{(SNR)_{p(1)}} \tag{5-77}$$

$$\langle N_e^2 \rangle = \frac{(T - 2T_p)^3}{3\sqrt{3}T_p} \exp\left(\frac{-(SNR)_{p(0)}}{8}\right) \tag{5-78}$$

and $(SNR)_{p(1)}$ is the ratio of peak pulse power to the rms noise power with the signal present, and $(SNR)_{p(0)}$ is the ratio of peak pulse power to the rms noise power with the signal absent.

$$(SNR)_p = \frac{(rGP_p)^2}{\langle i_n^2 \rangle}$$

If the system is operated above the error threshold, N_e is negligible, and, therefore, the SNR relationship becomes

$$(SNR)_{RMS} = \frac{\pi^2}{32}\left(\frac{T}{T_p} - 2\right)^2 \left(\frac{T}{T_p}\right)^2 \frac{(rGP_{avg})^2}{\langle i_{n(1)}^2 \rangle} \qquad (5\text{-}79)$$

For large T/T_p, the SNR is approximately proportional to $(T/T_p)^4$. In terms of decibels, the required SNR of the received pulse waveform, $(SNR)_r$, referenced to the required SNR of the baseband signal, $(SNR)_b$, becomes

$$(SNR)_{r,\,AVG/RMS} = (SNR)_{b,\,RMS} + 5.1 - 20 \log \left[\frac{T}{T_p}\left(\frac{T}{T_p} - 2\right)\right] \qquad (5\text{-}80)$$

$$(SNR)_{r,\,Pk/RMS} = (SNR)_{b,\,RMS} + 5.1 - 20 \log \left(\frac{T}{T_p} - 2\right) \qquad (5\text{-}81)$$

Bandwidth relationships.   From the previous, SNR is primarily a function of the relationship between T and T_p. T is, in turn, proportional to sampling frequency, and T_p is a function of the maximum transmission bandwidth permitted by the fiber link. For a Nyquist sampling rate, $1/T = f_s = 2(BW)_n$, and, assuming $(BW)_r = (BW)_n$, then for the cosine pulse, $1/T_p = (BW)_r$. Therefore:

$$T/T_p = \frac{(BW)_r}{2(BW)_b} \qquad (5\text{-}82)$$

$$(BW)_r = 2(BW)_b \frac{T}{T_p} \qquad (5\text{-}83)$$

Substituting Equation (5-82) into Equations (5-80) and (5-81), we have a required received $(SNR)_r$ of

$$(SNR)_{r,\,Avg/RMS} = (SNR)_{b,\,RMS} + 17.14 - 20 \log \left[\frac{(BW)_r}{(BW)_b}\left(\frac{(BW)_r}{(BW)_b} - 4\right)\right] \qquad (5\text{-}84)$$

$$(SNR)_{r,\,Pk/RMS} = (SNR)_{b,\,RMS} + 11.1 - 20 \log \left(\frac{(BW)_r}{(BW)_b} - 4\right) \qquad (5\text{-}85)$$

Application of PPM.   PPM is generally useful for short to moderate-length systems that require very high analog baseband SNR with very low received power. PPM cannot be multiplexed, and, therefore, it is intended for single-channel-per-fiber application.

Note that since the SNR is a very strong function of pulse width, T_p relative to T, it is strongly affected (degraded) by pulse dispersion in the fiber. SNR is approximately a function of $(1/T_p)^4$, and, therefore, SNR is degraded as a function of dispersion to the fourth power. For this reason, short transmission lengths are required when using multimode fiber. Single-mode fiber would permit very large improvement factors for PPM because of its very low dispersion.

The SNR improvement factor for PPM is impressive if the link bandwidth and component responses permit very narrow pulses (T_p). For example, if we assume that a bandwidth extension of 100 can be achieved, $(\text{BW})_r = 100\ (\text{BW})_b$, then the SNR improvement factor is

$$\frac{\text{SNR}_{b,\,\text{RMS}}}{\text{SNR}_{r,\,\text{Avg/RMS}}} = 20 \log (100 \times 96) - 17.14$$

$$= 79.64 - 17.14 = 62.5 \text{ dB}$$

5.6.4 Binary Pulse Encoding

The reasons for encoding digital signal are (a) to provide enough transitions that receiver regeneration circuitry can phase or frequency lock onto the signal timing, and (b) because the optical receiver is capacitively coupled and cannot respond to variations in the average DC signal level (low-frequency components) as signaling patterns change.

This capacitive coupling situation is shown in Figure 5.17. Here an NRZ signal is received by a DC-coupled detector preamplifier stage and presented as a voltage waveform at point (A). Even though the preamplifier can, in some cases, be DC-coupled, subsequent stages of amplification must be capacitively (AC) coupled so that the amplifier quiescent output level drift does not saturate the following stage. The average voltage level of the received signal (known also as the zero-frequency, or DC, component), and other low-frequency spectral components, do not pass through the capacitors. The result is signal "droop" at the input to the threshold detector, point (B), when there is a long chain of similar polarity pulses. When the threshold detector samples the signal in each time slot and compares it to the reference (V_{REF}), the "droop" results in decoding errors.

Even when long pulse chains do not exist, changes in signal patterns cause variations in the average DC component, and the signal baseline drifts up and down, causing SNR degradation (closing of the "eye" pattern) and thus increased BER and jitter.

The desired situation is for a constant-duty-cycle pulsed signal or one with constant or zero low-frequency components to be compatible with capacitive coupling and for clock recovery to have optical symbol transitions frequently or ideally within each bit period T. The desired conditions can be closely achieved with combinations of the following:

(a) Break up long-duration common optical symbol states by driving the signal to the baseline midpulse through return-to-zero (RZ) coding.

(b) Scramble the NRZ signal using a polynomial code that produces a random symbol pattern and ensures that chains of common optical symbol levels do not exist.

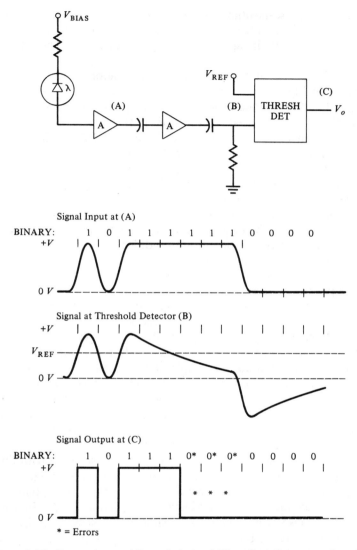

Figure 5.17 Errors due to AC-coupled signal "droop" at the regenerator with NRZ coding.

(c) Use phase-modulation codes, such as Manchester encoding, so that a signal is always present and at 50% duty cycle regardless of the input signal patterns.

(d) Bias the LED or laser at half power and transmit in a bipolar fashion (three levels) with a coding algorithm that ensures a constant average level of P_o, the midpower point.

(e) Transmit only the positive- and negative-going pulse transitions by biasing the source at midpower and driving it with the differentiated input signal, therefore, eliminating all low-frequency components.

(f) Use pulse-position modulation (PPM) employing very high-level very narrow pulses and integrate, or regenerate, just prior to the threshold detector, therefore, minimizing low-frequency components in the transmitted signal.

These signal formats and a representation of their frequency spectra are illustrated in Figure 5.18. Figure 5.19 illustrates some typical receiver designs used to detect and regenerate these signals, and provides an approximate comparison of the received SNR required of each from [13].

From Figure 5.18, note that both the NRZ and RZ codes have large DC and low-frequency components. For this reason, neither is acceptable for transmission without some sort of scrambling algorithm, except for the rare case of a DC-coupled system (short optical isolators). All other encoding approaches eliminate the low-frequency or DC component at the expense of doubling transmission bandwidth or reducing signal power in relation to threshold (trilevel codes). By scrambling the NRZ code, however, the original baseband bandwidth is maintained as well as the signal relationship to the decoding threshold. This is at the expense of equipment complexity and perhaps some overhead bits that increase bandwidth slightly.

Most high-performance optical trunking systems use scrambled NRZ. Although it may be more complex and costly than other approaches, the lower transmission bandwidth of NRZ is preserved along with superior SNR performance. This is important for long-distance applications.

Databus applications generally use a phase-modulated approach such as Manchester. Here bandwidth is generally not a constraint due to the shorter distances, and the coding is simple and inexpensive. With constant 50% duty cycles and differential receivers, detection and regeneration are simple and the SNR performance compares with NRZ. Signal bursts of multiple amplitudes from multiple locations can be accommodated as well since an "off" state is possible with such codes as Manchester. In MIL-STD-1773, a triple-pulse-width high state then low state marks the start of a transmission.

Bipolar encoding generally finds little application with fiber transmission, with the possible exception of transition (differential) encoding. The SNR performance of bipolar encoding is poor since the signal range is essentially divided in half to accommodate the two threshold levels (see Section 5.6.2). Transition encoding is often used, because of its simplicity, in short low-cost systems. Even though the SNR performance suffers, for short-distance interconnects, such as computer peripherals or for industrial applications, the low cost is the driver. When differential encoding is used, usually the transmitter sends refresh pulses if transitions are not frequent. This avoids compounding an error due to a noise burst into multiple bit errors.

Pulse-position modulation (PPM) is a unique method of transmitting a binary waveform. Most of the energy is at higher frequencies, where it can easily pass through capacitive coupling. The DC component is only $(T_p/T)P_{max}$. Once received, the bipolar signal can be reconstructed from the received pulse positions by decoding timing relationships, or, more simply, by integrating the received pulse waveform if the energy relationships are maintained. If the pulses are encoded such that the energy and time relationships within the transmitted pulses are adequate to provide the energy per pulse required of the received bipolar waveform (for proper

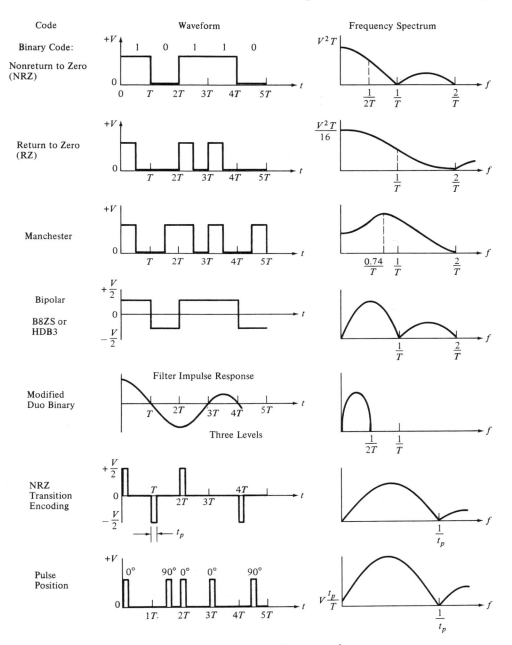

Figure 5.18   Binary encoding approaches.

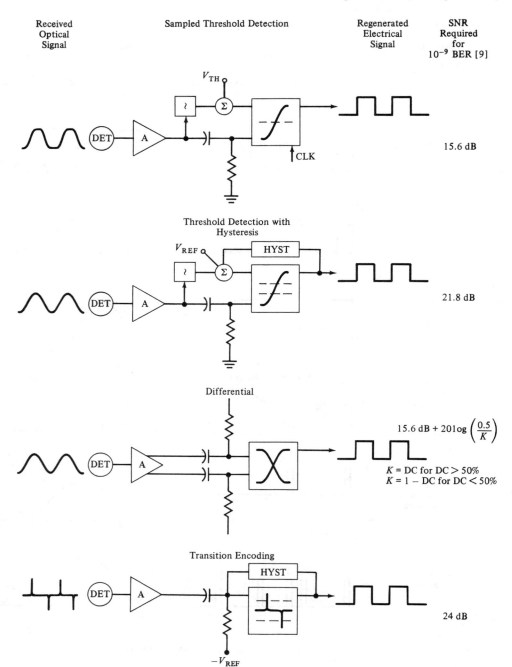

Figure 5.19   Receiver regenerator design considerations.

BER), then integrating the pulses re-creates the desired bipolar waveform. This PPM approach was introduced to the SAE A2K Committee by IBM for the draft MIL-STD-1773 optical databus standard [14] as an alternate means for Manchester code transmission. It simplifies threshold tracking since, with a high-frequency signal of little DC energy, the threshold circuitry can respond much faster. It eliminates the problem of transmitters having to be in a quiescent half-power state when "off" such as is required of the tristate Manchester.

5.7 WAVELENGTH-DIVISION MULTIPLEXING

Wavelength-division multiplexing (WDM) provides another level of multiplexing to individual or already multiplexed channels. With WDM, the individual transmission channels are carried on separate optical wavelengths and combined in a single fiber at the optical link layer. It is generally used when the number of fibers in a transmission link is inadequate for the capacity or becomes a critical cost item.

5.7.1 Operating Principle

Figure 5.20 shows the concept using four wavelengths as an example. Each transmission channel drives an individual laser or LED transmitter, each with a source operating at a different wavelength. The multiple-wavelength optical carriers are then combined onto a single fiber through a wavelength coupler. Since the coupler is wavelength-dependent, it can be designed so as not to suffer all of the loss associated with a standard N-port power splitter or combiner coupler.

The multiple optical carriers then propagate down the fiber and enter the wavelength demultiplexing coupler. This coupler is designed with a grating or series of

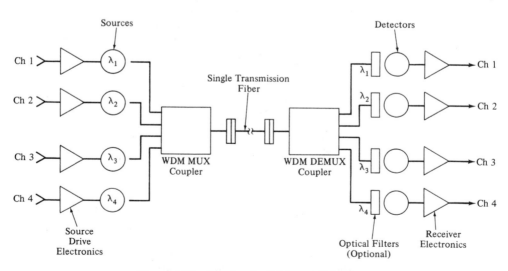

Figure 5.20   Wavelength-division multiplexing.

wavelength-selective coatings, such that the multiple optical carriers are separated spatially and coupled into separate receiving fibers. If the separation of the WDM demultiplexer coupler is not adequate to isolate crosstalk below the desired level, then wavelength-selective band-pass filters may be added (at the expense of some loss).

Once the optical carriers are separated, they are detected by photodetectors designed for the respective wavelengths of operation. Individual receivers then process the individual transmission channels.

5.7.2 Link-Design Considerations

WDM is, in essence, the operation of multiple optical links over one fiber. The design approach, therefore, includes performing the link analysis multiple times for the various optical carriers. The elements that the WDM devices add to the analysis is extra loss and the introduction of channel crosstalk. The analysis process generally consists of the following steps:

Power analysis for wavelength selection. The link budget analysis for a WDM system should be performed separately for each wavelength because most of the components in the link exhibit different performance at different wavelengths. Figure 5.21 shows a hypothetical case for a 44.7-Mb/s WDM system designed to operate at various wavelengths in the regions around 850, 1300, and 1550 nm. Even though equivalent launch power, coupler, and connector losses are assumed at all wavelengths, the wide variation in fiber loss and detector sensitivity limits such a broadband design to the performance constraints of the lower wavelengths. The

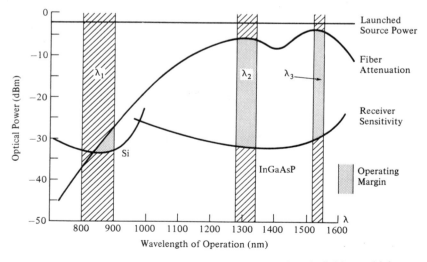

Figure 5.21 Link-power-budget considerations for wavelength-division multiplexing. This diagram illustrates the operating margin differences experienced when operating a WDM system in three bands assuming a constant fiber lengthof 10km and a signal bandwidth equal for all.

problem of power differential is not so great when the operating wavelengths are grouped within one window. Wavelength-separation limits of practical components, however, then become the critical factor.

Bandwidth analysis for wavelength selection.   The effect of dispersion in optical fibers with wavelength can be even more critical than that of the optical loss. Figure 5.22 shows the variation for typical classes of low-loss fibers over the same operating ranges described in Figure 5.21.

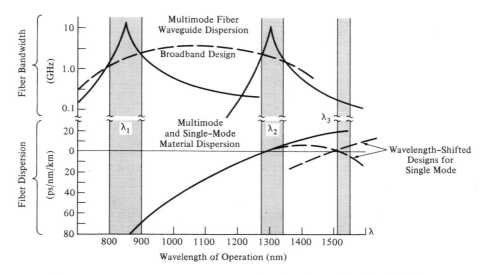

Figure 5.22   Bandwidth and dispersion considerations for wavelength multiplexing.

Channel-separation analysis.   Channel separation is discussed later. The decision on proper separation is a function of the selectivity of WDM couplers and filters, the spectral width of sources, and the tolerances and stability experienced with all these devices. Initial choice of separation can be based on avaliable coupler and source characteristics. A detailed worst-case power-budget analysis should then be performed to include the effects of wavelength drift and wavelength tolerance of the devices. Of greatest concern is the drift or tolerance of source wavelength with respect to the band edges of any optical filter or demultiplexer selectivity windows. Include the effects of time, temperature, and drive current on source wavelength.

Crosstalk analysis.   A good feel for expected crosstalk comes from the discussion of wavelength-separation analysis. A separate more refined analysis, however, should be performed where separations are close. The approach to a crosstalk analysis is to perform the link-budget power analysis again, under worst-case crosstalk conditions, but this time analyze the signal level received in each channel based on the power transmitted in the adjacent channel. For example, a 1220-to-1300-nm WDM crosstalk analysis might consist of analyzing received power and receiver sensitivity at some specific wavelength λ_x (say, 1290 nm), or band of wavelengths W_{λ_x} (say, 1280 to 1320 nm), within the 1300-nm receiver optical bandpass. The relationship for received crosstalk power is as follows:

Signal Path: λ_a to λ_b (for example, 1220-nm source to 1300-nm detector).

Wavelength Trial: (a) Perform multiple trials each at one of a series of wavelengths λ_x within the receiving channel optical passband W_{λ_b}, or (b) assume average power levels in a band of wavelengths representing the passband W_{λ_b}. The example that follows reflects the single-wavelength (λ_x) case.

Received Crosstalk Power:

$$P_t - L_c - L_s - L_{fa} - L_{wdma} - \mathrm{ISO}_{abx} \tag{5-86}$$

where P_t = Tx power (best-case power at λ_x nm)
L_c = mean connector loss = $N_c \times I_c$
L_s = mean splice loss = $N_s \times l_s$
L_{wdma} = WDM MUX loss at λ_x
$(\mathrm{ISO})_{abx}$ = WDM DEMUX isolation from λ_a to λ_b at λ_x
L_f = fiber attenuation at λ_x

Crosstalk noise level at the output of the receiver is the λ_a signal level based on detector sensitivity at λ_x.

5.7.3 WDM Coupler Design

The choice of WDM coupler design depends on:

(a) fiber compatability (single mode, multimode);
(b) channel separation and isolation;
(c) insertion loss per channel;
(d) cost versus performance trade-offs; and
(e) other practical considerations such as environmental performance.

      Couplers using wavelength-selective coatings can achieve 30- to 40-dB isolations with separations in the 100-nm range. Insertion losses between 2 and 4 dB are experienced with two-port multimode devices and 0.5 to 2 dB with single-mode. External filters can add another 20 to 40 dB or more isolation with added insertion loss in the range of 0.1 to 0.2 dB. Figure 5.23 shows a two-wavelength multimode system separated by 80 nm, where the WDM DEMUX coupler uses filter coatings. Isolation is 20 dB or better at the spectral skirts of the lasers and 30 to 40 dB at the center wavelengths.

      Grating couplers, shown in Figure 5.24, can be produced with 20-dB isolations between channels as close as 20 to 27 nm [15, 16] with from two to six channels. Coupler performance can be determined by the well-controlled characteristics of an optical grating.

      Although it may at first seem logical to use standard fiber splitters with filters attached to perform the function of WDM couplers, the approach is nonoptimum

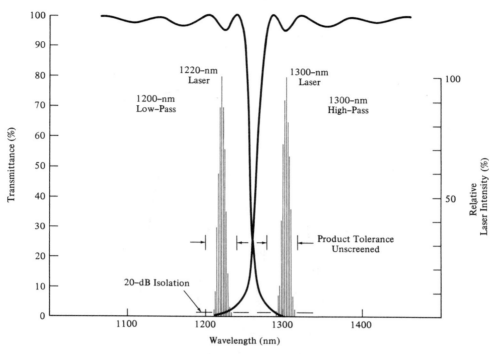

Figure 5.23 Typical transmittance characteristics of optical low-pass and high-pass filters used for WDM. Filter characteristics are shown for a hypothetical 1200- and 1300-nm system using multimode fiber.

and very lossy. The function of standard *N*-port couplers is to split power among *N* fibers. The splitting loss effects all optical carriers equally. Each optical wavelength channel at each demultiplexer output fiber has a throughput loss per channel of $10 \log (N)$ and excess loss. This loss is usually equalled as well at the multiplexing coupler since standard *N*-port couplers are actually fabricated as $N \times N$ port devices, therefore suffering a splitting loss as well.

WDM couplers, on the other hand, are fabricated in such a way as to route the optical signal through a path in the coupler depending on wavelength. These couplers do not split power, only route it. For this reason, at both the multiplexer and demultiplexer, the power in each channel remains undivided. The only loss experienced in a channel is the excess loss of the couplers. Although the excess loss may be higher in some designs, the results are significantly better. If we assume a 2-dB excess loss per WDM coupler, and no filtering required, the results from a four-channel system would be

$$L_{wdm} = 2L_e = 2(2 \text{ dB}) = 4\text{dB}$$

This is 10 dB better than the "brute-force" splitter and filter approach that would consume a 6-dB power splitting loss at either end and 1 dB for excess loss, or a total of 14 dB.

Care must be taken in considering the environmental effects on performance. Certain wavelength filter materials, if not properly sealed, are affected by humidity.

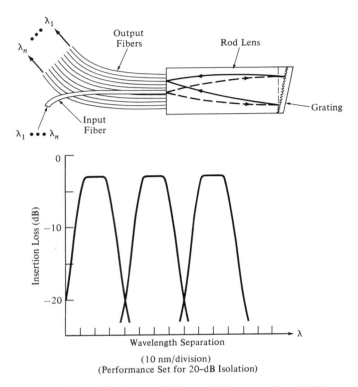

Figure 5.24 Wavelength-division multiplexing grating coupler.

Coupler designs that depend on an evanescent field have coupling ratios that are strong functions of wavelength. Temperature stresses all couplers since different materials are used in fabrication and mounting. The effect on channel isolation with temperature should be considered in the design.

5.7.4 Wavelength Selection

The selection of the wavelength ranges to be used generally follows the following analysis:

(a) Determine the ability of the wavelengths (or band) under consideration to meet the link budget for the capacity and distances required.

(b) Determine the ability of all wavelengths under consideration to meet link-bandwidth (or dispersion) requirements for the fiber selected.

(c) Determine the availability of product falling within the fiber "windows" of interest. It is not good practice to design around nonstandard or low-production-yield wavelengths.

(d) Determine practical channel separations that meet crosstalk and other parameters (discussed in the next section).

5.7.5 Channel Separation

Channel separation within a band of interest is selected on the basis of the following:

(a) The spacing must be wide enough such that the source spectral skirts do not reach into the wavelength range of the adjacent channel. If it does, crosstalk occurs.

(b) The spacing and bandwidths must be wide enough so that the drift with time and temperature of sources, couplers, and filters does not cause increased crosstalk or attenuation of power by drifting beyond the band edges. Sources should be temperature stabilized as well as other sensitive optical components. The 0.2 to 0.5 nm/°C drift of heterostructure lasers is unacceptable for WDM systems.

(c) Where spacing is narrow, attention must be given to the variation in laser wavelength with drive current and output power. An unexpected result may occur if the primary modes are too close to the "sharp cutoff" band edge of a filter or coupler. Wavelength jitter due to signal modulation can be converted to amplitude distortion within the slope of the filter band edge.

(d) With present multimode laser product, the source spectral width is wide (typically 2 to 5 nm) and the standard tolerances relatively loose on the center wavelength (5 to 20 nm). Stability with temperature or drive current is also rather poor. For WDM systems with multiple channels in the same fiber "window," screening and selection are a must. There is a substantial risk in designing a system that requires tight screening of components, however. Custom screening may be more expensive than is cost effective to the system. Vendors who do screen and offer a product line with multiple wavelengths in the same "window" are few and their product lines may change as demand dictates. Spares may become nonexistent or expensive. Spacing, therefore, should reflect the product environment.

(e) Eventually, distributed feedback (DFB) lasers with 50-MHz potential line widths or coupled cavity lasers (CCL) with 1- to 5-MHz potential line widths will become more commonplace [17]. These are single-mode or narrow-spectrum lasers where the wavelength and spectral width are controlled by a physical process (precisely etched physical gratings or external cavity lengths) instead of product yield and screening. Grating optics within the next 10 years is expected to offer 1-nm separations and from 5- to 50-channel multiplexing within a wavelength window. With nanometer-level control/selectivity on wavelength, sub-Angstrom-level line widths, and GaAs integrated optics-fabrication methods, WDM may become much more practical for widespread deployment in the future. Narrow line widths and coherent transmission techniques will eventually (10 to 20 years) permit gigahertz separations and 10- to 100-channel multiplexing within a wavelength window [15].

5.7.6 Crosstalk

Crosstalk is probably one of the most critical design considerations in WDM systems. Crosstalk is the presence of one channel's signal in an adjacent channel. It has the effect of raising the noise floor or reducing SNR. Crosstalk is caused by the following:

(a) The spectral skirts of one channel entering the demultiplexing and filtering passband of another cause crosstalk. When the optical carrier is modulated by its signal, power modulation on its spectral skirts is interpreted as "in-band" power modulation by the adjacent channel. For example, at the spectral edge of a multi-mode laser (± 2.5 nm), the power in the outer modes may only be 6 to 10 dB down. If these modes drifted into the 3-dB point of the adjacent band edge, the SNR in that band would be limited to SNR $= -(-6$ dBc $- 6$ dB$) = 12$ dB.

(b) Practical limits on selectivity and isolation cause crosstalk. For moderate separations (80 to 100 nm) and conventional filters, isolation levels achievable are in the 20-to-40-dB range. Grating-type couplers provide 20-dB isolations with separations in the 20-to-30-nm range.

(c) Nonlinear effects within the fiber at the high-power densities possible in single-mode systems can cause crosstalk or cross modulation. The mechanism is Raman scattering, which is a nonlinear stimulated scattering effect that allows the optical power at one wavelength to affect scattering and thus the optical power in another wavelength [18, 19, 20]. The optical power at one wavelength acts as a "pump" that interferes with and amplifies the signal power at another wavelength. The signal power in turn depletes the "pump" power. The phenomenon is such that the shorter wavelength is reduced in power while the longer wavelength is increased. The amount of interference, or crosstalk, is a function of signal power and wavelength separation. Tomita [18] measured the crosstalk in a 21-km single mode system to be -25 dB with the "pump"-wavelength power level at 0 dBm and a channel-wavelength spacing of 80 nm. This is acceptable for digital communications but not for analog. Iheda [19] has demonstrated, however, that this scattering effect drops rapidly as wavelength separations exceed 85 nm.

Crosstalk can be treated in the link-power budget as unwanted noise components of a received signal, in effect, a transmitted signal with a finite signal-to-crosstalk ratio (SCR). SCR causes increased receiver SNR requirements to a value of (SNR)$_r$ to compensate. For digital signals, we can derive a factor Q_r that represents the (SNR)$_r$ required for a given BER in the presence of crosstalk. Without crosstalk, Q is defined from Equation (4-49) as

$$Q = \frac{I_s}{2\langle i_n \rangle} \tag{5-87}$$

where I_s is the received signal current, which is roughly equal to the peak current received in a pulse. In the presence of digital crosstalk of peak level, I_{ct}, the peak signal is reduced to $I_s - I_{ct}$. Therefore, Q is reduced to

$$Q_{ct} = \frac{I_s - I_{ct}}{2\langle i_n \rangle} = \frac{I_s}{2\langle i_n \rangle}\left(1 - \frac{I_{ct}}{I_s}\right) \tag{5-88}$$

If we define the peak-signal-to-peak crosstalk ratio as (SCR)$_{pk/pk} = I_s/I_{ct}$, then Q in the presence of crosstalk is

$$Q_{ct} = Q\left(1 - \frac{1}{(\text{SCR})_{pk/pk}}\right) \tag{5-89}$$

For analog signals, we define SCR in terms of rms signal power to rms crosstalk power, $(SCR)_{RMS/RMS}$. The SNR in the presence of crosstalk of λ_2 into λ_1 is as follows:

$$(SNR)_{ct} = \frac{P_{sig(\lambda_1)}}{P_{n,rcvr} + P_{ct(\lambda_2)}} \tag{5-90}$$

$$(SNR)_{ct} = \frac{1}{1/(SNR)_r + 1/SCR} \tag{5-91}$$

where P_{sig} is the rms signal power, P_{ct} is the rms crosstalk power, and $(SNR)_r = P_{sig}/P_{n,rcvr}$ is the SNR for the receiver in the absence of all other noise.

5.7.7 Application of WDM

Most of the practical limits on WDM will dissapear in the future as integrated optics approaches begin to replace the bulk-device designs existing today. For example, Ito et al. [21] reports on a WDM design that uses monolithically integrated laser arrays and an integrated optic multiplexer and demultiplexer. The transmitter uses an array of five DFB lasers at wavelengths around 1310 nm separated by only 50 Angstroms. These devices were coupled to a single-mode fiber through a $LiNbO_3$ multiplexing waveguide optical coupler. The demultiplexer consisted of an aspherical geodesic lens on a silicon substrate glass waveguide that focused the received beam on a grating that separated the wavelengths. A second aspherical lens focused the beams onto the detectors. Insertion loss was less than 10 dB and crosstalk -17 dB.

As designs of this sort mature into low-cost production technologies, they will make WDM a very desirable approach. Single integrated sources, receivers, and multiplexers will replace the multiple bulk transmitters, receivers, and couplers required today.

With the advent of coherent optics, multiple-gigahertz data capacity on a single optical carrier compounded with 10- to 100-wavelength multiplexed optical carriers in a single wavelength window will provide astounding transmission capacities and networking options.

With today's technology, WDM is generally confined to applications where:

(a) the fiber in place is inadequate to handle the expansion in channel capacity;

(b) added wavelengths are reserved for potential future expansion;

(c) multiple high-quality or wideband channels required over a single fiber and other electrical multiplexing techniques are too expensive or will not maintain signal quality;

(d) transmitting vastly dissimilar signals over the same fiber and electrical multiplexing is too difficult or expensive;

(e) terminal redundancy and 1 : 1 protection switching exists, and it is desirable to use only a single fiber per transmit direction;

(f) systems are physically limited to a single fiber for duplex transmission; and

(g) a clear cost advantage exists for WDM over the cost of fiber, and reliable sources of components exist.

The use of WDM for two-way transmission over a single fiber has been limited by the availability of low-loss single-mode directional couplers that operate with a single wavelength, have insertion losses in the range of 0.5 dB, and isolation of 40 to 50 dB.

Figure 5.25 shows some common applications for WDM.

5.8 COHERENT OPTICS

Coherent optics is a subject deserving of an entire text. It implies the next generation of passive optical and electrooptical components and is, therefore, beyond the scope of this book. However, I deal with it briefly in order that an appreciation for the future performance advantages can be gained and so that the evolution of the technology can be put into perspective.

Coherent optics implies that we begin dealing with the optical carrier, as we do with electrical radio carriers, as a single-frequency sinusoid. With noncoherent optics, we treat the optical carrier as an incoherent noise source that we envelope modulate. Because the frequency of the envelope is in the 1- or 2-GHz maximum range (signal frequency) and the incoherent carrier frequencies are in the terahertz range (about 200,000 GHz), there is little interaction. Direct detection filters out the optical high-frequency components, leaving only those of the envelope modulating signal, which fall within the bandwidth response of the detector (in the gigahertz range).

By examining the carrier-power expression from Equation (5-52):

$$C(t) = 2P_o[1 + M_o \cos(\omega_s t)] \cos^2(\omega_o t)$$

where $\cos(\omega_s t)$ represents the modulating signal carrier, and $\cos(\omega_o t)$ a sinusoidal (coherent) optical carrier. After direct detection, we strip off the optical carrier power, leaving only the signal carrier as, from Equation (5-53):

$$i_d(t) = rGP_o[1 + M_o \cos(\omega_s t)]$$

Filtering the DC component and assuming a sinusoidal carrier of average power $= \frac{1}{2}$, we get an SNR for a PIN receiver of

$$\text{SNR} = \frac{(rP_o M_o)^2}{2erP_o(\text{BW})_s + 2eI_d(\text{BW})_s + N_{\text{rcvr}}} \tag{5-92}$$

Where receiver preamp noise dominates (as it usually does), the equation simplifies to

$$\text{SNR} = \frac{(rM_o P_o)^2}{N_{\text{rcvr}}} = \frac{P_{\text{sig}}^2}{N_{\text{rcvr}}} \tag{5-93}$$

SNR is, therefore, dictated directly by signal power and receiver noise power. If, on the other hand, the optical carrier is a sinusoid that is directly modulated by the signal carrier and then is detected within the band-pass of the detector response by heterodyning it down in frequency, the optical carrier is not lost. Under the conditions

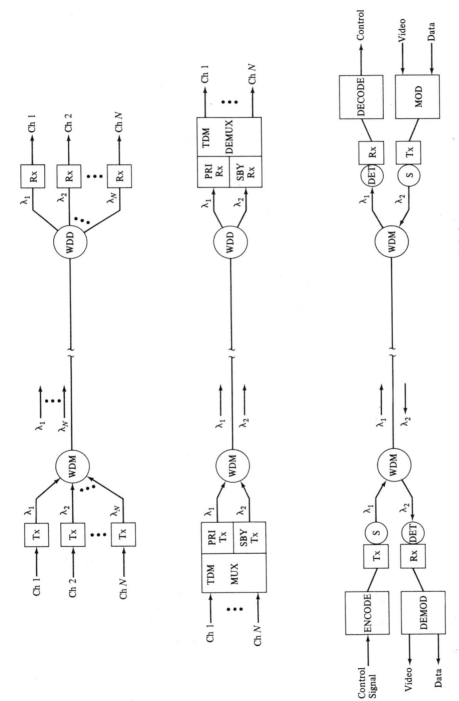

Figure 5.25 Applications of wavelength-division multiplexing.

of coherent detection, the following expression for received SNR is given following Kobrinski [15]:

$$\text{SNR} = \frac{rM_o P_o P_{lo}}{2er(P_o + P_{lo})(\text{BW})_s + 2eI_d(\text{BW})_s + N_{\text{rcvr}}} \qquad (5\text{-}94)$$

where P_{lo} is the power of the local oscillator. If we assume that the local oscillator power (which is the laser source at the receiver) is much greater than the signal power and that the resulting quantum noise dominates the receiver noise, we get

$$\text{SNR} = \frac{AP_{\text{sig}}}{B(\text{BW})_s} \qquad (5\text{-}95)$$

SNR becomes dominated by the signal power and signal bandwidth rather than receiver noise. The proportionality constants A and B and the expression vary with modulation technique, but in all cases, there is an improvement in sensitivity over that of noncoherent detection.

Figure 5.26 shows the process. A coherent laser is either directly modulated or its sinusoidal carrier output is modulated externally, as shown in the figure. Modulation can take many forms, amplitude, frequency, or phase, depending on the components used. The result is a sinusoidal carrier at optical frequencies (2 to 3×10^{14} Hz) varying at the signal rate in amplitude, phase, or frequency, depending on the modulation approach. Amplitude modulation is illustrated.

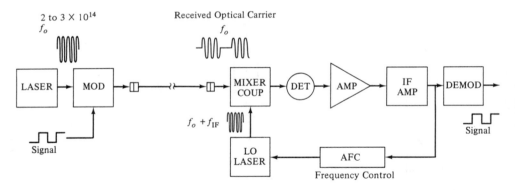

Figure 5.26   Heterodyne coherent transmission

If the system is a heterodyne, at the receiver, a local-oscillator laser is tuned to the received carrier at the frequency of the carrier (f_o) plus the IF frequency (f_{IF}). The two signals are optically mixed and the detected mixing product falls at the IF frequency (f_{IF}), which is generally in the gigahertz or less range, depending on such factors as signal bandwidth, laser spectral purity, and stability. A feedback control from the output of the detector maintains frequency of the LO laser so that the signal falls within the IF passband. The IF section is then followed by a signal demodulator that detects the phase-, frequency-, or amplitude-modulating signal. In homodyne systems, the LO-laser frequency and phase are the same as the signal. This approach is more sensitive than heterodyning. The modulator/demodulator can further improve sensitivity depending on the type of modulation and whether synchronous or nonsynchronous demodulation is used.

Improvements in sensitivity on the order of from 10 to 19 dB over direct detection using ASK, FSK, and PSK techniques have been reported by Kazovski [17], who compiled a survey of research from 1981 through 1985. Kazovski indicates the best reported DPSK (Differential Phase Shift Keying) heterodyne experiment running 8.5 dB below the best reported direct detection, with synchronous ASK heterodyne and nonsynchronous ASK homodyne running 4dB better (12.5 dB better than direct), asynchronous PSK heterodyne and synchronous ASK homodyne running another 3 dB better (15.5 dB better than direct), and the best is synchronous PSK homodyne running another 3 dB better (18.5 dB better than direct detection). In addition to sensitivity advantages, transmission capacity can be increased tenfold or more because the response characteristics of optical modulation components are in the gigahertz range.

Coherent optics requires a number of advances in component production capability. Line widths in the 10-kHz-to-1-MHz range are desired, which are only practical in external cavity-type lasers (although significant results have been obtained with broader DFB lasers). Low-loss external modulators or narrow-line lasers that can be directly modulated are required. Mixing couplers and polarization control are also required. External modulators of $LiNbO_3$ have been around for a long time, however, they have been lossy. Newer materials are evolving that should reduce this problem. Laser stability and frequency control are issues to be resolved. The performance advantages of coherent optics, as well as its flexibility, make the development worthwhile and will provide the next great evolutionary step in fiber optics.

REFERENCES

1. A. B. Carlson, *Communications Systems*, McGraw-Hill, New York, 1968.

2. D. R. Smith, *Digital Transmission Systems*, Van Nostrand Reinhold, New York, 1985.

3. R. L. Freeman, *Telecommunication Transmission Handbook*, Wiley, New York, 1981.

4. F. R. McDevitt and I. B. Slayton, *Optical Cable Communications*, RADC Contract No. F30602-74-C-0193, Harris Corp., Final Report RADC-TR-75–187 (A016846) (July 1978). Also available as *Fiber Optics Design Aid Package*, Vol. IV, Part II, Information Gatekeepers, Boston.

5. K. S. Shanmugam, *Digital and Analog Communications Systems*, Wiley, New York, 1979.

6. S. Haykin, *Communication Systems*, Wiley, New York, 1983.

7. J. Rieke and R. Graham, "The L3 Coaxial System Television Terminals," *Bell System Technical Journal* (July 1953).

8. Tran Muoi, "CATV Supertrunking Study", by PCO Inc., Unpublished.

9. T. D. Michaels, "Laser Diode Evaluation for Optical Analog Link," *IEEE Transactions CATV* **CATV-4,** 1 (January 1979): 30–45.

10. J. C. Daly, "Fiber Optic Intermodulation Distortion," *IEEE Transactions Communications* **COM-30,** 8 (August, 1982): 1954–1958.

11. W. M. Hubbard, "Efficient Utilization of Optical-Frequency Carriers for Low and Moderate Bandwidth Channels," *Bell System Technical Journal* **52** (May–June 1973): 731–765.

12. C. R. Patisaul, "Signal to Noise Ratio Improvement of Analog Pulse-Position Modulation Over Analog Intensity Modulation in Optical Communications Links," *Proceedings of the SPIE* **63** (August 1975).

13. AMP Corporation, *Designers Guide to Fiber Optics,* Harrisburg, PA, p. 160.

14. R. Betts, *An Aircraft 1553 Compatible Fiber Optic Interconnect System,* proposal to SAE A2K Standards Committee by IBM Federal Systems Division, Owego, NY.

15. H. Kobrinski and C. A. Brackett, "A Survey of Optical Frequency Multiplexing Techniques for Subscriber Loop Application," paper presented at the *Lightwave Journal* Symposium, "Fiber in the Subscriber Loop," Boston, 1987.

16. M. Seki et al., *Electronic Letters* **18** (March 1982): 257–258.

17. L. Kasovski, "Coherent Transmission Technology," paper presented at the *Lightwave Journal* Symposium, "Fiber in the Subscriber Loop," Boston, 1987.

18. A. Tomita, "Crosstalk Caused by Stimulated Raman Scattering in Single Mode Wavelength Multiplexed Systems," *Optical Letters* **8** (July 1983): 412–414.

19. M. Iheda, "Stimulated Raman Amplification Characteristics in Long Span Single Mode Silica Fibers," *Optical Communications* **39** (October 1981): 148–152.

20. A. R. Chraplyvy and P. S. Henry, "Performance Degradation Due to Stimulated Raman Scattering in WDM Optical Fiber Systems," *Electronics Letters* **19** (August 1983): 641–643.

21. T. Ito et al., "Waveguide Division Multiplexing System Using a Monolithically Integrated Laser Array and an Integrated-Optic Multiplexer/Demultiplexer," *Proceedings of the OFC* (1986).

6

Systems-Performance Analysis

6.1 ANALYSIS METHODOLOGY

Figure 6.1 shows a somewhat generic model for a fiber optic system. Although ECSA/ANSI terminology, borrowed from the SONET standard, is used to define and partition the functional layers of the transmission system, the terminology is adapted here to describe partitioning of a generic fiber transmission system (including asynchronous and analog transmission). The definitions are, therefore, not necessarily per the standard, but are conceptually similar. Likewise, the design methodology described here is unique, but conforms in principle to IEEE and ANSI standard methods of ANSI/ECSA T1X1/87-128R1 and EIA TSB 20 [1, 2]. The ANSI "joint engineering" method, which is described in Chapter 7, differs in that it engineers around a standard interface point and standard gain parameters.

The design is divided into stages for convenience. These include the following.

Stage 1: Service, path, and line-layer design. Service, path, and line-layer design is performed with the objective of delivering required end-to-end baseband service performance and of determining the resultant transmission performance of the fiber-optic section necessary to support these higher layers. This is achieved through a design and analysis of the signal processing (modulation, multiplexing, and encoding) required at the path and line-layer termination equipment as well as any preprocessing within the service layer.

Stage 2: Fiber-optic section design. The fiber-optic photonics-layer performance requirements are determined in order to achieve end-to-end single or cascaded (repeatered) section-layer performance as defined by Stage 1. This is achieved

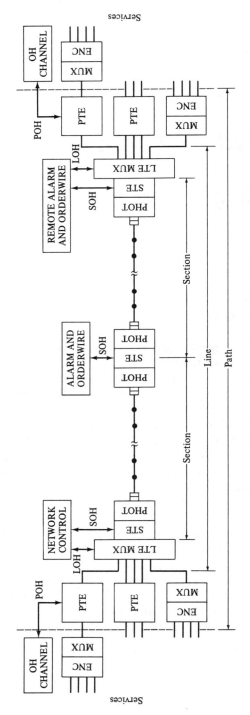

Figure 6.1 Systems-architecture functional block diagram.

by first selecting (a) candidate section-layer transmission modulation or encoding and overhead multiplexing (if any) approaches and (b) candidate fiber-optic technologies and component performance parameters that make up the photonics layer. A first-order analysis is then performed to determine the number of sections (repeater spans) required between line-layer termination points. Once a suitable design is achieved for the optical section, the transmission parameters for the photonics layer are defined. If a cascade of sections is required to achieve the distance requirement, the end-to-end section-layer transmission parameters are allocated to determine single-section performance.

Stage 3: Photonics-layer and optical-link refinement.   At this point, the technology and component choices made for the section-layer design are refined. Based on photonics-layer performance requirements, an optical link-budget analysis is made to determine whether the design and components chosen will achieve the transmission requirements defined in Stage 2.

The process is iterative in that it can be repeated as the design is refined in order to optimize performance, reliability, cost, and other design parameters.

Chapter 1 defined the performance parameters that applied at the various stages of the design. This chapter leads the designer through the three analysis stages in detail, providing some engineering aids and a worksheet approach to the analysis. The worksheets, provided as tables in each section, are not intended to be universal, but help summarize the analysis steps as used in practical applications. The analysis methodology is summarized in the next section and is illustrated in Chapter 1, Figure 1.11, and Table 6.1.

6.1.1 Design Stages

Stage 1: Service, path, and line-layer design

Step 1A: Define End-to-End Systems Performance

1A.1. Define service parameters $(SNR)_b$, $(BW)_b$, $(BER)_b$, $(BR)_b$, and jitter at the individual baseband channel input and output ports to the path-layer termination equipment as well as other systems-performance parameters.

1A.2. Calculate the end-to-end baseband channel service-performance limits required of the system in order to achieve required output performance based on the specified input parameters.

Step 1B: Design Systems Architecture.   Based on an understanding of the performance required, from Step 1A, design a functional systems architecture (all layers), as well as the multiplexing, modulation, encoding and overhead techniques to be used at the path, line, and section layers as well as any preprocessing within the services layer. This becomes the "candidate architecture" for performance analysis.

Step 1C: Design Services/Path Layer and Compute Line-Layer Performance.
The line-layer transmission requirements are defined based on an analysis of services and path-layer performance. This is achieved in the following steps:

TABLE 6.1   PERFORMANCE-ANALYSIS METHODOLOGY

Stage 1	1A	Define end-to-end systems performance
Service, path,		1A.1 Define service parameters
line design		1A.2 End-to-end baseband channel performance
	1B	Design systems architecture
	1C	Design path layer, determine line-layer performance
		1C.1 Design service, path layers—MOD, MUX, encode, POH
		1C.2 Signal unit conversion
		1C.3 Determine intermediate channel performance
	1D	Design line layer, determine section performance
		1D.1 Design line layer—multiplexing, LOH
		1D.2 Determine transmission channel performance
Stage 2	2A	Design section layer
Section design		2A.1 Design section—encoding, SOH
		2A.2 Determine preliminary photonics requirement
	2B	Photonics preliminary design
		2B.1 Functional design
		2B.2 Technology selection
		2B.3 Specify component performance
	2C	Evaluate design, determine number of section cascades
		2C.1 Link-bandwidth analysis
		2C.2 Link-power-budget analysis
	2D	Determine single-section/photonics performance
		2D.1 Cascaded-section performance
		2D.2 Revised photonics performance
Stage 3	3A	Link-bandwidth analysis—$(BW)_f$, $(BW)_r$
Photonics design	3B	Compute power penalties—P_p
	3C	Compute SNR penalties and $(SNR)_r$ or Q_r
	3D	Design receiver, determine MRP
	3E	Link-power-budget analysis—P_e
	3F	Systems-performance test

1C.1. Design a path layer by selecting specific path overhead (POH) and service/path-layer modulation, encoding, and low-level multiplexing (if required) approach or equipment.

1C.2. Convert SNR and BW baseband units to units common to the approach or equipment selected in Step 1C.1. The units used for defining baseband SNR and BW may be application-specific. Conversion to units that can be used with the tables provided in this text, or the specifications provided by an equipment supplier, may be required (peak to peak to rms conversion, for example).

1C.3. Determine the line-layer interface and performance parameters using the service/path to the line-termination-equipment transfer functions. This process differs depending on whether the path-termination equipment is to be analytically modeled, defined by a standard, or selected from product literature: (a) if a product is selected, then the termination-equipment output parameters are defined in the literature; (b) if a standard is selected, then the interface parameters are provided by that standard (DS-3 format or STS-1, for example); (c) if we choose to analytically model the services/path layer, then $(SNR)_c$ and

$(BW)_c$ [or $(BER)_c$, $(BR)_c$, and jitter] parameters are determined from the transfer functions for the selected modulation, encoding, and multiplexing approaches (per Chapter 5). Generally, the service/path layer consists of a signal-encoding or carrier-modulation function if the baseband signal is analog. It can also consist of some low-level multiplexing of analog or digital signals (including path-layer overhead) in order to achieve the desired format for line-layer interfacing.

Step 1D: Design the Line Layer and Determine Section-Layer Performance

1D.1. Design the line-layer, including specifics of multiplexing approach and line overhead (LOH). Although other processing could occur here, generally, the line layer consists of only an $N \times$ intermediate channel and a line-layer-overhead multiplexing function along with frame or output-signal formatting.

1D.2. Determine the section-layer interface parameters from the line-layer interface parameters (1C.3) and the line-to-section-layer transfer functions in a similar manner as in 1C.3.

Stage 2: Section-layer design.   The objective here is to define the section layer and determine the photonics-layer performance parameters for the worst-case span. This is achieved by designing the section layer, along with a rough design of the photonics layer, and determining whether repeaters are required in any of the spans. If repeaters are required due to span distance, then the section-layer performance parameters (from 1D) are allocated to an individual section to define the worst-case section performance. Based on section-layer design, these parameters are then converted to photonics-layer performance parameters.

Step 2A: Design Section Layer

2A.1. Define the signal format (encoding or remodulation) required for optical transmission by the photonics layer along with any section-layer overhead. Section-overhead multiplexing, framing, and error encoding are performed at this layer logically, but can be physically included in the line-termination-equipment multiplexing process. The same equipment conveniently performs both line and section functions if properly thought out, and, most often, the line- and section-termination equipment are combined in a modular shelf unit.

2A.2. Determine the new photonics-layer performance parameters based on format or signal changes (if any) imposed by the section layer in Step 2A.1.

Step 2B: Photonics-Layer Preliminary Design

2B.1. Select the basic technology to be used in the photonics layer.

2B.2. Select or specify candidate optical component performance.

Step 2C: Evaluate Design, Determine Number of Section Cascades (Repeaters).
Based on photonics transmission parameters from Step 2A.2 and a first-order worst-case link-bandwidth and power-budget analysis, determine the maximum span

length and the number of cascaded sections required to meet worst-case transmission-distance requirements between line-termination equipments. If the results are unacceptable, change the design choices in Steps 2A and 2B until the desired results are obtained. If the results are close, go on to complete Stage 3 to refine the photonics-layer design. If a gross problem still exists, Stage 1 choices may have to be revisited.

Step 2D: Compute Single-Section/Photonics-Layer Performance. If Step 2C indicates no repeatered sections, then the performance of a section is as defined in 1D.2, and the performance of the photonics layer is the same as defined from the preliminary analysis of 2A.2. From Step 2C, if repeaters are required, then:

> **2D.1.** For the worst-case span, allocate the section-layer transmission parameters from 1D.2 to the section cascades, so as to define the worst-case individual-section design parameters.
>
> **2D.2.** Based on the design of the section, recalculate the photonics-layer performance parameters (as was done in 2A.2) using the new section-layer parameters from 2D.1

Stage 3: Photonics-layer design analysis. Based on the fiber-optic components chosen in Stage 2, perform a more rigorous link-budget analysis to determine whether the systems gain and bandwidth are adequate to meet the performance defined for the photonics layer in Stage 2 under all conditions and with component tolerances taken into account. A rigorous link-budget analysis generally consists of six steps:

Step 3A: Link-Bandwidth Analysis.

> **3A.1.** Fixed bandwidth components: If transmission is analog and all component performance is fixed, including receiver bandwidth, compute the link-bandwidth analysis in order to determine (a) the adequacy of design to meet bandwidth requirements, or (b) the signal degradation experienced due to band limiting (to be converted to a power penalty in Step 3B).
>
> **3A.2.** Adjustable receiver bandwidth: If the receiver bandwidth is adjustable (generally, analog modulation), then perform the link-bandwidth analysis with preamplifier BW as an unknown. This determines the amount of bandwidth extension required of the receiver to compensate for fiber, source, and detector band limiting. The new $(BW)_r$ is used to calculate receiver MRP in Step 3D.
>
> **3A.3.** Digital band-limiting penalty: In a digital system, generally, the sensitivity of the receiver for a particular BER performance is related to fiber band limiting. Compute the bandwidth limits of the fiber and relate this to the bit rate for which the receiver was designed. If the ratio between electrical bandwidth and bit rate is approximately $(BW)_f < 0.7(BR)_t$, then the receiver must be derated more than 1 dB by applying a power penalty to the link-power budget in Step 3B.

Step 3B: Compute Power Penalties.   The sensitivity parameter (MRP) for the receiver is generally specified or computed assuming a noise- and distortion-free signal is received. Where noise and distortion mechanisms such as modal noise, band limiting, and source extinction ratio exist, the result is a reduction in signal-to-noise ratio at the receiver. In order to compensate, more optical signal power must be applied. The noise sources, therefore, can be accounted for as power penalties in the link-power budget.

Step 3C: Determine SNR Penalties and (SNR), or Qr.   Certain noise sources are a function of transmitted signal power and cannot be compensated by adding more signal power. Some of these include crosstalk, reflections, source modal noise, and the finite SNR of the transmitted signal. These become basic SNR limitations or penalties, $(SNR)_n$, imposed by the link. The only means of compensation is by improving the SNR (or sensitivity) of the receiver. This can be accomplished by specifying a better SNR parameter $(SNR)_r$ or Q_r for the receiver based on the $(SNR)_n$ or Q_n of the link. As a computational convenience, SNR penalties can be converted to equivalent power penalties. The option is up to the designer. If no degradation exists in the link, or it is treated as a power penalty, then $(SNR)_r = (SNR)_t$ or $Q_r = Q_t$.

Step 3D: Determine the Minimum Required Received Power (MRP).   Design the receiver (detector, preamp, and regenerator) and compute or determine the MRP required to achieve the $(SNR)_r$ or Q_r performance defined in 3C. For a purchased device, this is determined from supplier specifications. For an analytical model, it is calculated using appropriate noise bandwidth, $(SNR)_r$, $(CNR)_r$, or Q_r, and the noise characteristics of the detector and amplifier chosen (see Chapter 4).

Step 3E: Compute Optical Link-Power Budget.   A power-budget analysis is performed to determine that there is adequate optical power received and detected to achieve desired performance, once all optical losses, power penalties, and margins are accounted for. To be thorough, this is performed using as many statistical parameters as possible, and is computed for best- and worst-case conditions of component tolerances and environmental conditions.

Step 3F: Systems-Performance Test.   The "excess-power" results of worst- and best-case optical power-budget analyses are tested against receiver sensitivity and saturation parameters to determine dynamic range, performance margins, and statistical probability of operating within specifications.

If Step 3F shows the design to be inadequate or overdesigned, the most appropriate design changes are determined and the analysis is repeated.

6.1.2 Alternative Design Approach

The previous analysis approach assumes that we know the network parameters, that is, transmission distance and capacity, and are searching for a design to meet them. If, instead, we are solving for the transmission-distance capability of a proposed design, then the same steps are valid but Steps 2C and 2D would be eliminated and one would proceed directly from photonics design (Step 2B) to the link-budget analysis (Step 3). The link-budget analysis would also be performed in an alternate order, finding receiver MRP first (Step 3D) along with Step 3C and 3B (excluding band-limiting penalties). Use these to perform a preliminary link-power budget analysis (Step 3E) to find distance based on power limitation. Once basic distance limits are known, solve for band limiting (Step 3A) and resulting power penalties (Step 3B). Then go back and refine the link-power budget (Step 3E) and finally Step 3F.

6.2 DESIGN STEP 1A: DEFINING SYSTEMS END-TO-END PERFORMANCE

The initial steps in systems design are to define the system from the standpoint of baseband channel performance. Table 6.2 illustrates some of the typical parameters that might be used to define end-to-end baseband channel performance at the service and path layers. In the simplest sense, the designer is presented with the problem of transmitting N channels (with a specified bit rate or bandwidth) over distance D, and receiving each channel within a specified signal-to-noise ratio or bit rate. If input and output signal quality are specified, then the end-to-end systems performance can be calculated.

For a digital signal:

$$(\text{BER})_{b,\text{sys}} = (\text{BER})_{b,\text{out}} - (\text{BER})_{b,\text{in}} \tag{6-1}$$

TABLE 6.2   STEP 1A: END-TO-END PERFORMANCE SPECIFICATION

Step 1A.1: Define service parameters
  Transmission distance (D_s)
  Channel type (analog, digital)
  Number of channels (N)

Analog signaling performance
  Baseband signal-to-noise ratio in, $(\text{SNR})_{b,\text{in}}$
  Baseband signal-to-noise ratio out, $(\text{SNR})_{b,\text{out}}$
  End-to-End baseband channel bandwidth, $(\text{BW})_b$

Binary signaling performance
  Baseband channel bit rate, $(\text{BR})_b$
  Format (NRZ, RZ; bipolar, unipolar; BZ3S, etc.)
  Baseband channel bit-error rate in, $(\text{BER})_{b,\text{in}}$
  Baseband channel bit-error rate out, $(\text{BER})_{b,\text{out}}$
  Baseband channel jitter in, $t_{j,\text{in}}$
  Baseband channel jitter out, $t_{j,\text{out}}$

Step 1A.2: Calculate end-to-end per-channel performance
  $(\text{SNR})_{b,\text{sys}} = -10 \log (10^{-(\text{SNR})_{\text{out}}/10} - 10^{-(\text{SNR})_{\text{in}}/10}$
  $(\text{BER})_{b,\text{sys}} = (\text{BER})_{b,\text{out}} - (\text{BER})_{b,\text{in}}$
  Jitter: Define transform as a function of jitter versus signal frequency

For an analog signal, where SNR is expressed as rms power, $(SNR)_{RMS/RMS}$:

$$(SNR)_{b,sys} = \frac{1}{1/(SNR)_{out} - 1/(SNR)_{in}} \qquad (6\text{-}2)$$

or when SNR is expressed in decibels:

$$(SNR)_{b,sys} = -10 \log \left(10^{-(SNR)_{in}/10} + 10^{-(SNR)_{out}/10} \right) \qquad (6\text{-}3)$$

where $(SNR)_{out}$ = SNR at the output
$(SNR)_{in}$ = SNR at the input

6.3 STEP 1B: DESIGNING SYSTEMS ARCHITECTURE

The next step in the process is to develop the functional block diagram of the system, including the functional design of all layers as well as the systems topology, taking into account distance (probable repeaters), drop/insert points, overhead channels, and path redundancies. Figure 6.1 illustrates one example of such a diagram and Table 6.3 provides some common signal-processing options used at various layers. Such a design should include:

(a) defining all functional blocks and all interfaces between layers;
(b) specifying signal or frame formats at the interfaces to each layer to the degree possible;
(c) specifying interconnections between functional blocks as required by the topology and channel-interconnection requirements; and
(d) specifying other requirements, such as span distances, that may effect design.

The selection of transmission approach, analog versus digital, and the modulation, encoding, and multiplexing techniques to employ depends on many trade-off factors, industry standards, available product, and experience. Chapter 5 provides some of the trade-off parameters.

When multiple channels are to be transmitted, multiplexing is generally used to take advantage of fiber bandwidth when channel bandwidth is small relative to fiber bandwidth, and the cost of a multiplexer is less than the cost of multiple fibers, transmitters, and receivers. The multiplexing approach (time-division, frequency-division, wavelength-division) depends largely on the nature of the baseband channels to be transmitted. Fiber-optic systems product, generally, is mostly available for standard digital interfaces and, therefore, time-division multiplexing is most common. Video or voice signal, therefore, must be digitally encoded at the services or path layer. In some cases, however, signal format is not standard or is in a broadband analog format (such as video) and direct analog transmission with frequency-division or wavelength-division multiplexing is more economical.

Processing of analog signals at the service/path layer generally consists of analog-to-digital encoding or carrier modulation. This processing is performed for the following reasons:

TABLE 6.3 STEP 1B: SYSTEMS-ARCHITECTURE DESIGN OPTIONS

Channel type	Analog	Digital
Services/Path-layer processing common options		
Digital encoding:	PCM	NRZ
	DPCM	RZ
	ADPCM	DS-X Standard
	Deltamod	STS-X
Carrier modulation:		
Amplitude	AM	ASK
	AM/VSB	
Frequency	FM	FSK
Phase	PM	PSK
	DPM	QPSK
Subrate multiplexing:	FDM	TDM
Path overhead:	Subcarrier	Bit rob/stuff
	FDM	TDM
Line-layer multiplexing common options		
Channel MUX	FDM	TDM
Line OH	FDM Subchannel	Frame OH channel
	Subcarrier	Bit rob/stuff
Section-layer processing options		
Section OH	Subcarrier	Bit rob/stuff
	FDM	TDM
Reformat	Frequency translate	NRZ scramble
	Pulse encode	NRZI
		Manchester
		Biphase
Photonics-layer processing options		
Optical	Wavelength-division multiplexing (WDM)	

(a) Some SNR to CNR improvement is needed due to the limited SNR capability of the fiber transmission system (bandwidth is therefore traded for SNR).

(b) Photonics linearity is inadequate for direct intensity modulation.

(c) Line-layer multiplexing format demands that some form of carrier modulation or PCM be used.

6.4 STEP 1C: SERVICES/PATH-LAYER DESIGN

6.4.1 Step 1C.1: Design

Specific services and path-layer signal processing and line-layer formatting are designed based on the architectural choices of Step 1B. Path-layer overhead is determined and the means of multiplexing it into the information channels are chosen.

6.4.2 Step 1C.2: Baseband Unit Conversion

Based on the services and the design choices for the service/path-layer processing, the units specified for baseband performance may not be compatible with the interface-performance standards of the termination equipment. In most cases, the baseband signal parameters are specified in units that are common to the particular service, i.e., broadcast video, telephony, etc., as given in Chapter 2. These are probably not compatible with the rms power relationships provided in the tables in Chapter 5 or the specifications provided by a manufacturer. In order to analyze performance of the intermediate channels (at the path-to-line interface), we must first convert all parameters to units that are compatible with the transfer-function relationships for the services/path termination equipment.

In cases where analog signal processing (carrier modulation or digital encoding) is required, then conversion to the proper units must be performed in order to use the rms power relationships and tables in Chapter 5. The general relationship for SNR as expressed in decibels is

$$(SNR)_{b, \text{RMS/RMS}} = (SNR)_{b, \text{bb units}} - CF_{BW} - CF_{RMS} \tag{6-4}$$

where $(SNR)_{b, \text{bb units}}$ is the baseband SNR expressed in units unique to the service, CF_{BW} is a conversion factor for differences in noise-measurement bandwidth-filtering techniques between that of the service and that of the transmission-passband measurement standards, and CF_{RMS} is the conversion factor between the signal and noise units specified and that of the rms power units.

For example, in Chapter 2, we saw that a video baseband signal might be specified in terms of peak-to-peak white-to-blanking signal to rms noise with CCIR white-noise weighting, $(SNR)_{\text{ppwb/rms,}w}$. If we selected FM carrier modulation as our path-layer processing and were to use the tables in Chapter 5 to analyze SNR to CNR improvement, the units of SNR would have to be converted to rms signal power and rms noise power. The process would be to convert peak-to-peak voltage to rms power, and add the conversion factors (weighting) associated with the band limiting of the EIA weighting filter. The relationship given in Chapter 2 is as follows:

$$(SNR)_{\text{RMS/RMS}} = (SNR)_{\text{ppwb/rms,}w} - (CF)_{RMS} - WF$$

where CF_{RMS} = peak-to-peak voltage to rms power = 7.66 dB
WF = bandwidth weighting factor = 6.8 dB for EIA white noise

Therefore:

$$(SNR)_{\text{RMS/RMS}} = (SNR)_{\text{ppwb/rms,}w} - 14.46 \text{ dB}$$

For digital signals, bit rate is a standard unit, however, the signal format (bipolar, unipolar, RZ, NRZ) must be considered as well as any other characteristics dictated by standard (DS-0, DS-1, DS-3, DS4NA, STS-N, etc.) or nonstandard channel interfaces.

6.4.3 Step 1C.3: Service/Path to Line-Layer Interface Modulation and Encoding

In the case where standard interfaces are assumed or where equipment is a known product, the interface parameters are defined by the standards or the supplier's literature. In the case where the performance is to be analytically determined, however, the intermediate-channel interface performance must be computed from the "transfer functions" for the services/path termination equipment. Computation is particularly relevant to analog signal transmission. Table 6.4 outlines the process. Additional transfer functions for common modulation and encoding approaches are given in Chapter 5, Tables 5.3, 5.4, and 5.5.

The term "intermediate channel" is used to define the modulated carrier or digitally encoded state of any one of N channels between layers, that is, the channel state between the service baseband interface and the line-layer interface. The layer

TABLE 6.4   STEP 1C: SERVICES/PATH-LAYER PROCESSING

Step 1C.2: Unit conversion

For analog signals:

$$(SNR)_{RMS/RMS} = (SNR)_{b,\text{bb units}} - (CF)_{RMS} - (CF)_{BW}$$

For digital signals: Relate encoding standards

Step 1C.3: Calculate intermediate-channel performance

(a) No path processing:

Analog: $(SNR)_c = (SNR)_b$; $(BW)_c = (BW)_b$
Digital: $(BER)_c = (BER)_b$; $(BR)_c = (BR)_b$

(b) Analog carrier modulation (see Chapter 5, Table 5.3)

AM VSB: $(CNR)_c = (SNR)_b$
            $(BW)_c = (BW)_b$

FM: $(CNR)_{c,RMS} = (SNR)_{b,RMS} - 10 \log [\Delta f/(BW)_b]$
            $\qquad\qquad\quad - 10 \log \dfrac{(BW)_c}{(BW)_b} - 1.76$
            $(BW)_c = 2[\Delta f + (BW)_b]$

(c) PCM (see Chapter 5, Table 5.5)

Number of bits encoding: $n = \dfrac{(SNR)_{b,RMS} - 1.8 + QF}{6} =$

Channel bit rate: $(BR)_c = 2.35n(BW)_b$
Channel $(SNR)_{c,\text{pk/rms}} = 21.6$ dB at a BER $= 10^{-9}$

(d) Subrate multiplexing

Use Table 6.5 and Sections 5.4 and 5.5 of Chapter 5.

transfer functions provide the SNR and BW (or BER and BR) conversion factors between baseband to intermediate-channel performance or between intermediate channel at one layer to that at the next lower layer.

For example, with analog carrier modulation or analog-to-digital PCM encoding, bandwidth is traded for SNR, so there generally is some improvement in intermediate-channel $(SNR)_c$ or $(CNR)_c$ over that of the baseband parameter, $(SNR)_b$, and some increase in channel bandwidth, $(BW)_c$, over that of the baseband bandwidth, $(BW)_b$. This is illustrated in Chapter 5, Figure 5.2.

If the channels are digital in nature, and the encoding is changed at the path layer in order to map into a line-layer protocol, the new encoded signal bit rate, $(BR)_c$, and format (RZ, NRZ, etc.) must be determined.

6.4.4 Step 1C.3: Path-to-Line-Layer-Interface Subrate Multiplexing

Some baseband subrate channel multiplexing can be performed within or prior to the path-layer termination equipment in order to map the channels into formats that are "standard" for the path or line interfaces. For example, if we define the path layer as recommended in the SONET standard (see Chapter 7), PCM encoded voice channels would be multiplexed into a 24-channel group in a standard DS-1 format for presentation as a "service" to the SONET path-termination equipment. The path layer would, in turn, multiplex this along with multiple low-rate (DS-0, DS-1, etc.) channels and path overhead (POH) into a standard STS-1 line termination interface format. In the same manner, for a 565-Mb/s (12 DS-3) asynchronous system, we might consider the M13 (DS-1 to DS-3) multiplexer as a path-layer multiplexer for DS-1 service.

Channel-performance parameters, $(SNR)_c$, $(BER)_c$, $(BW)_c$, and $(BR)_c$, are modified by multiplexing and the associated signal reformatting.

In the absence of multiplexing (one baseband channel per line-layer channel), then path and line parameters are common:

$$(SNR)_{cl} = (SNR)_{cp} \qquad (BW)_{cl} = (BW)_{cp}$$

$$(BER)_{cl} = (BER)_{cp} \qquad (BR)_{cl} = (BR)_{cp}$$

where cl relates to line-layer intermediate channels, and cp to path-layer intermediate channels.

In the presence of subrate multiplexing, there is a transformation of intermediate-channel parameters between processing steps. The effect is generally to increase transmission bandwidth by approximately the channel bandwidth times the number of channels multiplexed plus some multiplexer efficiency and POH factor.

The SNR and BW transforms for multiplexing are described in Sections 5.4 and 5.5 in Chapter 5, and are summarized in Table 6.5.

TABLE 6.5 MULTIPLEXING TRANSFORMS

Calculate channel-performance parameters after multiplexing (subscripts represent line-layer multiplexing):

(a) No multiplexing:

$$(SNR)_t = (SNR)_c; \qquad (BW)_t = (BW)_c$$

$$(BER)_t = (BER)_c; \qquad (BR)_t = (BR)_c$$

(b) Frequency-division multiplexing:

Transmission bandwidth:

$$(BW)_t = (BW)_z + (BW)_{LOH} + N[(BW)_c + (BW)_{POH} + (BW)_g]$$

where $(BW)_g$ = guard band (half on either side)

$(BW)_z$ = zero-frequency band

$(BW)_{LOH}, (BW)_{POH}$ = line and path overhead channel subcarrier bandwidths, respectively

Per channel transmitted SNR:

$$(SNR)_{ct} = (SNR)_c$$

(c) Time-division multiplexing:

Transmission bit-error rate:

$$(BER)_t = (BER)_c$$

Transmission bit rate:

$$(BR)_t = \frac{N(BR)_c}{FEF} = N(BR)_c + (BR)_{FOH}$$

$$FEF = \frac{N(BPC)}{N(BPC) + FOH}$$

$(BR)_c$ = information-channel bit rate

BPC = information bits per channel

$(BR)_{FOH}$ = multiplexer frame overhead plus layer-overhead composite bit rate

FOH = multiplexer frame overhead plus layer-overhead bits per N channel group

6.5 STEP 1D: LINE-LAYER DESIGN

Typically, the line-layer function is simply one of multiplexing N standard line formatted channels and line overhead (LOH) into a single section-layer channel and formatting the signal for that layer. We restrict the definition to just that. Examples include the SONET-defined standard line layer that multiplexes N STS-1 channels and LOH into a STS-N channel, or a 560-Mb/s fiber-optics terminal that multiplexes a 12 DS-3 channel and overhead into a 560-Mb/s channel for subsequent optical transmission.

The multiplexing transform relationships are provided in Chapter 5 and summarized in Table 6.5. The objective of this analysis is the definition of the optical section-layer end-to-end performance parameters by transforming intermediate-channel performance values to transmission-channel performance values.

6.6 STEP 2A: OPTICAL SECTION-LAYER DESIGN

The optical subsystem can be designed once the section-layer end-to-end transmission parameters are determined from Step 1D, since the choice of components depends on these parameters. Section-layer design consists of the following:

(a) designing a functional section and photonics layer;

(b) performing a first-order analysis to determine approximately how many repeaters may be required for the worst-case span and refining the initial design as appropriate;

(c) computing final per-section-and-photonics-layer end-to-end performance parameters based on the number of repeater cascades.

Section-layer design is an iterative process. The result is a refinement of the photonics design to minimize the number of repeater spans required and optimize performance. This stage of the process, therefore, uses a rough worst-case link-bandwidth and power-budget analysis approach to arrive at section-layer design limits and estimate performance. Once it appears that the design is correct, a more refined analysis of the photonics-layer performance is achieved in Stage 3.

6.6.1 Step 2A.1 Section-Layer Functional Design

The function of the section layer is to provide transmission-signal formatting prior to optical transmission by the photonics layer. In a digital transmission system, this includes framing, scrambling, and section error monitoring, and creates the transmission-symbol formatting required for proper signal reception and regeneration. The section layer can also provide overhead (SOH) such as local orderwire. At the terminal location, the function can be physically performed within the line multiplexing equipment. Within a repeater, the section layer is present to serve the photonics layer, but requires no higher layers to be present. Section-layer design and preliminary analysis is summarized in Table 6.7.

Functional design of the section layer should include the following steps:

(a) Select the section-layer transmission format, including the final transmission-frequency plan or pulse encoding for analog transmission or pulsed symbol format for digital transmission (Chapter 5, Section 5.6).

(b) Define the section overhead (SOH), if any, and the multiplexing approach to be used.

(c) Merge the SOH into the frequency plan (if analog) or time-slot position (if digital) within the photonics transmission format.

6.6.2 Step 2A.2 Determine Photonics-Layer Performance

Photonics-layer transmission parameters are determined by change in spectral occupancy, $(BW)_t$ and $(BR)_t$, and signal performance, $(SNR)_t$ and $(BER)_t$, that the section-layer transformations impose on the transmitted signal at the interface to the section layer as defined in Step 1D.2. Such changes might include the following:

(a) SOH time-division multiplexing: The same multiplexing rules as described in Step 1C.3 apply to the insertion of SOH into a serial data stream:

$$(BR)_t = (BR)_{ts} + (BR)_{SOH} \qquad (6-5)$$

where $(BR)_{SOH}$ is the SOH bit rate including added framing bits, $(BR)_{ts}$ is the bit rate of the transmission signal at the line/section interface, and $(BR)_t$ is the resulting transmission bit rate for the photonics layer.

(b) SOH frequency multiplexing: In analog transmission, the SOH can be carried as a subcarrier or multiplexed in as another channel. Follow the same rules as described in Chapter 5, Section 5.6, in determining the new $(CNR)_t$ and $(BW)_t$ at the photonics layer.

$$(BW)_t = (BW)_{ts} + (BW)_{SOH} + (BW)_{gs} \qquad (6-6)$$

where $(BW)_{ts}$ is the transmission bandwidth of the section end to end, $(BW)_{SOH}$ is the bandwidth required of the overhead subcarrier, and $(BW)_{gs}$ is any added guard band between the SOH subcarrier and the information channels.

(c) Digital encoding for transmission: Changes in encoding frame format can affect the transmission bit rate, $(BR)_t$. Changes in the transmitted-symbol format from, say, NRZ to an RZ, Manchester, or scrambled NRZ format may not change bit rate, but may change spectral occupancy of the signal as well as the requirements on signal regeneration at the receiver. Symbol format, therefore, has a direct effect on receiver sensitivity. See Chapter 5, Section 5.6.4, for a discussion.

6.7 STEP 2B: PHOTONICS-LAYER PRELIMINARY DESIGN

The objective of a preliminary design is to select candidate technologies and components and roughly test the choices against the performance criteria of Step 2A in order to optimize performance and reduce or eliminate repeaters within practical constraints of cost and other factors. The steps include:

1. functional design;
2. technology selection; and
3. component-performance definition.

6.7.1 Step 2B.1 Functional Design

Functional design of the photonics layer should include such elements as:

(a) optical link topology;
(b) wavelength multiplexing, if any;
(c) coupler configuration and interconnection;
(d) connector, patch panel, and other interface points;
(e) splices, where identifiable; and
(f) other physical parameters such as span distance.

6.7.2 Step 2B.2 Technology Selection

Technology is selected on the basis of estimated performance. Choices may include:

(a) single-mode versus multimode;
(b) wavelength of operation;
(c) fiber core size, attenuation, and bandwidth range;
(d) source type (LED, ILD);
(e) detector and receiver (APD, PIN, PIN/FET, etc.); and
(f) coupler technology (fused, GRIN lens, optical grating, integrated optics).

These choices are generally made from experience or are aided by engineering design charts or tables, such as those in Chapter 3. Figure 3.1 can be used as a convenient aid at this stage of the design.

6.7.3 Step 2B.3: Component-Performance Specification

Once the basic technology and component types are selected, then component-performance parameters are specified or determined from supplier specifications. Where supplier specifications are not available, some of the tables presented in Chapter 3 can be used as an aid. Table 6.6 provides a guide to the component parameters needed for a complete link-budget analysis.

TABLE 6.6 STEP 2B: TECHNOLOGY SELECTION AND OPTICAL-COMPONENT
PERFORMANCE SPECIFICATION

Basic parameters required for a systems design

Transmitter:
>   Type (ILD, LED)
>   Center wavelength (nm)
>   Spectral width (nm)
>   Coupled power (dBm)
>   Derating for temperature, lifetime (dB)
>   Rise/fall time (ns)
>   Modulation index versus linearity (analog)
>   Extinction ratio (digital)

Receiver:
>   Type (PIN or APD, FET or bipolar, TXP or IFE)
>   MRP at specified BER and BR or SNR and BW
>   Bandwidth (Hz) or response time (ns)
>   Dynamic range (dB)
>   Saturation power (dBm)
>   Derating for band limiting (dB)
>   Derating for temperature (dB/°C)

Connector:
>   Insertion loss (dB), mean and standard deviation
>   Derating for mating lifetime and temperature (dB)

Splice:
>   Insertion loss (l_s) (dB), mean and standard deviation

Fiber [measured at specified wavelength(s)]:
>   Attenuation coefficient (dB/km), mean and standard deviation
>   Temperature derating (dB/km)
>   Multimode dispersion (ns/km), typical and worst case
>   Multimode concatenation factor
>   Material dispersion (ps/nm/km) at worst-case λ
>   Reel length (km)

Where analytical receiver design is involved

Detector:
>   Type and material (silicon APD, GaAs PIN, etc.)
>   Rise/fall time (ns)
>   Responsivity (A/W)
>   Optimum gain and F (excessive noise factor) at mean P_d
>   Dark and leakage currents (A)
>   Capacitance (F)

Preamp:
>   Type (bipolar or FET, IFE or TXP)
>   Input resistance (ohms)
>   Input capacitance (F)
>   Transconductance (mhos) or Beta
>   FET channel noise factor (τ)

6.8 STEP 2C: EVALUATE DESIGN AND ESTIMATE THE NUMBER OF REPEATERS

Before detailed analysis is performed, component choice should be tested with a first-order analysis to determine (1) whether it performs within the desired range, and (2) the number of repeater spans. If repeaters are required, the transmission parameters must be improved on a per-section span basis in order to account for cascading degradation.

Simple tests are performed on the basis of bandwidth and power budget over the total transmission distance D. If the result is that the span length is too long for repeaterless operation, then (a) fiber loss or dispersion, or other component performance, must be changed, or (b) repeaters must be added and the per-section performance adjusted accordingly.

6.8.1 Step 2C.1: Link-Bandwidth Section Analysis

The objective at this point is to determine the effect of link band limiting on span distance (SD). The bandwidth of the optical link, $(BW)_l$, or total photonics, $(BW)_p$, for the longest span should be evaluated to ensure that it is greater than that required by the transmitted spectrum, $(BW)_t$, by an amount that induces no more than a 1-dB power penalty on the received signal (see Section 6.12). Typically, the following are required for less than a 1-dB power penalty, from Figures 6.9 and 6.13:

$(BW)_{l,\text{elect}} > 0.65$ to 0.70 $(BR)_t$ for digital transmission, depending on the receiver.

$(BW)_{p,\text{elect}} > 1.3$ to 1.9 $(BW)_t$ for analog transmission, depending on the receiver.

For convenience, the analysis is generally performed in terms of component response time and then converted to bandwidth using [3]:

$$(BW) = \frac{0.350}{t_r} \tag{6-7}$$

The relationship between distance and bandwidth for analog signals is derived by setting receiver $(BW)_r = (BW)_t$ and solving for $(BW)_p/(BW)_t = X$. If factor X is greater than X (1 dB), then the design is satisfactory, where the 1-dB P_p factors X (1 dB) from Figure 6.9 are

X (1 dB) = 1.3 for thermal-noise-dominant receiver

X (1 dB) = 1.67 for a bipolar receiver or APD

X (1 dB) = 1.9 for a PIN/FET

If X is less than these figures, then the resulting P_p can be determined from Figure 6.9 or distance must be reduced accordingly. The relationships are given in Table 6.7.

TABLE 6.7   STEP 2C: PRELIMINARY DESIGN EVALUATION OF THE SECTION
CASCADE

Step 2C.1: Link-bandwidth section analysis
Effect of band limiting on span length or power penalty
Analog signals (see Figure 6.9):

$$(BW)_l > X(BW)_t, \text{ bipolar for a 1-db } P_p$$

where $X = 1.67$ for a bipolar amplifier or APD

$\quad\quad\quad = 1.9$ for an FET amplifier

$$D = \sqrt{\frac{0.111/[X(BW)_t]^2 - t_{rs}^2 - t_{rd}^2}{1.182(t_{dw}^2 + t_{dm}^2 \, \Delta\lambda^2)}}$$

where t_{rs}, t_{rd} = rise time of source and detector, respectively

$\quad\quad t_{dw}$ = multimode dispersion in ns/km

$\quad\quad t_{dm}$ = material dispersion in ps/nm/km

$\quad(BW)_t$ = transmission bandwidth = $0.35/t_r$ (photonics)

$\quad(BW)_l$ = link bandwidth (source, fiber, detector)

Digital signals (see Figure 6.13):

$$(BW)_e \text{ (Tx, fiber, detector)} > 0.7 \, BR \text{ for a 1-dB } P_p$$

$$D = \frac{\{[0.7785 - t_{dx}^2(BR)^2]/(t_{dw}^2 + t_{dm}^2 \, \Delta\lambda^2)\}^{1/2}}{BR}$$

Step 2C.2: Link-power-budget section analysis
With D as a known, solve the power budget for span distance (SD) and, the number of spans (NS)
using the worst-case conditions.

$$NS = INT\left(1 + \frac{D}{SD}\right)$$

$$SD \text{ (km)} = \frac{P_t - MRP - M - P_p - N_c l_c - N_s l_s}{l_f(dB/km)}$$

If the distance between repeaters is inadequate, then refine the component selection and repeat this
stage.

For digital signals, the relationship between bit rate and distance is based on an
approximation for a 1-dB power penalty following Midwinter [14]. The derivation is
given in Section 6.12.4 in the discussion on intersymbol interference. The relationship is provided in Table 6.7 as well.

If only the fiber is involved in the analysis, then use the following by substituting (4-41) into (6-7):

$$(BW)_{f,\text{elect}} = 0.7(BW)_{f,\text{opt}} = \frac{0.322}{t_{df,\text{FWHM}}} \tag{6-8}$$

6.8.2 Step 2C.2: Link-Power-Budget Section Analysis

With D as a known, solve the power budget for span distance (SD) and number of
spans (NS) using a worst-case approach. The relationships are

$$\text{SD (km)} = \frac{P_t - \text{MRP} - M - P_p - N_c l_c - N_s l_s}{l_f (\text{dB/km})} \qquad (6\text{-}9)$$

where P_t = worst-case coupled optical power

MRP = worst-case minimum required received optical power for specified SNR or BER at the receiver

M = power margins allowed for component drift and aging (1 to 3 dB), added splices for repair (1 to 2 dB), and miscellaneous operating margin (3 dB)

P_p = any known power penalties or those derived from the bandwidth budget

$N_c l_c$ = number of connectors times the maximum loss per connector, or if $N_c > 2$ use the mean loss plus derating factor (if any) times N_c

$N_s l_s$ = number of splices times the mean loss per splice

l_f = worst-case mean filter attenuation over the temperature range in dB/km

If span length (SD) is less than the total transmission distance (D) to the point where it does not appear that practical changes in the components can make a difference, then assume that repeaters are required:

$$\text{NS} = \text{INT}\left(1 + \frac{D}{\text{SD}}\right) \qquad (6\text{-}10)$$

6.9 STEP 2D: ALLOCATION TO A SINGLE SECTION

When it is determined from Stage 2C that repeaters are required in the optical transmission subsystem (cascaded sections), the end-to-end section-layer performance from Step 1D.2 must be allocated to a single section. Allocation to each section in a cascade results in the performance requirements in each section being greater than the end-to-end requirements. In practice, all section spans may not be the same and, therefore, the parameter allocation concentrates on the worst-case span. For simplicity, in this section, we assume equal spans, leaving any adjustments to the designer on a case-by-case basis.

Three situations are addressed here: (a) SNR and bandwidth allocation for analog signal transmission with linear nonregenerative repeaters, (b) BER allocation for digital transmission with regenerative and nonregenerative repeaters, and (c) jitter allocation for digital transmission with regenerative and nonregenerative repeaters and multiplexer effects.

Figures 6.2 and 6.3 show an optical subsystem with a series of cascaded links (repeaters) and the effects the cascading has on SNR and bandwidth for analog signals and on BER and jitter for digital signals. The design procedure, Step 2D.1, is summarized in Table 6.8. From a physical sense, each functional block represents a section (transmitter, fiber span, and receiver). Each block, therefore, cuts the repeater in half at the section-layer interface.

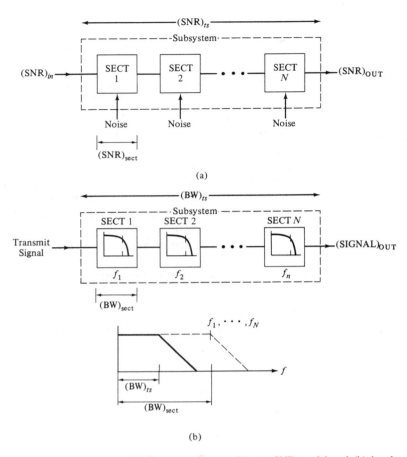

Figure 6.2 Analog network transmission model. (a) SNR model and (b) bandwidth model.

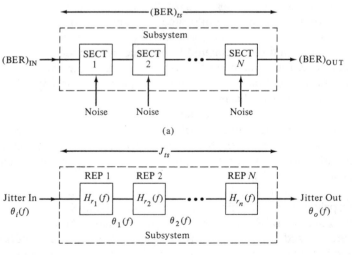

Figure 6.3 Binary network transmission model. (A) Error-performance model and (b) jitter-performance model.

TABLE 6.8 STEP 2D.1: TRANSMISSION PARAMETER ALLOCATION
FOR CASCADED SECTIONS

Analog transmission (Figure 6.2):

$$(CNR)_{sect} = (CNR)_{ts} + 10 \log (N)$$

$$(BW)_{sect} = 1.1(BW)_{ts}\sqrt{N}$$

where N is the number of section spans in cascade, and $(CNR)_{ts}$ and $(BW)_{ts}$ are the end-to-end
requirements of the total cascade (CNR in dB).

Digital transmission (Figure 6.3):

$$(BER)_{sect} = \frac{(BER)_{ts}}{N}$$

$$Q_{sect} = \text{depends on } (BER)_{sect}; \text{ see Figure 4.9}$$

$$(BR)_{sect} = (BR)_{ts}$$

If nonregenerative repeaters:

$$t_{r,sect} = \frac{t_{r,ts}}{1.1\sqrt{N}}$$

$$Q_{sect} = Qt_s\sqrt{N}$$

6.9.1 Step 2D.1 Analog Transmission SNR Allocation

In the transmission of a signal of $(SNR)_{in}$ through the cascaded sections shown in
Figure 6.2(a), noise is introduced at each individual section. This degrades the
signal-to-noise ratio to a level $(SNR)_{out}$ at the output. The result is a value for
section-layer subsystem performance, $(SNR)_{ts}$. $(SNR)_{ts}$ must be such that it meets the
performance requirements of the line layer as determined in Step 1D.2.

$(SNR)_{ts}$ is a combination of the individual section signal-to-noise ratios,
$(SNR)_1$ through $(SNR)_N$. We can combine SNR by normalizing all noise terms and
assuming unity signal. If noise terms are expressed as rms current, $(SNR)_{rms/rms}$, then
they combine as the square root of the sum of the squares:

$$(SNR)_{ts} = \frac{\text{unity signal}}{\left(\dfrac{\langle i_{n1}\rangle^2}{\langle i_{s1}\rangle^2} + \cdots + \dfrac{\langle i_{nN}\rangle^2}{\langle i_{sN}\rangle^2}\right)^{1/2}} \tag{6-11}$$

If noise is expressed as noise power, however, $(SNR)_{RMS/RMS}$, it is already in
the form displayed and, therefore, as a term, combines directly. If we assume
$(SNR)_{RMS/RMS}$ is expressed in decibels, we get

$$(SNR)_{ts} = -10 \log (10^{-(SNR)_1/10} + \cdots + 10^{-(SNR)_N/10}) \tag{6-12}$$

If the section spans are identical, then, $(SNR)_{sect}$ in decibels is

$$(SNR)_{sect,RMS/RMS} = (SNR)_{ts,RMS/RMS} + 10 \log (N) \tag{6-13}$$

and when SNR is not expressed in decibels:

$$(SNR)_{sect,RMS/RMS} = (SNR)_{ts,RMS/RMS} \times N \qquad (6-14)$$

For example, assume that an analog signal is to be transmitted over a fiber network 60 km in length. Assume the rms carrier-to-noise ratio, $(CNR)_{ts}$, calculated for the end-to-end section layer, from Stage 1D.2, is 20 dB, and that two repeaters were estimated from Stage 2C, $N =$ three cascades. Calculating the $(CNR)_{sect}$ requirement of an individual section in the cascade gives

$$(CNR)_{sect} = (CNR)_{ts} + 10 \log (N) = 20 + 10 \log (3) = 24.77 \text{ dB}$$

6.9.2 Step 2D.1 Continued
Analog Transmission Bandwidth Allocation

When transmitting analog signals, successive stages (cascaded band-limiting elements) degrade the available bandwidth of the transmission channel. Figure 6.2(b) illustrates an optical subsystem containing N section spans with individual passbands $(BW)_1$ through $(BW)_N$. Although the sections are shown as low-pass components, the analysis is applicable as well to band-pass characteristics. The design goal is that the band-pass of the total section cascade is equivalent to and, therefore, passes the required transmitted signal bandwidth, $(BW)_{ts}$, determined in Step 1D.2.

The individual section transmission bandwidth, $(BW)_{sect}$ for a network of N cascaded sections, where each is identical, is for cascades of two to five sections [3]:

$$(BW)_{sect} = 1.1(BW)_{ts} \sqrt{N} \qquad (6-15)$$

and for six or more, this is more accurate [3]:

$$(BW)_{sect} = \frac{(BW)_{ts}}{\sqrt{2^{1/N} - 1}} \qquad (6-16)$$

When the bandwidths of elements in a system are not identical, the following relationship can be used as an estimate:

$$(BW)_{ts} = \frac{1}{\sqrt{\dfrac{1}{(BW)_1^2} + \dfrac{1}{(BW)_2^2} + \cdots + \dfrac{1}{(BW)_N^2}}} \qquad (6-17)$$

6.9.3 Step 2D.1 Continued
Digital Transmission BER Allocation

The functional block diagram of a digital transmission subsystem of N cascaded sections is shown in Figure 6.3. The quantity of received data errors over a stated period of time is the defining factor for signal quality, rather than SNR. A number of terms are generally used to describe error performance. Two of the most common are

(a) bit-error rate (BER): the number of errored bits received divided by the total number of bits transmitted, per unit of time; and

(b) error-free seconds (EFS): the probability or percentage of 1-second time intervals that are free of errors.

The relationship between the two for statistically independent errors is [4]

$$\text{EFS (\%)} = (1 - \text{BER})^{\text{BR}} \times 100 \tag{6-18}$$

If we assume an error-free signal at the input to the network shown in Figure 6.3 and that each cascaded section in the network contributes a bit-error rate of $(\text{BER})_{\text{sect}}$, the total bit-error rate at the output is [4]

$$(\text{BER})_{ts} = \frac{1 - [1 - 2(\text{BER})_{\text{sect}}^N]}{2} \tag{6-19}$$

When $N \times (\text{BER})_{ts} \ll 1$, then:

$$(\text{BER})_{\text{sect}} = \frac{(\text{BER})_{ts}}{N} \tag{6-20}$$

$$(\text{BER})_{ts} = (\text{BER})_1 + (\text{BER})_2 + \cdots + (\text{BER})_N \tag{6-21}$$

where $(\text{BER})_N$ is the BER of individual section number N.

The relationship for EFS is simply 100% less the sum of the individual contributions of each section:

$$(\text{EFS})_t \, (\%) = 100 - \{[100 - (\text{EFS})_1] + \cdots + [100 - (\text{EFS})_N]\} \tag{6-22}$$

These relationships assume regenerative repeaters between stages in the cascade. For nonregenerative (linear) repeaters, another relationship exists. As discussed in Chapter 4, BER is a function of received $(\text{SNR})_{\text{pk/rms}}$ or Q at the threshold detector (regenerator). In a system with linear repeater cascades (no regeneration), the resultant Q from a series of N cascades is

$$\frac{1}{Q_{ts}^2} = \frac{1}{Q_1^2} + \cdots + \frac{1}{Q_N^2} \tag{6-23}$$

and when all sections are identical:

$$Q_{\text{sect}} = Q_{ts} \sqrt{N} \tag{6-24}$$

6.9.4 Step 2D.1 Continued
Digital Transmission Rise-Time Degradation

In most digital transmission systems, regenerative repeaters are used. At each repeater, the received signal is sampled, decoded, and retimed. The result is a series of new transmission pulses with proper pulse shape and regenerated timing. No rise-time degradation exists from repeater to repeater since the pulses are reshaped. If, however, linear repeaters (nonregenerative) are used, then pulse shape and rise time degenerate from section to section.

The following relationship [3] applies within approximately 10% accuracy for a series of nonidentical links or series elements of rise times $t_{r(1)}$ through $t_{r(N)}$:

$$t_{r,\text{sys}} = 1.1\sqrt{t_{r1}^2 + \cdots + t_{rN}^2} \qquad (6\text{-}25)$$

An approximation for an individual section rise time of a subsystem with N identical cascaded sections is, therefore,

$$t_{r,\text{sect}} = \frac{t_{r,\text{ts}}}{1.1\sqrt{N}} \qquad (6\text{-}26)$$

6.9.5 Step 2D.1 Continued
Digital Transmission Jitter Accumulation

Jitter is defined as short-term timing or phase variations from ideal in a received re-generated signal pulse. Jitter is generated in (a) repeaters and receivers, wherever threshold detection and signal regeneration takes place (see Chapter 4), and (b) multiplexers or frame-formatting circuitry where bits are buffered, stuffed, and retimed. The timing uncertainty caused by jitter can result in (a) bit errors, (b) synchronization slips, (c) crosstalk from adjacent signal patterns, and (d) distortions in digital-to-analog conversions (PCM, PDM, or PPM) where phase- to amplitude-noise conversion can take place.

In most digital systems, the terminal equipment is specified in terms of the amount or percentage of timing jitter that it can tolerate. It is important, therefore, for the systems designer to understand jitter buildup and/or attenuation within the transmission system. Jitter requirements for a transmission system are generally specified in terms of the following parameters:

(a) Jitter accommodation (input jitter tolerance): This is the amount of sinusoidal jitter (peak to peak) at the input to the system that can be tolerated without producing errors.

(b) Jitter transfer characteristics: This is the amount of jitter enhancement or attenuation of the input jitter that the system provides.

(c) Jitter generation (intrinsic jitter): This is the amount of jitter the system produces with no jitter applied at the input.

Figure 6.4 illustrates the form that these specifications can take. Note that the jitter functions are frequency-dependent. This is because the retiming functions within regenerators, receivers, and demultiplexers contain local clocks or oscillators that attenuate or reduce the effects of high-frequency phase variations (see Section 4.6). Jitter specifications vary, depending on equipment type and application. Values that might accommodate Figure 6.4 can be found in ANSI/EIA TSB-19-1986 [8]. An example for DS-3 transmission is given in Table 6.9.

Standards for jitter are contained in the following:

(a) ANSI/EIA TSB-19-1986, *Optical Fiber Digital Transmission Systems Considerations for Users and Suppliers;*

(b) CCITT recommendations, G.700 series;

(c) ANSI standards for DS1, DS1C, DS2, DS3, and DS4NA levels of the digital hierarchy;

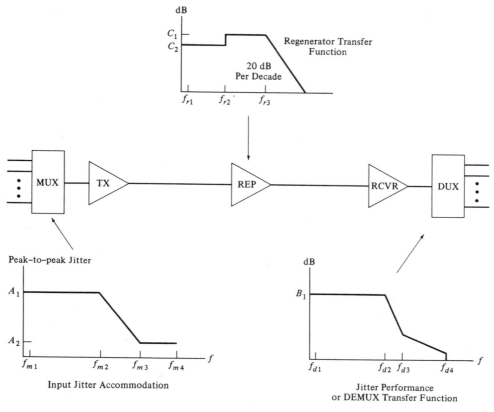

Figure 6.4 Format for jitter specification in an optical communication system.

TABLE 6.9 JITTER REQUIREMENTS FOR DS-3 TRANSMISSION PER ANSI/EIA TSB-19-1986[a]

Multiplexer jitter accommodation					
A_1 Unit Interval (pk-pk)	A_2 Unit Interval (pk-pk)	f_{m1} (Hz)	f_{m2} (Hz)	f_{m3} (Hz)	f_{m4} (Hz)
5	0.1	10	2.3×10^3	60×10^3	300×10^3
M3-1 demultiplexer transfer function					
B_1 (dB)	f_{d1} (Hz)	f_{d2} (Hz)	f_{d3} (Hz)	f_{d4} (Hz)	
0.1	10	350	2.5×10^3	15×10^3	
Regenerator transfer function					
C_1 (dB)	C_2 (dB)	f_{r1} (Hz)	f_{r2} (Hz)	f_{r3} (Hz)	
0.05	0.25	5	$5f_{d2}$	$BR \times 3.14 \times 10^{-3}$	

[a] Refer to Figure 6.4.

271

(d) ECSA/ANSI T1X1 SONET standard;

(e) Fiber-optic terminal and multiplexer equipment supplier specifications and other associated standards.

Jitter values are specified for a specific network model and measurement method. Generally, jitter is measured for a terminal or transmission link by looping back the transmitted signal, at the highest level, at either a repeater or terminal, as the case may be.

As indicated in Chapter 4, jitter can be classified in three categories:

1. Systematic jitter: This is pattern-dependent jitter accumulated in a systematic predictable manner, such as might be related to a predominant spectral component or average DC-level drift in the transmitted signal, intersymbol interference, local oscillator offsets, and such.

2. Nonsystematic jitter: This caused by uncorrelated jitter sources, such as timing variations between repeaters and waiting-time jitter introduced by multiplexing.

3. Random jitter: This is uncorrelated noise-related jitter, such as might be caused by amplitude noise converted to phase noise at the sampling instant.

Jitter typically accumulates in a system as a function of the number of regenerators or MUX/DEMUX stages (N) that the binary signal passes through. Accumulation is a complex function and, generally, the mechanisms are not easily isolated. The relationships that follow provide some approximations useful in first-order systems design.

Systematic timing jitter. The model for the introduction and accumulation of systematic jitter is given in Figure 6.3(b). The output jitter spectrum $\Theta_o(f)$ at each repeater is related to the input phase jitter spectrum $\Theta_i(f)$ and the repeater transfer function $H_r(f)$:

$$\Theta_o(f) = \Theta_i(f) \times H_r(f) \qquad (6\text{-}27)$$

If we assume that jitter adds linearly from repeater to repeater, jitter can be shown to accumulate:

$$\Theta_o(f) = \Theta_i(f)[H_r(f) + H_r(f)^2 + \cdots + H_r(f)^N] \qquad (6\text{-}28)$$

Assuming that each repeater is regenerative, transfer function $H_r(f)$ takes the form similar to a low-pass filter, as seen in Figure 6.4, with a roll-off frequency that is a function of the "loop" bandwidth, $(BW)_l$, of the oscillator circuit or phase-locked loop. The accumulation at low frequencies, defined as timing jitter, is [4]

Power Density:

$$J_n = J_i \times N^2 \qquad (6\text{-}29)$$

Amplitude:

$$t_{jn} = t_{ji} \times N \qquad (6\text{-}30)$$

This power accumulation is reduced by 20 dB per decade as frequency increases past f_{r3} (per Figure 6.4).

Random data patterns.   The previous relationships assume that the jitter introduced at each repeater is frequency- (and thus pattern-) dependent. If the data pattern is random, then the power density of the jitter introduced is constant, analogous to the white-noise spectrum. Under this circumstance, Smith [4] and Byrne et al. [5] have determined the rms jitter accumulation to be related by

$$\langle t_j^2 \rangle = \frac{J_i(\text{BW})_i N'}{2} \qquad (6\text{-}31)$$

where

$$N' = N - \frac{1}{2}\left(\frac{(2N-1)!}{4^{N-1}[(N-1)!]^2}\right)$$

The rms accumulation is directly proportional to N' and the jitter amplitude to $\sqrt{N'}$. For long repeater chains ($N > 100$), N' approximates N. These long chains, without intervening multiplexers, are not normally found in fiber-optic systems, however.

Repetitive data patterns.   According to Smith [4], jitter at the end of N repeaters due to repetitive changes in data patterns accumulates as a function of

$$t_{j,\text{tot}} = t_{ji} \times N \qquad (6\text{-}32)$$

Uncorrelated jitter sources.   Rowe [6] has shown that rms jitter created by random uncorrelated jitter sources accumulates:

$$\langle t_{jn}^2 \rangle / \langle t_{ji}^2 \rangle = (N)^{1/4} \qquad (6\text{-}33)$$

For this reason, systematic jitter sources predominate.

Regenerator sampling time (or frequency) alignment.   According to Smith [4], the maximum alignment jitter does not increase with the number of cascades, but is bounded by the maximum jitter injected at each repeater.

Multiplexer jitter.   Most fiber transmission systems use asynchronous multiplexers, that is, multiplexers that combine a multitude of nontime synchronous channels. To ensure that no data bits are dropped due to timing differences, the multiplexer samples at a rate slightly greater than the sum of the input rates. Pulses are therefore "stuffed" periodically into the input data streams to make up the timing differences. These pulses are then removed at the demultiplexer. Over a longer period, known as the "waiting time," the timing catches up and the pulse is deleted, resulting in a phase jump. In networks with long chains of multiplexer/demultiplexer stages, waiting-time jitter has been shown [7] to accumulate in proportion to

$$t_{j,\text{tot}} = t_{ji} \times \sqrt{N} \tag{6-34}$$

The demultiplexer retimes the output signal with a "smoothing loop" that attempts to produce a jitter-free output clock at the average rate of the input data. The demultiplexer, therefore, acts as a jitter reducer in a system. The jitter tolerance and transfer characteristics of the multiplexers dominate the jitter performance of the transmission system. For a more in-depth analysis of multiplexer jitter mechanisms, see Smith [4].

6.10 STAGE 3: PHOTONICS-LAYER DESIGN EVALUATION

The objective of the Stage 3 analysis (using link-budget-analysis methodology) is to determine whether the proper components and design parameters were chosen in Stage 2, such that (a) the signal power reaching the detector/receiver is in the proper range and (b) the system's bandwidth is adequate, to achieve performance requirements. As described in Section 6.1, the analysis consists of seven steps:

3A Link-bandwidth analysis
3B Compute power penalties
3C Compute SNR penalties and $(SNR)_r$ or Q_r
3D Design receiver, determine MRP
3E Link-power-budget analysis
3F Systems-performance test

6.10.1 Consider Operating Conditions and Tolerances

In order that the analysis represents practical conditions, component-parameter variations and tolerances must be taken into account. This can be achieved by analyzing the design for three cases:

1. Nominal case: All environmental conditions are nominal; all component parameters are at their specified mean or typical values; and operating time is midlife.

2. Best case: Environmental conditions are those within the operating range that produce the maximum received power and fastest responses; loss components are at the minimum end of their tolerance range; source power and receiver sensitivity are at the high end of the tolerance ranges; power penalties are minimum; and operation is at start of life.

3. Worst case: Environmental conditions are those within the operating range that produce the minimum received optical power and slowest responses; loss components are at the maximum end of their tolerance range; source power and receiver sensitivity are at the low end of the tolerance ranges; power penalties are maximum; and operation is near end of life.

The results of such an analysis provide three necessary pieces of information:

(1) whether the system will operate under all conditions;
(2) how marginal the design is (minimum power and saturation); and
(3) how much dynamic range is needed within the receiver to accommodate variations in tolerances and operating conditions.

6.10.2 Statistical Analysis Approach

Practical components and their interactions in a system never exhibit a best-, worst-, or nominal-case value 100% of the time regardless of how they are specified. Given a number of like components, their performance from component to component is generally grouped around some mean value, but varies about that mean often and to quite a significant extent. Peformance from component to component often follows what is called a "normal distribution." This distribution is illustrated in Figure 6.5. It is generated by plotting the function [15, 16]

$$P = \frac{1}{\sqrt{2\pi}\,\sigma} \exp - \frac{X^2}{2\sigma^2} \qquad (6\text{-}35)$$

The P axis in the figure shows the probability that a component will be measured at a particular value on the X axis. The percentage probability that a component will be measured less than a particular value on the X axis is given on the lower X-axis scale.

The mean is the "nominal" around which the components are expected to perform, specifically, the mean value is that for which 50% of the measured values fall below. 84.13% of the measured values are contained within one standard deviation (1σ) above mean, 97.73% within 2σ, and 99.87% within 3σ. In specifying component tolerances, generally, the 2σ value is considered the worst-case tolerance [8] although some standardize on 3σ.

The difference between the mean value and the advertised performance specification depends on whether the supplier is specifying a "maximum" value or a "typical" value. For example, fiber-optic cable specified at a maximum loss of 0.5 dB/km is almost always delivered at a mean loss of around 0.3 to 0.4 dB/km to ensure the specification is not exceeded. Some small percentage of the fibers will be actually at 0.5 dB/km. Likewise, connectors that are specified at a "typical" insertion loss of 0.5 dB perform in this range on the average, but specific matings can measure 1.5 dB or more.

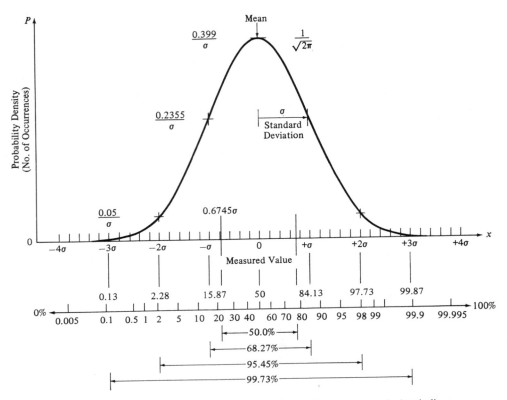

Figure 6.5 Normal distribution. The normal distribution curve is marked to indicate mean and standard-deviation relationships, and the percentage of measured values that can be expected to fall between designated points on the curve.

Ignoring the "statistical" nature of component performance can create extremely overconservative designs. It can also result in much time wasted after a system is installed trying to solve seemingly "out-of-spec" conditions that are not really problems but simply a result of ignoring the statistics of the situation. If, for example, worst-case tolerance specifications for all components in series are simply summed, the result is a design for a condition that would virtually never occur in practice.

Where multiple like components operate in series, such as splices, sections of cable, or couplers and connectors in the case of a databus, their performance must be treated on a statistical basis whenever the statistical data is available.

When components can be specified in terms of mean performance with a standard deviation in tolerance, summation is as follows:

$$\text{Total mean value} = \sum_{i=1}^{n} \overline{X}_i \tag{6-36}$$

$$\text{Total standard deviation} = \sqrt{\sum_{i=1}^{n} \sigma_i^2} \tag{6-37}$$

This approach assumes that component tolerances follow a normal distribution, which is not always the case, but is the most practical assumption we can make.

6.11 LINK-BANDWIDTH ANALYSIS

The fiber-optic link-bandwidth model is shown in Figure 6.6. A link-bandwidth analysis is generally performed in order to determine one of the following: (1) the frequency response of the photonics in order to evaluate the adequacy of the total design; (2) the bandwidth of a particular component in the link (such as the receiver) required to achieve end-to-end bandwidth performance when cascaded with other components; or (3) the magnitude of the band limiting is in the optical portion of the link so that it can be compensated with the appropriate power penalty.

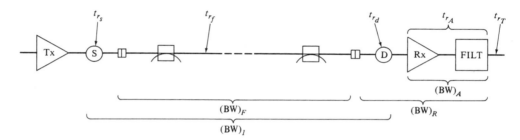

Figure 6.6   Link-bandwidth-analysis parameter definition.

The response of a component or system can be characterized in the time domain in terms of rise time or in the frequency domain in terms of bandwidth. Analysis can be performed from either standpoint, however, it is often more convenient to analyze performance in terms of rise time. Conversion to bandwidth can be performed at the conclusion of the analysis using the approximation [3] from Equation (6-7):

$$(BW)_{3dB} = \frac{0.35}{t_r}$$

Relationships for the bandwidth and rise-time effects of concatenated systems elements were given in Equation (6-25) as

$$t_r = 1.1\sqrt{t_{r1}^2 + t_{r2}^2 + \cdots + t_{rn}^2}$$

Rise-time analysis can be aided by using Table 6.10, which is designed as a worksheet. Generally, only a worst-case analysis is adequate for bandwidth unless there is reason to do otherwise. The worksheet can be used with variations depending on the problem to be solved.

TABLE 6.10 STEP 3A: LINK-BANDWIDTH BUDGET WORKSHEET

Case: _____ ; Conditions: _____ ; Solve for: _____

Required link results:

Transmission Bandwidth, $(BW)_t$ = _____ Hz

Bit Rate, $(BR)_t$ = _____ bps; Encoding = _____

Tx Length, $D_t = D_s(1 + \mu) + D_r$ = _____ = _____ km

Equivalent rise time:

$t_r = 350/\text{bandwidth (MHz)}$ = _____ ns

$t_r = 700/\text{BR(Mbps)}$ for NRZ; $350/\text{BR(Mbps)}$ for RZ = _____ ns

Component Parameter	Total t_r	Total t_r^2
Source response: rise time =	_____	
Fiber multimode dispersion:		
$\quad t_r = 1.087 \times t_{dw}(\text{ns/km}) \times D_t^\gamma(\text{km})$		
$\quad t_r = 1.087 \times (\quad) \times (\quad)^{(\quad)} =$	_____	
Fiber-material dispersion:		
$\quad t_r = 1.087 t_{dm}(\text{ps/nm/km}) \times \Delta\lambda(\text{nm}) \times D_t(\text{km}) \times 10^{-3}$ ns		
$\quad t_r = 1.087(\quad) \times (\quad) \times (\quad) \times 10^{-3} =$	_____	
Receiver response:		
$\quad$ Detector Rise Time (ns) =	_____	_____
$\quad$ Receiver BW Filter: $t_r = 0.35/(BW)_r =$	_____	_____
$\quad$ Sum of the Squares =		=====
Total rise time:		
$\quad t_{r,\text{tot}}(\text{ns}) = 1.1\sqrt{\text{sum of squares}} =$		=====
$\quad (BW)_{\text{tot}}(\text{MHz}) = 350/t_{r,\text{tot}}(\text{ns}) =$		=====

If a single component's response is to be determined in order to achieve end-to-end performance, simply leave that item, t_{rx}, blank on the worksheet and compile the sum of the squares of the remaining component, t_{rn}. Then, knowing the total end-to-end requirement for the photonics t_{rP}, apply the following formula to find that component's required rise time:

$$t_{rx} = \sqrt{\frac{t_{rP}^2}{1.1} - \sum_{1}^{N} t_{rn}^2} \qquad (6\text{-}38)$$

Note that the rise-time/bandwidth relationships assume that all component response can be represented by simple band-pass networks and the fiber optical/electrical bandwidth assumes Gaussian response. This is not always the case. Figure 6.7 shows the range that sources and fiber fall within. A critical analysis would have to adjust to this.

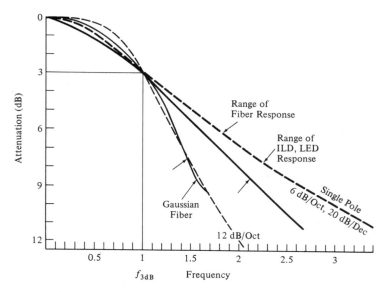

Figure 6.7   Typical ranges of optical-component response.

Case 1: Evaluating end-to-end response.   If the resultant band-pass of the entire photonics and $t_{r,tot}$ from optical source to receiver output are desired, then do the following:

(a) Use the worksheet as is, filling in the data for all components in the link. Solve for total rise time, $t_{r,tot}$, resulting from the series cascade of source, fiber detector, and receiver (or receiver preamp).

(b) Convert the resultant total rise time to bandwidth using Equation (6-7) and compare with the required bandwidth.

(c) Reflect deficiencies as a power penalty (Step 3B), or select new components, or adjust receiver bandwidth to achieve desired results.

Case 2: Determining required receiver bandwidth.   In order to achieve the required photonics bandwidth, $(BW)_t$, the receiver bandwidth can be adjusted (increased beyond $(BW)_t$ to compensate for band limiting of the other optical components. The bandwidth budget is performed in order to solve for optimum receiver-noise bandwidth $(BW)_r$. This parameter is then used in the receiver SNR calculations to find the minimum required detected power (MDP).

(a) Compute the link-bandwidth budget (Table 6.10) for the rise time of the optical and electrooptical link components, t_{rL}, leaving the receiver/preamp rise time blank.

(b) Solve Equation (6-38) for the receiver rise time:

$$t_{rR} = \sqrt{\frac{t_{rP}^2}{1.1} - t_{rL}^2} \tag{6-39}$$

(c) Convert rise time to bandwidth (BW)$_r$ using Equation (6-7).

(d) Assume that this is receiver-noise bandwidth (unless better data is available) and use it to calculate receiver-noise current (Step 3D) or equate the difference between receiver bandwidth (BW)$_r$ and (BW)$_t$ as a power penalty in Step 3B.

Case 3: Determining fiber band-limiting penalty: Receiver with fixed bandwidth. When the receiver bandwidth is fixed, the bandwidth budget is performed on the optical link to solve for the resultant power penalty required to compensate for band limiting. Generally, with a purchased receiver, the receiver is designed and specified slightly wider than the transmitted signal passband, say, (BW)$_r$ = (BW)$_t$/0.7, in order to accommodate some pulse dispersion or fiber band limiting. Any band limiting beyond this point is reflected as a power penalty.

If source and detector responses are negligible or included in the terminal performance, and, therefore, fiber band limiting is the only factor, then simply solve for the two dispersion elements, leaving the rest of Table 6.10 blank. Apply the resultant total to the curves (given by the receiver manufacturer) for power penalty versus signal band limiting (Step 3B).

(a) Solve for total fiber dispersion using Table 6.10 or as follows:

Multimode Dispersion:

$$T_{dw} = t_{dw} \text{ (ns/km)} \times D^\gamma \text{ (km)} \tag{6-40}$$

Material Dispersion:

$$T_{dm} = t_{dm} \text{ (ps/nm/km)} \times \Delta\lambda \text{ (nm)} \times D \text{ (km)} \tag{6-41}$$

$$T_{df} = \sqrt{T_{dw}^2 + T_{dm}^2} \tag{6-42}$$

(b) Convert to the appropriate bandwidth function from Chapter 4, where T_{df} is expressed as FWHM:

$$(BW)_{f,\text{opt}} = \frac{441}{T_{df}} \text{ MHz/ns} \tag{4-40}$$

$$(BW)_{f,\text{elect}} = \frac{312}{T_{df}} \text{ MHz/ns} \tag{4-39}$$

(c) Use this bandwidth with the power penalty curves provided by the receiver manufacturer, or see Section 6.12, to compute the power penalty (Step 3B).

6.12 STEP 3B: POWER PENALTIES

The signal-to-noise performance defined for the photonics layer in Steps 2A.2 or 2D.2 can be applied directly as the receiver SNR requirement only if there is no noise generated elsewhere in the photonics. If noise exists in the form of signal distortion at the transmitter, modal noise in the fiber, crosstalk in couplers, reflections in passive components, or from any other source besides the receiver, then it must either be accommodated by (1) adding signal power or (2) increasing SNR requirements of the receiver to compensate. Table 6.11 summarizes the power penalty design procedure. The remainder of this section provides the details.

6.12.1 SNR Penalties

When we can conveniently relate the noise level directly as received noise power that can be compensated for by adding more signal, then we can add a power penalty (P_p) to the link-power budget to compensate.

When the noise source is a function of signal power (such as a reflection or harmonic distortion) and is presented or measured in terms of a signal-to-noise limit, $(SNR)_n$, a signal-to-distortion ratio (SDR), or a signal-to-crosstalk ratio (SCR), it is more convenient to treat it as an SNR imitation or penalty. The SNR penalty is most accurately accommodated by increasing the $(SNR)_t$ requirement of the receiver to a higher value, $(SNR)_r$, or Q_r in the case of a digital receiver. The method is discussed in Section 6.13.

As a shortcut approximation, we can also treat SNR penalties as power penalties by making some assumptions regarding the MDP relationships. From Chapter 1, Figure 1.9, the total photonics SNR requirement, $(SNR)_t$, was related as a function of receiver $(SNR)_r$ and the noise contribution of the optical link $(SNR)_n$. The relationship for noise power $(SNR)_{RMS/RMS}$ is, from Equation (6-11):

$$\frac{1}{(SNR)_t} = \frac{1}{(SNR)_n} + \frac{1}{(SNR)_r} \qquad (6\text{-}43)$$

Solving for $(SNR)_r$, we get

$$(SNR)_r = \frac{(SNR)_t(SNR)_n}{(SNR)_n - (SNR)_t} \qquad (6\text{-}44)$$

Power penalty is a function of the increase in MDP resulting from changing the SNR requirement from its design value of $(SNR)_t$ to the new value $(SNR)_r$ in order to accommodate link noise related by $(SNR)_n$. The relationship is

$$P_p = 10 \log \frac{(MDP)_r}{(MDP)_t} \qquad (6\text{-}45)$$

where $(MDP)_r$ is the new requirement, and $(MDP)_t$ is the original requirement when $(SNR)_t$ is assigned as the receiver SNR. Referring to the equations for MDP in Section 4.10, we see from Equation (4-101) that MDP varies as a direct function of $(SNR)_{RMS/RMS}$ when quantum noise is dominant and as the square root of $(SNR)_{RMS/RMS}$ when amplifier and detector noise are dominant. Therefore:

$$P_p = Z \log \frac{(SNR)_r}{(SNR)_t} \qquad (6\text{-}46)$$

where $Z = 10$ for quantum-noise dominance, and $Z = 5$ for amplifier-noise dominance. Substituting Equation (6-44) for $(SNR)_r$ in Equation (6-46) and simplifying, we get

TABLE 6.11   STEP 3B: POWER PENALTIES

Analog transmission

(a) Bandlimiting effect on MDP when we set $(BR)_r = (BW)_i$:

$$P_p = -X \log \left[1 - \frac{(BW)_i^2}{(BW)_i^2} \right]$$

$\dfrac{(BW)_{e,\text{sect}}}{(BW)_t} > Y$ for a 1-dB P_p; see Figure 6.9

Dominant Noise:	Thermal	Quantum or Bipolar Shott	FET Channel
X:	2.5	5.0	7.5
Y:	1.3	1.67	1.9

(b) SNR penalties when we set $(SNR)_r = (SNR)_{t,\text{RMS/RMS}}$:

$$P_p = -Z \log \left[1 - \frac{(SNR)_t}{(SNR)_n} \right]$$

$Z = 5$ for amplifier- or detector-noise dominance

$Z = 10$ for quantum-noise dominance (generally APDs)

(c) Band-limiting equalization for FDM (see Section 5.4.2):

High-frequency channels effected: $P_p = \dfrac{\text{Channel } N \text{ power decrease}}{N}$

All channels effected: $P_p \quad = \dfrac{\text{Channel } N \text{ power decrease}}{2}$

Digital transmission:

(a) Optical-link band-limiting:

$\dfrac{(BW)_{e,\text{link}}}{BR} > 0.65$ for a 1-dB P_p; see Figure 6.13

$$P_p = 4.15 \left(\frac{\text{FWHM}}{T} \right)^4 \qquad \text{(approximation)}$$

(b) Compensation penalties when Q_t must remain fixed:

$$P_p = -5 \log \left[1 - \left(\frac{Q_t}{Q_n} \right)^2 \right]$$

(c) Extinction-ratio penalty (see Figure 6.12):

$$ER = 0.1 \text{ approximately for 1-dB } P_p$$

(d) Partition modal noise: Use P_p in (b) above

$$Q_n = \frac{\sqrt{2}}{(\pi (BR) D \sigma_\lambda t_{dm})^2}$$

$$P_p = -Z \log \left(1 - \frac{(\text{SNR})_t}{(\text{SNR})_n} \right) \qquad (6\text{-}47)$$

For digital signaling, Q is given as a function of peak signal to rms noise, $(\text{SNR})_{\text{pk/rms}}$, and, therefore, sums as the square root of the sum of the squares, as indicated in Equation (6-23). From Section 4.10.3, MDP is generally a function of Q since amplifier noise generally dominates. Therefore, from Equation (6-45):

$$P_p = 10 \log \frac{Q_r}{Q_t} \qquad (6\text{-}48)$$

Solving Equation (6-23) for Q_r in the presence of link-noise limitation Q_n is:

$$Q_r = \left(\frac{Q_t^2 Q_n^2}{Q_n^2 - Q_t^2} \right)^{1/2}$$

Substituting this relation for Q_r in Equation (6-48), we get

$$P_p = -5 \log \left[1 - \left(\frac{Q_t}{Q_n} \right)^2 \right] \qquad (6\text{-}49)$$

6.12.2 Analog Modulation Penalties

Figure 6.8 shows the process of electrical-to-optical conversion at the source and the associated signal-to-power relationships when using linear intensity modulation (analog transmission). The bias conditions and linear characteristics of the source cause two sources of signal degradation from the ideal: (1) harmonic distortions in the transmitted signal, and (2) modulation-index limitations on the amount of total power transmitted as signal power. Due to these limitations, some of the power transmitted is not signal power and, therefore, can be considered unwanted noise power.

Harmonic-distortion penalties.   Harmonic distortion is most conveniently treated as a signal-to-distortion ratio (SDR) limit and as such can be related approximately as in Equation (6-47), where SDR is substituted for $(\text{SNR})_n$. Refer to Section 5.4.3 for the derivation of SDR.

Modulation index.   In Figure 6.8, the source is biased with forward current I_b at some midpoint optical power P_o. Signal current $i_s(t)$ modulates the optical power around P_o with a magnitude of P_s peak to peak. The bias point and magnitude of I_s is chosen such that the signal is not affected by the nonlinear regions of the source. At the detector, the composite received signal appears as a photocurrent modulated around some average value corresponding to P_o. Since some of the optical power transmitted is not signal, we can define an optical carrier-modulation index M_o:

$$M_o = \frac{P_s}{P_{\max}} \qquad (6\text{-}50)$$

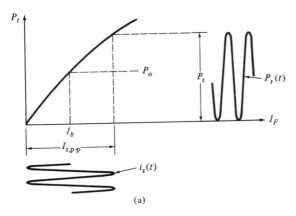

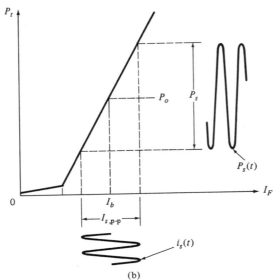

Figure 6.8   Source drive relationships for analog modulation. (a) LEDs and (b) laser diodes.

For analog signaling, this relationship is included directly in the receiver MDP analytical relationships. Therefore, no power-penalty compensation is required if the relationships in this book are used. For a purchased receiver, the specifications may handle M_o differently. From Equation (4-101), MDP is a direct function of $1/M_o$ for amplifier- and leakage-noise dominance, and $1/M_o^2$ for quantum-noise dominance. Therefore, the power-penalty allowance is

$$P_p \text{ (dB)} = X \log \frac{1}{M_o} \tag{6-51}$$

where $X = 10$ for amplifier- and leakage-noise dominance, and $X = 20$ for quantum-noise dominance.

6.12.3 Analog Band-Limiting Penalties

The response limitations of electrooptical components (source and detector) and the dispersion effects within optical fiber create a reduction in signal amplitude with modulation frequency. When we specify MDP (Section 4.10.1), we assume a specific receiver (BW$_r$) intended to permit the signal to pass undistorted. If receiver (BW$_r$) is set at the transmission signal bandwidth (BW)$_t$, then the optical elements would have to offer infinite bandwidth so as not to influence signal power. This is because the finite bandwidth of the optical components cascades with receiver bandwidth to reduce the total signal passband, thus reducing signal level. In order to compensate, receiver bandwidth must be widened or power added at the upper band edge (equalization). Either condition can be represented as a power penalty.

Case 1: Receiver bandwidth extension. From Equation (6-17), the bandwidth of cascaded photonics elements (source, fiber, detector, receiver filter) creates a total bandwidth limit (BW)$_p$ for the photonics:

$$\frac{1}{(\text{BW})_p^2} = \frac{1}{(\text{BW})_s^2} + \frac{1}{(\text{BW})_f^2} + \frac{1}{(\text{BW})_d^2} + \frac{1}{(\text{BW})_r^2} \tag{6-52}$$

If we combine all elements with the exception of the optical receiver filtering as link bandwidth (BW)$_l$, set the photonics-bandwidth requirement as equal to (BW)$_t$, (BW)$_p$ = (BW)$_t$, and solve for the required receiver bandwidth (BW)$_r$, we get

$$(\text{BW})_r = \sqrt{\frac{(\text{BW})_l^2 (\text{BW})_t^2}{(\text{BW})_l^2 - (\text{BW})_t^2}} \tag{6-53}$$

Knowing (BW)$_r$, we can take two approaches to compensate for band limiting:

1. recalculate MDP with the revised receiver bandwidth requirements, (BW)$_r$, and make $P_p = 0$;
2. use the original MDP calculated with (BW)$_r$ = (BW)$_t$, and add a power penalty that approximates the results of added noise due to bandwidth extension.

From the power-penalty (P_p) method, P_p is a function of the increase in MDP from the design value of (MDP)$_t$ to the new value (MDP)$_r$. This is as a result of the increase in receiver bandwidth, from Equation (6-45):

$$P_p = 10 \log \frac{(\text{MDP})_r}{(\text{MDP})_t} \tag{6-45}$$

In observing the MDP equations, Equation (4-101), we see that MDP varies differently as a function of bandwidth depending on which noise term dominates. If quantum noise q dominates, then MDP is a direct function of BW, and, as before:

$$P_p = -5 \log \left[1 - \left(\frac{(\text{BW})_t}{(\text{BW})_l} \right)^2 \right] \tag{6-54}$$

If amplifier noise A dominates, then we can simplify Equation (4-101) to

$$\text{MDP} = \frac{\langle i_{nA} \rangle}{rM_o} \sqrt{\text{SNR}} \qquad (6\text{-}55)$$

Referring to the amplifier-noise equations in Section 4.9, Chapter 4, we see that for the bipolar case, MDP is a direct function of BW and, therefore, P_p is equivalent to Equation (6-54):

$$P_{p,\text{BIP}} = -5 \log \left[1 - \left(\frac{(\text{BW})_t}{(\text{BW})_l} \right)^2 \right] \qquad (6\text{-}56)$$

For the FET case, however, noise power increases as the third power of BW. MDP is, therefore, a function of BW to the $3/2$ power. P_p, therefore, becomes:

$$P_{p,\text{FET}} = -7.5 \log \left[1 - \left(\frac{(\text{BW})_t}{(\text{BW})_l} \right)^2 \right] \qquad (6\text{-}57)$$

Where amplifier thermal noise dominates, MDP varies as the square root of BW and, therefore, the power penalty is

$$P_{p,\text{therm}} = -2.5 \log \left[1 - \left(\frac{(\text{BW})_t}{(\text{BW})_l} \right)^2 \right] \qquad (6\text{-}58)$$

This relationship is plotted in Figure 6.9. Note that for the bipolar case, for example, a 1-dB penalty requires that the link electrical bandwidth be 1.67 times that of the overall requirement of $(\text{BW})_t$ if the receiver is set at $(\text{BW})_t$.

Case 2: Band equalization. Within a signal-frequency spectrum, the higher frequencies are reduced in power by passband band limiting (see Figure 5.9). The higher frequencies of the spectrum can be compensated by adding power at the expense of reducing power to the lower. In Chapter 5, Section 5.4.2, this was discussed with respect to frequency-division multiplexed channels. The approximate power penalties for compensation were determined from Equations (5-28) and (5-29) as

$$P_p = \frac{p_{m(-)}}{N} \text{ for only one of } N \text{ channels affected}$$

$$P_p = \frac{p_{m(-)}}{2} \text{ for all channels affected}$$

where $p_{m(-)}$ is the power reduction due to passband attenuation between upper and lower band edges. If channel m is within the passband, use Figure 6.9 passband characteristics. If it falls beyond, use Figure 6.7 to estimate power reduction.

6.12.4 Extinction Ratio and Intersymbol Interference

As shown in Figure 6.10 for binary signaling, the source is generally biased at a zero or low-power state $P_{(0)}$ and pulsed to a maximum pulse power for the alternate state $P_{(1)}$. With an LED, the value of drive current at $P_{(0)}$ can be zero since the linear

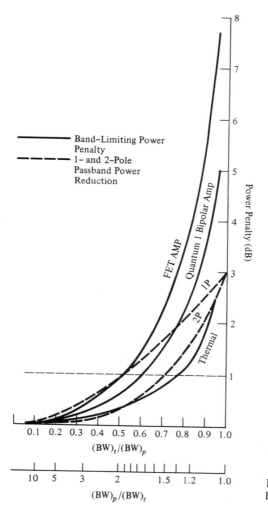

Figure 6.9 Analog transmission band-limiting power penalty.

operating region of an LED extends to zero forward current. With a laser diode, however, the linear operating region only extends down to a region just above the lasing threshold. The value at $P_{(0)}$ depends, therefore, on a bias point that is chosen based on the operating characteristics desired.

When the source is biased at some value other than zero power with LEDs, or above the lasing threshold with laser diodes, then a portion of the received optical power is not signal. The result is similar to the effect of the modulation index in that the signal-to-noise ratio at the receiver is affected by the division of total received optical power into its signal and nonsignal components. The term used for binary transmission, however, is the extinction ratio, which is defined during a signal pulse as

$$ER = \frac{\text{power in low state}}{\text{power in high state}} = \frac{P_{(0)}}{P_{(1)}} \qquad (6\text{-}59)$$

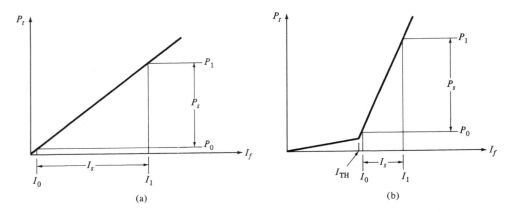

Figure 6.10 Source drive relationships for digital transmitters. (a) LEDs and (b) laser diodes.

At the receiver this becomes

$$\text{ER} = \frac{P_{d,\min}}{P_{d,\max}} = \frac{I_{d(0)}}{I_{ds} + I_{d(0)}} \tag{6-60}$$

where $I_{d(0)}$ = detected photocurrent at the bias point of source, $P_{(0)}$

I_{ds} = peak-to-peak the value of signal photocurrent during a signal pulse

Since the relationships for MDP versus Q (Chapter 4) are based on average received power P_m during a pulse, we have to relate the extinction ratio in these terms and take into account encoding, where pulse width T_p can be smaller than the full bit period T. Signal power per pulse is

$$P_s = P_{(1)} - P_{(0)} \tag{6-61}$$

By taking into account pulse period T_p, mean power P_m is

$$P_m = P_{(0)} + \frac{P_s}{2}\frac{T_p}{T} \tag{6-62}$$

If we define mean energy E_m as

$$E_m = P_m(T) \tag{6-63}$$

the extinction ratio can be written in terms of mean power or energy by substituting Equations (6-63) and (6-62) into Equation (6-59):

$$\text{ER} = \frac{P_{(0)}}{P_{(1)}} = \frac{E_m - E_s/2}{E_m + E_s/2} = \frac{P_m T - (P_s T_p)/2}{P_m T + (P_s T_p)/2} \tag{6-64}$$

In terms of signal power P_s and energy per pulse E_p, signal energy per pulse can be related as

$$E_p = P_s T_p \tag{6-65}$$

and, therefore, the extinction ratio is

$$ER = \frac{E_{p(0)}}{E_{p(1)}} = \frac{P_{(0)}T}{P_{(0)}T + P_s T_p} \tag{6-66}$$

Power penalty due to the extinction ratio can be compared to signal reduction in relation to mean power. A larger mean power contributes to both quantum noise at the detector and modal noise within the source and fiber. The effect of extinction ratio on modal noise is discussed in Section 6.12.6. The magnitude of quantum noise at the detector is a direct function of mean detected power (Chapter 4, Section 4.8). The relationships for receiver MDP in Chapter 4 assumed $P_m = P_s/2$, and, therefore, ER = 0. When using these relationships, the effects of extinction ratio must be treated separately as a power penalty.

Alternatively, we could include the results of the extinction ratio and other pulse-distortion phenomena such as intersymbol interference within the analytical relationships for MDP. Intersymbol interference is caused by pulse dispersion and band limiting, which spreads the power in transmitted pulses time slots into the adjacent time slots. Figure 6.11 illustrates the effect along with that of extinction ratio. Using this and Figure 6.10, we can determine the signal reduction and noise power contribution of ER and ISI.

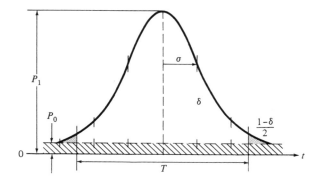

Figure 6.11 Gaussian pulse with intersymbol interference (ISI) and finite extinction ratio. ISI shown at $T = 5\sigma$.

$$P_s = P_{(1)} - P_{(0)} \tag{6-67}$$

$$P_{(0)} = P_m - \frac{P_s}{2} \tag{6-68}$$

$$P_{(1)} = \frac{P_{(0)}}{ER} = \frac{P_m - P_s/2}{ER} \tag{6-69}$$

Substituting Equations (6-68) and (6-69) into Equation (6-67), we get

$$P_s = 2P_m \frac{1 - ER}{1 + ER} \tag{6-70}$$

In the presence of intersymbol interference (ISI), the energy per bit is diminished by the fraction δ and, therefore, the signal power becomes

$$P'_s = \delta P_s = 2P_m \left(\frac{1 - ER}{1 + ER} \right) \delta$$
$$= 2k_1 P_m \tag{6-71}$$

where $k_1 = [(1 - ER)/(1 + ER)]\delta$.

The noise contribution to the signal in pulse period T due to ISI is

$$\langle i_{nl} \rangle = rGP_s(1 - \delta) = 2rGP_m k_2 \tag{6-72}$$

where $k_2 = [(1 - \text{ER})/(1 + \text{ER})](1 - \delta)$.

If we wish to maintain Q_t in the presence of ISI and ER, then

$$Q_t = \frac{I_s'}{2\langle i_n' \rangle} = \frac{rGP_s'}{2\sqrt{\langle i_{nD}^2 \rangle + \langle i_{nA}^2 \rangle + \langle i_{nl}^2 \rangle}} \tag{6-73}$$

$$Q_t = \frac{2sP_m}{2\sqrt{qP_m + l + a + i}} \tag{6-74}$$

where $s = rGk_1$
$q = 2erG^2F(\text{BW})_n$
$l = 2e(G^2FI_{dm} + I_{du})(\text{BW})_n$
$a = \langle i_{nA}^2 \rangle$
$i = (2rGk_2P_m)^2$

Solving the quadratic for P_m, which we define as the minimum required detected power (MDP), we get

$$\text{MDP} = \frac{Q}{2s^2}\left(Qq + \sqrt{\frac{(Qq)^2}{4} + s^2(l + a + i)} \right) \tag{6-75}$$

Substituting the values from Equation (6-74), assigning J_1 and J_2 as noise bandwidth factors, eliminating negligible factors, and simplifying, we get

$$\text{MDP} = \frac{Q}{rk_1}\left(\frac{QeF(\text{BR})J_1}{k_1} + \sqrt{\frac{a + l + i}{G^2}} \right) \tag{6-76}$$

Extinction-ratio penalty.   If there is no ISI, then $\delta = 1$, $k_2 = 0$, and k_1 is

$$k_1 = \frac{1 - \text{ER}}{1 + \text{ER}}$$

Substituting into Equation (6-76) and solving for various values of ER, we can obtain the power penalty versus ER. The solution for a typical PIN/FET and an APD, with $G = 100$ and $F = 4$, is plotted in Figure 6.12. See Garrett and Midwinter [9] and Hooper and White [10] for additional discussions on extinction-ratio penalties.

Intersymbol interference (ISI).   Power penalty (reduction in MDP) at the receiver as a function of intersymbol interference can be obtained by solving the MDP equation, Equation (6-76), for a constant Q, while varying δ and setting ER to zero. If ER = 0, then $k_1 = \delta$ and $k_2 = 1 - \delta$. This case is plotted in Figure 6.13 as the dashed line for a typical PIN/FET case where amplifier noise dominates.

An alternate means of determining approximate power penalty due to ISI is to observe the reduction in energy within pulse period T and the corresponding increase in interference in adjacent pulse periods as rms pulse width increases. This is shown in Figure 6.11 for the case of $\sigma = T/5$ with a Gaussian pulse. If we look at

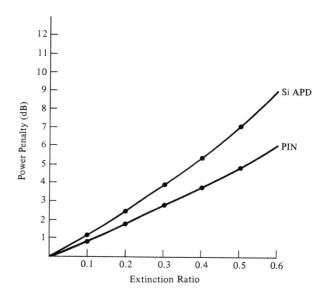

Figure 6.12 Power penalty as a function of extinction ratio.

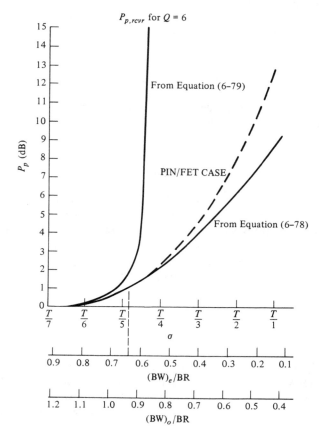

Figure 6.13 Power penalties due to intersymbol interference. Note: $(BW)_e$ and $(BW)_o$ are based on NRZ format.

TABLE 6.12 FRACTION OF ENERGY IN PULSE
WITH PERIOD T DUE TO INTERSYMBOL
INTERFERENCE[a]

$x = \dfrac{T}{\sigma}$	Fraction of energy
0.5	0.1974
1.0	0.3830
1.5	0.5468
2.0	0.6826
2.5	0.7888
3.0	0.8664
3.5	0.9198
4.0	0.9546
4.5	0.9756
5.0	0.9876
5.5	0.9940
6.0	0.9974

[a] $T = X\sigma$

the fraction of energy within a pulse period T as δ (see Table 6.12) and that within the adjacent period as $1 - \delta$, we get an energy-to-noise ratio that (in the absence of other noise) relates to Q as follows:

$$Q = \frac{\text{average signal}}{\text{rms noise}} = \frac{\delta P_{(1)}/2}{(1 - \delta)P_{(1)}/2} = \frac{\delta}{1 - \delta} \tag{6-77}$$

If we define the reduction in SNR at the receiver from the design value of Q_t to a reduced value Q_n as the power penalty, then we have

$$P_p = +5 \log \left[1 + \left(\frac{Q_t}{Q_n} \right)^2 \right] \tag{6-78}$$

A plot of Equation (6-78) for different values of T/X is given in Figure 6.13 for P_p as a function of Q^2. It compares closely to the example for a PIN/FET.

It is often desired to represent power penalty as the amount of penalty incurred by increasing the receiver Q to compensate for Q_n in order to reestablish the design value Q_t for the end-to-end photonics performance. This is also plotted in Figure 6.13 as $P_{p,\text{rcvr}}$. Note that T cannot be much less than $T/4.5$ or the $Q = 6$ condition for a 10^{-9} BER cannot be achieved. The relationship is

$$P_p = -5 \log \left[1 - \left(\frac{Q_t}{Q_n} \right)^2 \right] \tag{6-79}$$

The scales in Figure 6.13 related P_p to bit rate and bandwith using the Gaussian pulse-shape assumptions $(\text{BW})_e = 0.133/\sigma$ and $(\text{BW})_o = 0.187/\sigma$ and the NRZ assumption $\text{BR} = 1/T$ from Section 4.4.2. Note that for a 1-dB power penalty, the following conditions can be used as a conservative rule of thumb for Gaussian pulse shapes at the receiver:

$$(BW)_e > 0.65(BR) \tag{6-80}$$

$$(BW)_o > 0.9(BR) \tag{6-81}$$

A further discussion of intersymbol interference can be found in Gower [11] and Personick [12]. Midwinter [14] determined a simple curve-fitting relationship that gives power penalty directly in decibels based on a $T/2$ rectangular pulse for practical receivers:

$$P_p \, (\text{dB}) = 2\left(\frac{t_e}{T}\right)^4 \tag{6-82}$$

where t_e is the pulse width at the $1/e$ amplitude point of the received pulse (see Figure 4.6). Using the rms width of the rectangular $T/2$ pulse as $\sigma = 0.144T$ and the relationship for a Gaussian received pulse of $\sigma = 0.354t_e$, Midwinter derives the relationship for t_e/T for the received signal pulse as 0.408. Adding the elements of fiber multimode (t_{ew}) and material (t_{em}) dispersion at the $1/e$ points, we get (sum of squares):

$$\left(\frac{t_e}{T}\right)^2 = (0.408)^2 + \left(\frac{t_{ew}}{T}\right)^2 + \left(\frac{t_{em}}{T}\right)^2 \tag{6-83}$$

From Equation (6-82), we can derive the 1-dB P_p as $t_e/T = 0.8409$. If we assume that the concatenation factor on the fiber is 1 for simplicity, substituting the formulas in Table 6.10 for fiber dispersion into Equation (6-83), using the ratio $t_e = 1.2 \, t_{d,\text{FWHM}}$, and solving for D, we can derive the maximum distance:

$$D_{\max} = \frac{1}{\text{BR}} \sqrt{\frac{0.7785 - [t_{dx}(\text{BR})]^2}{t_{dw}^2 + (\Delta\lambda t_{dm})^2}} \tag{6-84}$$

where dispersions t_{dw}, t_{dm}, and t_{dx} are FWHM, and t_{dx} relates to nonfiber band-limiting elements such as the source or detector, if applicable.

6.12.5 Source-Bias Effects on Lasing Delay

When transmitting binary signaling with a laser diode, the source is generally biased at the low- or zero-power state $P_{(0)}$ and pulsed to a maximum power for the alternate state $P_{(1)}$. The linear operating region of a laser, however, only extends down to a region just above the lasing threshold. If the bias is below threshold, the resulting output power waveform is distorted due to a delay in lasing action as the current pulse drives through the nonlasing region below threshold. The effect is shown in Figure 6.14. The delay is characterized as a reduction in pulse width from the ideal value T_p to a value T_p' by a lasing delay time T_d:

$$T_p' = T_p - T_d \tag{6-85}$$

A reduction in pulse width T_p, within a bit period T, can be directly translated into a reduction in power or energy per bit. This reduction, in turn, can be translated into a power penalty ($-\text{dB}$):

$$P_p = 10 \log \frac{\text{actual energy/pulse}}{\text{desired energy/pulse}} \tag{6-86}$$

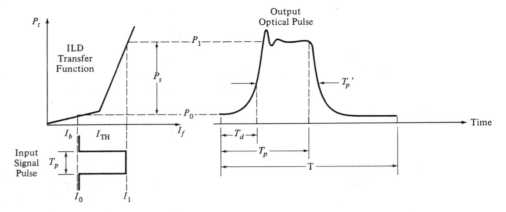

Figure 6.14 Effects of lasing delay on pulse shape.

By substituting Equation (6-65) into Equation (6-86), we get

$$P_p = 10 \log \frac{P_s(T_p - T_d)}{P_s T_p} = 10 \log \frac{T_p - T_d}{T_p} \tag{6-87}$$

In Chapter 3, Section 3.3, we saw that source delay time T_d in the spontaneous emission region (LED) was a function of $\tau \ln$ (drive current ratio). For the laser, this can be characterized by

$$T_d = \tau \ln \frac{\text{total signal current}}{\text{signal current in lasing region}}$$

From Figure 6.16, this becomes:

$$T_d = \tau \ln \frac{I_s}{I_s + I_b - I_{TH}} \tag{6-88}$$

where I_s = peak (zero to maximum) pulse signal current driving the laser
   I_b = bias point (zero signal point) of the laser forward current, i.e., $P_{(0)}$
   I_{TH} = value of the drive current at the lasing threshold
   Δ = slope of the laser transfer function in the lasing region = $\Delta P_s / \Delta I_s$
   τ = carrier lifetime of the device

In addition, T_d changes on a pulse-by-pulse basis, depending on the state of the previous bit and the size of T with respect to decay time between pulses while the laser is in the lower state $P_{(0)}$. This variation in pulse width with varying binary signals is known as "patterning." The degree of patterning depends on not only the characteristics of the laser (slope, threshold), but on the design of the transmitter, whether it maintains constant mean power or constant bias. Transmitter design, at this level of detail, is beyond the scope of this book. See Garrett and Midwinter [9], who have given an in-depth analytical treatment to the subject of patterning penalties and transmitter design.

   In nearly all transmitters, the parameters of bias and signal level are controlled by an optical feedback loop to maintain constant mean power. This ensures near-constant operation as the laser varies with time and temperature. Power penalties occur as a result of nonoptimum control and signal patterns. Figure 6.15 shows the general effect that laser bias along with changes in threshold and signal magnitude might have on power penalty.

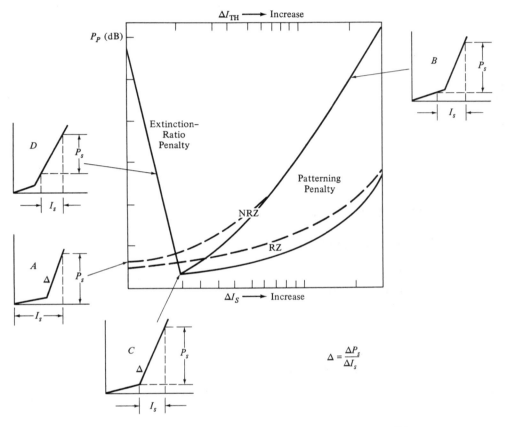

Figure 6.15   Effect of source-bias conditions on power penalty [9].

Case 1: Zero bias.   In some designs, pulse-signal current level is controlled to maintain constant mean power, and I_b is fixed at zero. For this case, power penalty is represented as condition A in Figure 6.15. If either laser threshold (I_{TH}) or laser slope (Δ) increase, turn-on delay and patterning penalties increase. This is also apparent from Equations (6-87) and (6-88).

Case 2: Constant drive current.   In some designs, pulse current I_s is held constant by the drive electronics, and the mean output power P_m is held constant by the drive and feedback circuitry. In this case, the power penalty is represented as a function of slope Δ and pulse current I_s. For low P_m and high I_s, the value for $P_{(0)}$ may fall below threshold, illustrated as condition B in Figure 6.15, creating a lasing delay similar to the case for zero bias.

If the laser is driven such that $I_b = I_{TH}$, shown as condition C in Figure 6.15, then any turn-on delay approaches zero ($T_d = 0$). At this point, the power penalty due to delay or patterning also approaches zero ($P_p = 0$). There can be some minor extinction-ratio penalty due to a nonzero value of $P_{(0)}$ at I_{TH}, however.

If P_m is large and I_s is relatively small, then I_b may exceed I_{TH}, as in condition D. In this case, there is no turn-on delay or patterning penalties, but extinction-ratio penalties increase with decreasing I_s.

The effect that photonics design and operating environment have on power penalty can be summarized as follows:

(a) Laser threshold: A low I_{TH} is desirous for low power penalty.

(b) Time: As lasers age, I_{TH} increases more than slope decreases, causing increased power penalty.

(c) Temperature: I_{TH} increases as a strong function of temperature, causing an increase in power penalty. Thermoelectric cooling helps to reduce this effect.

(d) Signal format: Reducing pulse width (half-width RZ, for example) reduces the effects of patterning since it gives the laser time between pulses to recover.

6.12.6 Modal Noise

Modal noise in a fiber is generally caused by the following:

(a) Speckle noise: This is an interaction at an interface, such as a connector, between wavefronts of multiple modes traveling with different propagation times in a fiber.

(b) Partition noise: This is the effect of increasing modal wavefront interference with distance due to intramodal or material dispersion.

Speckle noise.   With speckle noise, if all of the modes are coupled by the receiving fiber, there is no resultant noise. However, if partial (mode-selective) coupling occurs, then amplitude-modulated noise occurs as a function of the number of modes interacting and the contrast between the resultant patterns. The resultant noise is a function of contrast between patterns at the mode-selective joint. The more modes transmitted and propagating, the lower the contrast and, likewise, the noise. Modes can be increased by choosing a multiline laser or, in binary systems, biasing the laser around threshold, where multiple modes are stimulated.

A large extinction ratio decreases SNR due to speckle noise since it increases total power (and, therefore, noise) while reducing the percentage of signal power. Figure 6.16 assumes a single-mode laser coupled to a multimode fiber and shows the trend in optical power penalty resulting from increasing number of spectral sources (increasing number of modes) and extinction ratio. The plots were derived from data given by Rawson, Goodman, and Norton in [17].

Partition noise.   Partition noise in single-mode fibers can be characterized as well by its effect on power penalty. In a binary system, the limitation on the ability to maintain a received peak-signal-to-rms-noise ratio (characterized by factor Q_t), due to partition-noise contribution, is from (4-75):

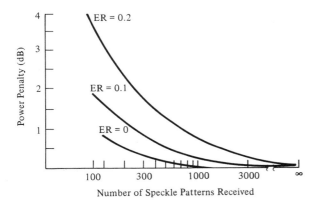

Figure 6.16 Increase in power penalty due to model "speckle" noise in a multimode binary transmission system.

$$Q_n = \frac{\sqrt{2}}{[\pi\,(\mathrm{BR})D\sigma_\lambda t_{dm} \times 10^{-12}]^2} \tag{6-89}$$

where BR = bit rate, bps

   D = length of fiber, km

   σ_λ = rms-source spectral width, nm

   t_{dm} = material dispersion, ps/nm/km

Knowing the required Q_t and the SNR limit of partition noise Q_n, we can calculate power penalty P_p from

$$P_p = -5 \log \left[1 - \left(\frac{Q_t}{Q_n} \right)^2 \right] \tag{6-90}$$

where Q_t is that required to achieve the desired received bit-error rate (for example, $Q_t = 6$ for a 10^{-9} BER).

If the transmission format is analog, then, since $2Q = i_{s,\mathrm{p\text{-}p}}/\langle i_n \rangle$ and BW $\approx$ BR/2, the relationship is

$$\frac{i_{s,\mathrm{pp}}}{\langle i_n \rangle} = \frac{0.5\sqrt{2}}{[\pi\,(\mathrm{BW})D\sigma_\lambda t_{dm} \times 10^{-12}]^2} \tag{6-91}$$

To find the power penalty, use Equation (6-47) if the previous SNR is converted to rms. If the SNR is compared in terms of the peak-to-peak signal, then Equation (6-90) can be used where $(\mathrm{SNR})_{\mathrm{p\text{-}p/rms}}$ is substituted for Q.

6.13 STEP 3C: SNR PENALTY AND RECEIVER (SNR)$_r$

Noise sources or signal degradation (or distortion) can be treated directly as power penalties, as discussed in the last section. Another method of handling these noise sources, particularly when they are measured in relation to signal level, is to treat them as an SNR reduction. If a noise source is related in terms of signal-to-noise ratio, signal-to-distortion ratio, or such, it is more accurate to treat it as an SNR relationship than to use the power-penalty approximations.

Noise sources that are more properly handled in this fashion include the following:

(a) harmonic distortion generated at the source and measured in terms of decibels below the carrier level or carrier-to-distortion ratio (CDR);

(b) modal noise that is a function of signal power and measured as signal-to-modal-noise ratio (SMR);

(c) crosstalk in WDM or directional couplers measured in terms of isolation (decibels below signal level) or signal-to-crosstalk ratio (SCR); and

(d) reflections in passive optical components measured with reference to forward signal power as signal-to-reflection ratio (SRR).

Since $(SNR)_t$ for the photonics layer is fixed by the higher layers, we must increase the required SNR for the receiver to a value $(SNR)_r$ in order to compensate for the SNR limit, $(SNR)_n$, imposed by noise in the link. The relationship, assuming rms noise power, is

$$\frac{1}{(SNR)_r^2} = \frac{1}{(SNR)_t^2} - \frac{1}{(SNR)_n^2} \tag{6-92}$$

where $(SNR)_n$ results from the sum of all transmitted noise components, SMR, SDR, SRR, SCR, and so forth:

$$\frac{1}{(SNR)_n^2} = \frac{1}{(SMR)^2} + \frac{1}{(SDR)^2} + \cdots + \frac{1}{(SCR)^2} \tag{6-93}$$

Table 6.13 represents these expressions in terms of decibels.

TABLE 6.13   STEP 3C: SNR PENALTIES[a]

Analog transmission:

$$(SNR)_r = -10 \log \left(10^{-(SNR)_t/10} - 10^{-(SNR)_n/10}\right)$$

$(SNR)_t$ = required transmission SNR of phototonics

where $(SNR)_n$ is the SNR limitation due to distortion and intermodulation (SDR), modal noise (SMR), reflections (SRR), crosstalk (SCR), etc.

$$(SNR)_n = -10 \log \left(10^{-SDR/10} + 10^{-SMR/10} + 10^{-SRR/10} + 10^{-SCR/10}\right) \text{ in dB}$$

Digital transmission:

$$Q_r = \frac{Q_n Qt}{\sqrt{Q_n^2 - Qt^2}}$$

where Q_t = Q required for transmission

Q_n = Q degradation from source and link noise similar to the analog case = $(SNR)_{n,avg/rms}$, $(SCR)_{avg/avg}$, or $(SRR)_{avg/avg}$

[a] $(SNR)_r$ or Q_r is the revised receiver performance necessary to compensate for link-introduced noise related as $(SNR)_n$ or Q_n.

In a digital system, the signal-to-noise ratio is measured in terms of Q from Equation (4-48):

$$Q = \frac{\text{peak signal}}{2(\text{rms noise})} = \frac{i_{s,\text{pk}}/2}{\langle i_n \rangle}$$

where $i_{s,\text{pk}}$ is the difference between peak signal and threshold. As with SNR, the relationship of received Q_r to Q_t in the presence of noise-source limitation Q_n is

$$\frac{1}{Q_t^2} = \frac{1}{Q_n^2} + \frac{1}{Q_r^2} \qquad (6\text{-}94)$$

Solving for the resultant receiver Q_r to compensate for Q_n, we get

$$Q_r = \frac{Q_n Q_t}{\sqrt{Q_n^2 - Q_t^2}} \qquad (6\text{-}95)$$

where Q_n could be the result of crosstalk, reflections, or modal noise defined as

$$Q_n = (\text{SNR})_{n,\text{av/rms}}$$

6.14 STEP 3D: MINIMUM REQUIRED RECEIVED OPTICAL POWER

In order to complete the link-budget analysis, receiver sensitivity must be determined under the operating conditions specified. Receiver sensitivity is defined in terms of the minimum received power (MRP) required to achieve the $(\text{SNR})_r$, $(\text{CNR})_r$, or $(\text{BER})_r$ for a particular signal format (as determined by the section encoding) and under specific environmental conditions. MRP is measured and defined as the power coupled into the optical aperture of the receiver. This is generally referenced to the input aperture of the connector attached to the receiver.

The minimum received power (MRP) required is defined in terms of the minimum detected power (MDP) required and any power loss (L_d) incurred in coupling from the aperture into the detector itself:

$$\text{MRP}\,(-\text{dBm}) = \text{MDP}\,(-\text{dBm}) + L_d\,(\text{dB}) \qquad (6\text{-}96)$$

The loss mechanism (L_d) between the receiving end of the optical fiber (or fiber pigtail) and the detector is generally a reflection and an area mismatch loss. When the detector contains an integral optical pigtail, this loss is generally very small and is often accommodated in the detector responsivity (r) specification as a lower value for responsivity than expected for the detector alone. Detector coupling loss (L_d) was covered briefly in Chapter 3.

MDP was derived in Chapter 4 along with the contribution of detector and receiver noise and the performance characteristics of various receiver designs. A summary is provided in Tables 6.14 and 6.15.

TABLE 6.14 STEP 3D: DETERMINE MINIMUM REQUIRED RECEIVED OPTICAL POWER (MRP)

Methodology

1. If the receiver is purchased, determine MRP specifications using $(SNR)_r$, $(CNR)_r$ or Q_r and $(BW)_r$ or $(BR)_t$.
2. If the receiver MRP is analytically derived: (a) determine detector and amplifier noise current, then (b) compute MDP, then (c) MRP assuming fiber-to-detector coupling loss (L_d)

$$MRP = MDP + L_d$$

MDP calculation

Analog intensity modulation:

$$MDP = \frac{eF(SNR)_r(BW)}{rM_o^2\langle s^2\rangle}\left(1 + \sqrt{1 + \frac{M_o^2\langle s^2\rangle\{\langle i_{nA}^2\rangle + \langle i_{nL}^2\rangle\}}{[eFG(BW)]^2(SNR)_r}}\right)$$

where BW is the receiver noise bandwidth, generally set at $(BW)_t$ for single-channel transmission and $(BW)_c$ for FDM.

Carrier modulation:

$$MDP = \frac{2eF(CNR)_c(BW)_c}{r}\frac{N^2}{M_o^2}\left[1 + \sqrt{1 + \frac{M_o^2}{N^2}\frac{\{\langle i_{nA}^2\rangle + \langle i_{nL}^2\rangle\}}{2[eFG(BW)_c]^2(CNR)_c}}\right]$$

where $(BW)_c$ is the channel bandwidth of each of N channels.

Digital signaling:

$$MDP = \frac{Q_r}{r}\left[Q_r eF(BR)J_1 + \sqrt{\frac{\langle i_{nA}^2\rangle}{G^2} + 2eI_{dm}F(BR)J_2}\right]$$

where	NRZ	RZ
$J_1 =$	0.5	0.5
$J_2 =$	0.563	0.403

6.15 STEP 3E: LINK-POWER BUDGET

The objective of the optical link-power-budget analysis is to determine whether the proper components and design parameters were chosen, such that the signal power reaching the detector/receiver is in the range to achieve performance requirements. In order for the analysis to represent practical conditions, component-parameter variations and tolerances must be taken into account. This can be achieved by (a) using statistical data to the extent possible, and (b) analyzing the design for best-case and worst-case conditions.

6.15.1 Performance Relationships

The analysis is basically a matter of adding power gains and losses as the signal propagates down the optical link. The design meets performance requirements if (a)

TABLE 6.15   STEP 3D: AMPLIFIER NOISE

Thermal-noise dominant (voltage amplifier):

$$\langle i_{nA}^2 \rangle = \frac{4kT\,(\text{BW})}{R_i}$$

Bipolar preamplifier:

Analog signaling:

$$\langle i_{nA}^2 \rangle = \frac{4kT}{R_i}\text{BW} + \frac{8\pi kTCo}{\sqrt{\beta}}(\text{BW2})^2$$

Binary signaling:

$$\langle i_{nA}^2 \rangle = \frac{4kT}{R_i}(\text{BR})J_2 + \frac{8\pi kTCo}{\sqrt{\beta}}(\text{BR})^2\sqrt{J_2 J_3}$$

FET Preamplifier:

Analog signaling:

$$\langle i_{nA}^2 \rangle = \frac{4kT}{R_i}\text{BW} + \frac{16\pi^2 kT\tau C_i^2}{g_m}(\text{BW3})^3(1 + fc/f)$$

Binary signaling:

$$\langle i_{nA}^2 \rangle = \frac{4kT}{R_i}(\text{BR})J_2 + \frac{16\pi^2 kT\tau C_i^2}{g_m}(\text{BR})^3 J_x J_3$$

where BR = transmission bit rate, $(\text{BR})_t$, bps

BW = noise BW, use signal BW, $(\text{BW})_t$ or $(\text{BW})_c$, Hz

$$(\text{BW2})^2 = \frac{f_h^2 - f_l^2}{2}$$

$$(\text{BW3})^3 = \frac{f_h^3 - f_l^3}{3}$$

R_i = amplifier input or feedback resistance, ohms

C_t = total capacitance (detector + stray + transistor), farads

f_c = $1/f$ noise corner frequency = 10 to 50 MHz

	J_f	J_2	J_3
NRZ	0.184	0.563	0.0868
RZ	0.0984	0.403	0.0361

g_m = transconductance $\approx$ 0.03

$$J_x = 1 + \frac{J_f f_c}{(\text{BR})J_3}$$

k = Boltzmann constant = 1.38×10^{-23} J/K

T = temperature, degrees Kelvin = 273 + °C

β = transistor current gain (h_{fe})

τ = channel-noise factor = 1.8 for GaAs MESFETs

systems gain G_s is greater or equal to systems loss and (b) received power is less than that which creates saturation and distortion at the receiver.

Figure 6.17 identifies the component-performance parameters that make up the link-power budget. Figure 6.18 graphically represents the relationship between the component power and loss parameters. The controlling parameter on the transmitting end is the optical source power coupled into the fiber (P_t). This power varies with device type and coupling efficiency, as shown for the LED and ILD. The controlling parameter on the receiving end is the minimum required received power (MRP) necessary to achieve received signal performance. This also varies with

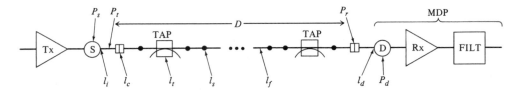

Figure 6.17   Link-power budget parameters.

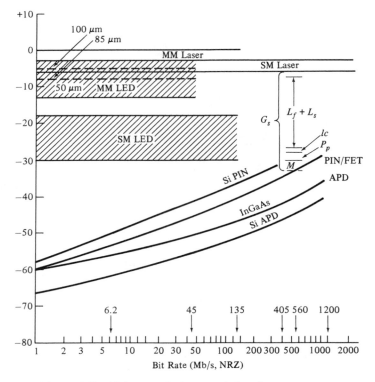

Figure 6.18   Link-power-budget practical performance ranges.

device type (PIN or APD) and wavelength of operation. Since receiver noise increases with bandwidth, MRP also increases (sensitivity decreases) with bandwidth or data rate.

Systems power gain G_s is the difference in optical power between transmitted coupled power P_t and required received optical power MRP.

$$G_s = P_t - \text{MRP} \tag{6-97}$$

An alternate definition excludes the source and detector to fiber coupling losses and defines source power P_s and required detected power MDP at the semiconductor device interface:

$$G_s' = P_s - \text{MDP} \tag{6-98}$$

Systems gain, therefore, defines the amount of optical power loss L_o that can be tolerated and how much margin is available for drift and degradation. Optical power margin is the excess power remaining when the optical power losses (expressed as positive values) are subtracted from the systems gain:

$$M = G_s - L_o \tag{6-99}$$

Some of this margin, called excess margin (P_e), must be reserved for unexpected variations in component performance.

$$P_e = G_s - L_o - M - P_p \tag{6-100}$$

Optical power losses L_o result from:

(a) fiber attenuation L_f;
(b) connector insertion loss L_c;
(c) splice attenuation L_s;
(d) coupler port-to-port insertion loss L_t;
(e) source-power insertion loss L_i and coupling loss L_d at the detector for unterminated devices.

By combining all margin components as one factor M and detailing the loss components, the total link-power-budget relationship becomes

$$P_e = (P_t - \text{MRP}) - (L_c + L_t + L_f + L_s) - M - P_p \tag{6-101}$$

Or for noncoupled sources and detectors:

$$P_e = [(P_s - L_i) - (\text{MDP} + L_d)] - (L_c + L_t + L_f + L_s) - M - P_p \tag{6-102}$$

The optical loss equation and, therefore, the link-power budget can be applied to almost any link architecture with some variation in its form. Figure 6.19 shows the form that the optical loss relationship might take for various common architectures.

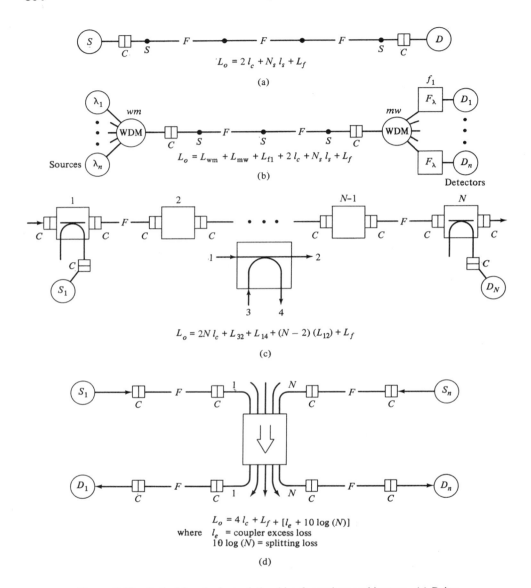

Figure 6.19 Optical loss-budget relationships for various architecture. (a) Point-to-point simplex link, (b) point-to-point wavelength multiplexed, (c) Tee bus, and (d) star bus.

6.15.2 Statistical Relationships

As mentioned in Section 6.10, statistical parameters should be used to the degree possible in the analysis. Figure 6.20 shows how the statistical relationships between component performance interact in the link-budget analysis.

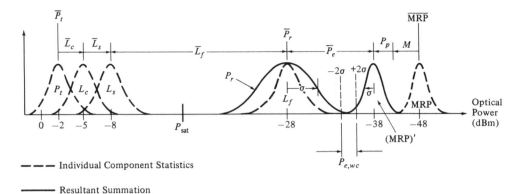

Figure 6.20   Statistical representation of link-power budget.

Worst-case power performance for a component can be defined as

$$P_{wc} = \overline{P}_{\text{mean}} - 2P_{\text{std dev}} \qquad (6\text{-}103)$$

Total mean power performance is calculated as the sum of the mean values for each component measured at the operating conditions being analyzed.

$$\overline{P}_{\text{tot}} = \overline{P}_1 + \overline{P}_2 + \cdots + \overline{P}_n \qquad (6\text{-}104)$$

Total standard deviation is the square root of the sum of the squares of the 1σ values:

$$\sigma_{tot}^2 = (\sigma_1^2 + \sigma_2^2 + \cdots + \sigma_n^2)^{1/2} \qquad (6\text{-}105)$$

For an individual component, the standard deviation for performance generally considers not only normal manufacturing tolerances (σ_{tol}), but variations with environment and aging (σ_{env}) and measurement uncertainties (σ_{meas}):

$$\sigma_{tot}^2 = (\sigma_{tol}^2 + \sigma_{env}^2 + \sigma_{meas}^2)^{1/2} \qquad (6\text{-}106)$$

Worst- and best-case conditions are generally defined as the 2σ (97.73% probability) or 3σ (99.87% probability) values, depending on the convention used. Refer to Figure 6.5.

Substituting Equation (6-101) in Equation (6-103) for gain and loss components, a worst-case (2σ) analysis becomes

$$P_{e,wc} = [(\overline{P}_t - \overline{\text{MRP}}) - \overline{L}_o - \overline{M} - \overline{P}_p] - 2[\sigma_{P_t}^2 + \sigma_{\text{MRP}}^2) + \sigma_{L_o}^2]^{1/2} \qquad (6\text{-}107)$$

and, likewise, a mean ($+2\sigma$) best-case analysis becomes

$$P_{e,bc} = \{(\overline{P}_t - \overline{\text{MRP}}) - \overline{L}_o\} + 2[\sigma_{P_t}^2 + \sigma_{\text{MRP}}^2) + \sigma_{L_o}^2]^{1/2} \qquad (6\text{-}108)$$

Table 6.16 provides a worksheet for the link-power-budget analysis. One such analysis is performed for each anticipated operating condition (worst, best, and nominal).

TABLE 6.16   STEP 3E: LINK-POWER-BUDGET WORKSHEET

Operating Conditions: _____

Cable Length D: Span D_s: _____   Uncertain μ: _____   Margin D_r: _____

               $D_s(1 + \mu) + D_r =$ _____   +   _____   =   _____ km

Couple Power P_t	$\bar{P}_s$	$-$	$L_{coupler}$	$=$	$\bar{P}_t$		$\bar{P}_t$	σ_{P_t}
Nominal (dBm)	_____	$-$ _____		$=$ _____		$=$	_____ *	_____ *
Minimum (dBm)	_____	$-$ _____		$=$				
	$-$ _____	Temperature	$-$ _____	Lifetime $=$ _____				

Power Loss L_o	Type	N Units	Mean l_o	Std Dev	$N \times$ Mean	$N\sigma^2$
L_c connector	_____	_____	_____	_____	_____ *	_____ *
		_____	_____	_____	_____ *	_____ *
	Derating	_____	_____	_____		
L_t coupler		_____	_____	_____	_____ *	_____ *
		_____	_____	_____	_____ *	_____ *
L_s splice	$D_s(1 + \mu)/D_f + 1 =$ _____	_____	_____	_____	_____ *	_____ *
	N_r repair	_____	_____	_____		
	Derating	_____	_____	_____		
L_f fiber	$D_s(1 + \mu) =$ _____	_____	_____	_____	_____ *	_____ *
	D_r repair	_____	_____	_____		
	Temperature	_____	_____	_____		
	Wavelength	_____	_____	_____		
	Measurement	_____		_____		_____ *

	Mean	1 Standard Deviation
Loss total (dB):		
$\bar{L}_{o,wc}$, $1\sigma_{wc} =$	_____	_____
*$\bar{L}_{o,bc}$, $1\sigma_{bc} =$	_____ *	_____ *
Received power (dBm):		
$P_{r,min} = \bar{P}_{t,min} - \bar{L}_{o,wc}$, $1\sigma_{wc} =$	_____	_____
Note: sum * for best case $P_{r,max} = \bar{P}_{t,max} - \bar{L}_{o,bc}$, $1\sigma_{bc} =$	_____ *	_____ *

Receiver MRP (dBm):	$\overline{MDP}$	$+$	L_{coup}	$=$	$\overline{MRP}$		$\overline{MRP}$	σ_{MRP}
Nominal:	_____	$+$ _____		$=$ _____		$=$	_____ *	
		Temperature			Wavelength			
Worst Case: $\overline{MRP}$ +	_____	Derating	$+$ _____		Derating	$=$	_____	

Margin + penalty (dB):		M, P_p
	Operating Margin M_{op}	_____
	Band-Limiting Penalty	_____
	Modal-Noise Penalty	_____
	Extinction Ratio and Other P_p	_____
	Total $M + P_p =$	_____

Excess Margin P_e (dB):	Mean $\bar{P}_{e,wc}$	Standard Deviation $\sigma_{Pe,wc}$

$$\text{Worst-Case Mean} = \bar{P}_{t,min} - \bar{L}_{o,wc} - (MRP)_{wc} - M - P_p = \underline{\qquad}$$

$$\text{Standard Deviation} = 1\sigma_{P_e} = (\sigma_{P_t}^2 + \sigma_{L_{o,wc}}^2 + \sigma_{MRP}^2)^{1/2} = \underline{\qquad}$$

6.15.3 Component-Performance Relationships

Chapter 3 provides an in-depth discussion of the individual parameters that make up the individual component characterization. A summary is provided here for worst- and best-case conditions (mean ± 2 standard deviations). If statistical parameters do not exist (and worst-/best-case do) for some components, use worst/best case in lieu of mean ± 2 sigma for those, but continue with the statistical approach for other components. Note that although the statistical component relationships are given independently in what follows, when applied to the link-power budget, all standard deviations for all components are summed together as the square root of the sum of their squares, as shown in Equation (6-105).

Coupled power (P_t). $P_{t,wc}$ is the worst-case end of life (calculated on an MTBF basis) and $P_{t,bc}$ is the best-case average source power coupled into the transmitting aperture before the connector (generally, a pigtail fiber) under the range of environmental conditions specified. P_t is based on a random 50% duty cycle pulse for NRZ or format specific for others. In any case, it must be consistent with MDP relationships. If statistical values are provided, they can be used here, but be sure to include environmental effects ($\sigma_{P_t,T}$) as well.

Minimum required received power (MRP). $(MRP)_{wc}$ is the maximum worst-case and $(MRP)_{bc}$ is the minimum best-case, mean optical power required to be coupled into the optical aperture of the receiver (generally, the pigtail after the connector) for specified operation at the worst-/best-case source center wavelength and within the range of environmental conditions specified. MRP is based on the same signal format as P_t and must be consistent with MDP relationships. $(MRP)_{wc}$ is defined for $(SNR)_r$ or Q_r and, therefore, includes the adjustments for all SNR penalties, $(SNR)_p$. $(MRP)_{bc}$ may exclude $(SNR)_p$ if appropriate. Note that power penalties (P_p) are handled as a separate parameter. Statistical parameters can be used here if available, but be sure to include environmental effects ($\sigma_{MRP,T}$) as well.

Connector loss (L_c).   Connector loss is specified as follows, where N_c is the number of connectors, $\bar{l}_c$ is the mean loss per connector, and σ_c is the standard deviation for connector loss. For best-/worst-case analysis, these are the best-/worst-case values over the allowable range of environmental conditions. Worst case is

$$L_{c,wc} = N_c \bar{l}_c + 2(N_c \sigma_c^2)^{1/2} \qquad (6\text{-}109)$$

The best-case relationship has the same form except that the 2-sigma term is subtracted.

Coupler (tap) loss (L_t).   Insertion loss is specified similarly to that of the connector, where N_{pp} is the number of transitions between similar coupler input and output ports, $\bar{l}_{pp}$ is the mean port to port loss, and σ_{pp} is the standard deviation of port-to-port loss. The relationship should include the best-/worst-case source center-wavelength variations ($\bar{l}_{pp,w}$) and temperature variations ($\bar{l}_{pp,T}$), if applicable. Generally, such tolerance parameters as uniformity are specified with the mean loss parameter (see Chapter 3). Worst case is

$$L_{t,wc} = N_{pp}\bar{l}_{pp} + 2(\bar{N}_{pp}\,\sigma_{pp}^2)^{1/2} \tag{6-110}$$

Best case reflects a minus 2-sigma relationship.

Splice loss (L_s).   Splice loss is a function of the number of splices (N_s) that can be derived from the total length:

$$N_s = \text{number of cable spans} + 1 \tag{6-111}$$

This accounts for the pigtail splices at the ends at the terminal or patch panel. A worst-case analysis would include an allowance for an estimated number of repair splices (N_r) based on route vulnerability and determined in splice pairs. N_r is a minimum of 2. Loss parameters include unit mean loss ($\bar{l}_s$) at room temperature, and standard deviations for intrinsic and extrinsic losses (σ_s) and for temperature effects ($\sigma_{s,T}$), if any. Worst-case loss is

$$L_{s,wc} = (N_s + N_r)\bar{l}_s + 2[\sigma_s^2 + \sigma_{s,T}^2)(N_s + N_r)]^{1/2} \tag{6-112}$$

Best-case loss assumes N_r is not present and has a minus 2-sigma relationship.

Fiber loss (L_f).   Fiber mean loss ($\bar{l}_f$) and standard deviation loss (σ_f) are generally specified in terms of decibels/kilometer measured at room temperature on the reel for a specific wavelength. Total loss is, therefore, a function of total length for all spans spliced together within a section:

$$D = (1 + \mu)D_s + D_r \tag{6-113}$$

where μ is the measurement uncertainty between route distance and actual cable length, D_s is the route distance between terminals or repeaters, and D_r is the cable-length reserve assumed for repair.

The total cable loss parameter must also take into effect the mean ($\bar{l}_{f,w}$) and standard deviation ($\sigma_{f,w}$) change in cable loss (dB/km) due to a different operating wavelength (worst-case source variation), the mean ($\bar{l}_{f,T}$) and standard deviation ($\sigma_{f,T}$) change in loss (dB/km) due to temperature, and the uncertainty in cable loss measurement by the supplier ($\sigma_{f,l}$) expressed in decibels/kilometer. The total worst-case loss is

$$L_{f,wc} = [(1 + \mu)D_s + D_r](\bar{l}_f + \bar{l}_{f,w} + \bar{l}_{f,T})$$
$$+ 2\{[(1 + \mu)D_s + D_r](\sigma_f^2 + \sigma_{f,w}^2 + \sigma_{f,T}^2 + \sigma_{f,l}^2)\}^{1/2} \tag{6-114}$$

The best-case relationship would use -2 sigma, omit D_r and μ, and use only the best-case mean variations with wavelength and temperature.

Power penalties (P_p).   Power penalties are generally derived as mean values although, where the data exists, they could just as well be derived as best- and worst-case values. For worst-case analysis, use worst-case penalties. For best-case analysis, either omit all P_p or use best-case P_p.

Margins (M).   Margins are used for all power or loss variations that were not accounted for in the individual component parameters. Although the previous relationships accounted for variations, often the specific data are not available, and, therefore, a margin must be assumed to account for them. The size of the margian depends on the degree of uncertainty expected in the calculations. If not accounted for before, margins might include:

(a) nonquantified component drift with temperature (M_T);

(b) nonquantified component degradation with age (M_a);

(c) performance variation of fiber attenuation, detector responsivity, and coupler isolation with wavelength (M_w);

(d) normal operating margin (M_{op}) generally specified at 3 dB; and

(e) repair margin for future maintenance action (M_r).

For a worst-case analysis, use worst-case mean values, derating factors, and include an operating margin (M_{op}) as a minimum (usually 3 dB), and any other margins to account for uncertainties in the analysis. For a best-case analysis, exclude all margins and use best-case mean values.

6.16 STEP 3F: SYSTEMS-PERFORMANCE TEST

The following performance tests determine whether the design is adequate. The tests are basically a calculation of performance margins. The statistical representation of the test parameters given here is shown in Figure 6.21. If any of the tests indicate inadequate or marginal design, the most appropriate component parameter(s) should be adjusted (Steps 2A, 2B, and 3D) and the analysis performed again. Table 6.17 provides a summary of this analysis in a worksheet format.

(a) Excess power (P_e) is the power remaining after all losses, power penalties, and operating margins are taken into account. It is derived directly from the worksheet of Table 6.16.

$$P_e = \overline{P}_e \pm 2\sigma_{P_e} \tag{6-115}$$

Assuming statistical parameters exist for P_t and MRP, the relationship is

$$P_e = (\overline{P}_t - \overline{MRP} - \overline{L}_o - M - P_p) \pm 2(\sigma_{P_t}^2 + \sigma_{MRP}^2 + \sigma_{L_o}^2)^{1/2} \tag{6-116}$$

(b) Saturation margin (M_{sat}) is defined as the power margin between best-case received power (P_{max}) and the specified receiver saturation power level (P_{sat}).

$$M_{sat} = \overline{P}_{sat} - \overline{P}_{r,max} \tag{6-117}$$

where $P_{r,max} = \overline{P}_{t,max} - \overline{L}_{o,bc} + 2\sigma_{bc}$.

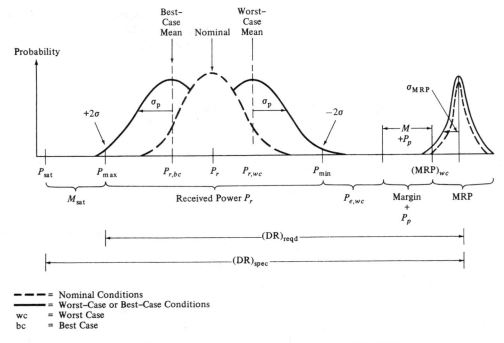

Figure 6.21   Statistical representation of received optical power.

(c) Dynamic range (DR): The required dynamic range is defined as the range of received optical power over which the receiver must operate without degradation.

$$(DR)_{reqd} = \overline{P}_{rmax} - \overline{P}_{rmin} + 4\sigma_{P_{r,wc}} \qquad (6\text{-}118)$$

The specified dynamic range for the receiver is the difference between the saturation power and worst-case MRP required for proper operation.

$$(DR)_{spec} = P_{sat} - (\overline{MRP})_{wc} \qquad (6\text{-}119)$$

(d) Operational probability factor (OP) defines the degree of confidence (probability) for satisfactory operation. It can be performed only when statistical data is used in the link budget. The percent probability is determined by calculating the number of standard deviations between a mean performance and some limit, such as minimum power or saturation.

OP expressed in the number of standard deviations for minimum performance is

$$(OP)_{min} = \frac{\text{worst-case mean } P_e}{P_e \text{ standard deviation}} \qquad (6\text{-}120)$$

Knowing $(OP)_{sat}$, in $N\sigma$, and using Figure 6.5, we can determine the percent probability that all systems gain values remain equal to or greater than all losses, margins,

TABLE 6.17   STEP 3F: LINK-PERFORMANCE TEST[a]

Worst-case excess power margin:

$$P_{e,\text{wc}} = \overline{P}_{e,\text{wc}} - 2\sigma_{P_{e,\text{wc}}}$$

$$P_e = (P_{t,\min} - \overline{L}_{o,\text{wc}} - (\text{MRP})_{\text{wc}} - M - P_p) - 2\sigma_{P_{e,\text{wc}}}$$

Saturation margin (M_{sat}):

$$M_{\text{sat}} = P_{r,\text{sat}} - P_{r,\max}$$

$$M_{\text{sat}} = P_{r,\text{sat}} - (\overline{P}_{r,\max} + 2\sigma_{P_r,\text{bc}})$$

Dynamic range (DR):

Required:

$$(\text{DR})_{\text{reqd}} = (\overline{P}_{r,\max} - \overline{P}_{r,\min}) + 4\sigma_{P_r,\text{wc}}$$

Specified:

$$(\text{DR})_{\text{spec}} = P_{\text{sat}} - (\overline{\text{MRP}})_{\text{wc}} + 2\sigma_{\text{MRP}}$$

Operational probability (OP):

Use Figure 6.5 to determine %OP.

For minimum performance:

$$(\text{OP})_{\min} = \frac{\text{mean } P_{e,\text{wc}}}{1\sigma_{P_e,\text{wc}}} = \underline{\quad} \text{ standard deviations}$$

$$(\text{OP})_{\min} = \underline{\quad} \text{ sigma} = \underline{\quad} \text{ \% probability of success}$$

OP for Nonsaturation:

$$(\text{OP})_{\text{sat}} = \frac{P_{r,\text{sat}} - \text{mean } P_{r,\max}}{1\sigma_{P_r,\text{bc}}}$$

$$(\text{OP})_{\text{sat}} = \underline{\quad} \text{ sigma} = \underline{\quad} \text{ \% of probability of nonsaturation}$$

[a] Parameter values listed can be obtained from Table 6.16.

and penalties.

OP for nonsaturation becomes

$$(\text{OP})_{\text{sat}} = \frac{P_{r,\text{sat}} - \text{mean } \overline{P}_{r,\max}}{P_r \text{ standard deviation}} \tag{6-121}$$

Knowing $(\text{OP})_{\text{sat}}$ (in $N\sigma$) and using Figure 6.5, we can determine the percent probability that all received power remains at or below saturation limits.

6.17 SYSTEMS DESIGN EXAMPLE: PCM/TDM CASE

In order to demonstrate the methodology, a design example will be performed. The case to be selected will be a high quality video transmission trunk, using PCM encoding and TDM multiplexing. The use of a complex signal such as video will fully demonstrate the technique. The trunking system is illustrated in Chapter 7, Figure 7.19. See Chapter 2 for a discussion of the video signal.

6.17.1 Stage 1: Define End-To-End Systems Performance (Table 6.2)

Step 1A.1: Define service parameters.

Transmission distance = 60 km
Video $(SNR)_{b,in}$ = 70 dB (ppwb/rms, w)
Video $(SNR)_{b,out}$ = 60 dB (ppwb/rms, w)
                              CCIR white-noise weighting
Video baseband channel BW, $(BW)_b$ = 4.2 MHz
Number of channels = 8
Stereo audio $(SNR)_{RMS/RMS}$ = 60 dB (RMS/RMS)
Audio bandwidth = 20 kHz per channel

Step 1A.2: Calculate end-to-end performance [Equation (6-3)].

$$(SNR)_{b,sys} = -10 \log (10^{-(SNR)_{out}/10} - 10^{-(SNR)_{in}/10})$$
$$= -10 \log (10^{-60/10} - 10^{-70/10}) = 60.46 \text{ dB}$$

Step 1B: Design systems architecture (Table 6.3). The functional block
diagram of all layers is shown in Figure 6.1 and a list of processing choices is given in Table 6.3. The signal processing chosen for this example is as follows:

(a) Path-layer modulation/MUX:

Pulse-code modulation (PCM)
Stereo sound PCM encoded and multiplexed with video
Path overhead = 2.4 kb/s bandwidth data channel

(b) Line-layer multiplexing:

Time-division modulation (TDM)
number of channels = 8
No line overhead except multiplexer framing

(c) Section layer:

NRZ Scrambling only, no overhead

Step 1C: Design path layer, determine intermediate-channel performance.

Step 1C.1: Design Path Layer.

PCM encoding: Linear quantizing for video and audio separate sound PCM encoding

TDM MUX of video and two stereo

Frame overhead is 100 kb/s including 2.4 kb/s data

Step 1C.2: Unit Conversion (Table 6.4). Convert the baseband signal parameters to terms that can be used with the PCM transfer-function formulas. See Chapter 5 for tables and Chapter 2 for video conversion factors. From Equation (6-4), Table 2.3, and Equation (2-9), for the picture portion only, we get

$$(\text{SNR})_{\text{RMS/RMS}} = \text{SNR (baseband term)} - (\text{CF})_{\text{RMS}} - (\text{WF})_{\text{BW}}$$
$$\text{CF} = \text{baseband-to-rms conversion} = 7.66 \text{ dB}$$
$$\text{WF} = \text{CCIR white-noise weighting improvement} = 6.2 \text{ dB}$$
$$(\text{SNR})_{\text{RMS/RMS}} = (\text{SNR})_{\text{ppwb/rms},w} - 7.66 \text{ dB} - 6.2 \text{ dB}$$
$$= (\text{SNR})_{\text{ppwb/rms},w} - 13.86 \text{ dB}$$
$$= 60.46 \text{ dB} - 13.86 \text{ dB} = 46.6 \text{ dB}$$

For the audio portion, $(\text{SNR})_{\text{RMS/RMS}}$ was defined as 60 dB.

Step 1C.3: Intermediate-Channel Performance (Table 6.4). Using Table 6.4 or 5.5 (Chapter 5), calculate the channel parameters required at the line layer.

$$\frac{\text{Number of bits}}{\text{sample}} = n = \frac{(\text{SNR})_{b,\text{RMS/RMS}} - 1.8}{6}$$

For the video, we get

$$n = \frac{46.6 - 1.8}{6} = 8 \text{ bits encoding}$$

$$(\text{BR})_c = 2.35n(\text{BW})_b = 2.35(8)(4.2 \text{ MHz}) = 78.96 \text{ Mb/s}$$

and for the audio, we get, per channel:

$$n = \frac{60 - 1.8}{6} = 10 \text{ bits encoding}$$

$$(\text{BR})_c = 2.35n(\text{BW})_b = 2.35(10)(20 \text{ kHz}) = 470 \text{ kb/s}$$

Multiplexing one video and two audio plus a 100-kb/s frame overhead that contains a 2.4-kb/s data channel results in a data rate per channel of

$$(\text{BR})_c = 78.96 + 2(0.47) + 0.1 = 80 \text{ Mb/s}$$

The channel BER desired at the line-layer $(\text{BER})_c = 10^{-9}$.

Step 1D: Design line-layer multiplexing (Table 6.5), determine optical section performance.

Step 1D.1: Design Multiplexing.

Number of Channels: 8 video/audio/data channel groups
Multiplexer: TDM

Line overhead: 64-kb/s voice channel

Frame overhead: 6 bits per 8-channel group to include 64-kb/s data channel and section scrambling/error detection

Output coding: NRZ

Step 1D.2: Determine Section Performance.   From Table 6.5, or Chapter 5, section 5.5.2, the required bit-error rate and bit rate at the section layer are

$$(\text{BER})_{ts} = (\text{BER})_c = 10^{-9}$$

$$(\text{BR})_{ts} = \frac{N(\text{BR})_c}{\text{FEF}}$$

$$\text{FEF} = \frac{N(\text{BPC})}{N(\text{BPC}) + \text{FOH}}$$

$$= \frac{8(8)}{8(8) + 6} = 0.91429$$

$$(\text{BR})_{ts} = \frac{8(80)}{0.91429} = 700 \text{ Mb/s}$$

6.17.2 Stage 2: Section Design

Step 2A: Design optical section layer, determine photonics performance (Section 6.6.1).

Step 2A.1 Design Section Layer.

Section overhead: None

Frame encoding: Scrambled NRZ with parity-error detection coding to be achieved in the line-termination-equipment multiplexer

Step 2A.2 Determine Preliminary Photonics Performance.   Since the section-layer encoding remained NRZ, and all scrambling and error detection was performed in the frame overhead of the line-layer multiplexer, photonics performance is the same as that defined for the section layer in Step 1D.2:

$(\text{BR})_t = 700 \text{ Mb/s}$

$(\text{BER})_t = 10^{-9}$

Encoding: NRZ scrambled for zero DC component

Transmission distance $D_s = 60 \text{ km}$

If it is determined that this performance cannot be achieved without a repeater, then individual-span photonics performance is revised in Steps 2C and 2D.

Step 2B: Photonics-layer preliminary design.

Step 2B.1: Functional Design.   Simple point-to-point link.

Step 2B.2: Select Technology (Figure 3.1)

Fiber: Single-mode
Wavelength: 1300 nm
Source: ILD
Receiver: PIN/FET

Step 2B.3: Component-Performance Specification (Table 6.6).   See Table 6.18.

Step 2C: Preliminary performance estimate, estimate number of repeaters (Table 6.7).   A first rough estimation of repeater spacing is done to determine whether design and component choices are adequate and to determine how many repeaters may be necessary. If satisfactory, the more detailed link-budget analysis follows. If not, then the design can be revised before continuing.

(a) Fiber bandwidth: For proper operation (1 dB or less compression), the total transmission distance (from Table 6.7) for an NRZ signal, assuming fiber dispersion dominates ($t_{dx} = 0$), must be less than or equal to

$$D = \frac{[0.7785/(t_{dw}^2 + t_{dm}^2 \Delta\lambda^2)]^{1/2}}{BR}$$

where $t_{dw} = 0$ for single-mode
$t_{dm} = 3 \text{ps/nm/km}$ for the worst case
$\Delta\lambda = 2.5$ nm for the worst case
$BR = 700 \times 10^6$

$$D = \frac{1}{7 \times 10^8} \sqrt{\frac{0.7785}{5.625 \times 10^{-23}}} = 168 \text{ km}$$

There does not appear to be a bandwidth limitation.

(b) Link-power budget: Use worst-case component specifications for the systems-gain calculation ($P_t - MRP$) and the worst-case mean values for optical losses (mean plus deratings). Add a large operating margin (6 dB or so) and assume a 1-dB P_p by definition. If receiver MRP must be calculated, either do so at this time (Step 3D) or use design curves (Figure 3.3) or vendor data to estimate MRP for this initial calculation. From Table 6.7 or Equation (6-9), and by using values from Table 6.18, total span distance SD is

TABLE 6.18 DEVICE PARAMETERS

Source:	Nominal	Standard deviation	Worst
Wavelength (nm)	1300	±10	
Coupled Power (dBm)	0		−3
Derating	−1 dB for time and temperature		
Rise time (ns)	0.2		0.3
Spectral width (nm)	2		2.5 (1 nm rms)
Extinction ratio	0.05		0.1

Detector: PIN	Nominal		Worst
Rise time (ns)	0.2		0.3
Responsivity	0.6 at 1300 nm		
Gain, F	1		
I_{du} (A)			10^{-9}
C_{det} (pF)			1.0
Coupling loss (dB)			0.5

Preamp: GaAs FET			
R_1(ohms)	10^6		
Noise Factor	1.8		
C_a (pF)	1.0		
g_m (mhos)	0.03		
Saturation (dBm)	−10		

Connector:	Mean	Standard deviation	Worst
l_c (dB) sm/mm	0.5	0.33	1.0
sm/sm	0.7	0.33	1.5

Splice:	Mean	Standard deviation	
l_s (dB)	0.15	0.1	

Fiber:	Mean	Standard deviation	Worst
l_f (dB/km)	0.35	0.1	
Temperature derating	0.05 dB/km at −10°C		
Measurement uncertainty (dB/km)		0.1	
Reel length (Df)	3 km	±10% max	2.7 km
Dispersion (ps/nm/km)			
1300 nm	2.0		3.0
1500 nm	17.0		20.0

$$SD = \frac{P_t - \text{MRP} - N_c l_c - N_s l_s - P_p - M}{l_f}$$

$$= \frac{(-3 - 1) - (-30) - 2(1.5) - 21(.15) - 1 - 6}{0.35 + 0.05}$$

$$= \frac{12.85}{0.4} = 32.125 \text{ km}$$

If this distance was short by a few kilometers, we would go forward with the detailed analysis assuming that it might be successful. In this case, however, the difference is too great. The other option is to redesign using a different set of components or a different technology (1550 nm, for example). In this case, we use the chosen design.

$$N = \text{INT}\left(1 + \frac{D_{\text{tot}}}{D_{\text{span}}}\right)$$

$$= \text{INT}\left(1 + \frac{60}{32.125}\right) = 2 \text{ section spans required}$$

Step 2D: Calculate single-section performance.

Step 2D.1: Cascaded-Section Performance (Table 6.8).   From Table 6.8, Equation (6-20), and Figure 6.3, we get

$$(\text{BER})_{\text{sect}} = \frac{(\text{BER})_{ts}}{N}$$

$$= \frac{10^{-9}}{2} = 5 \times 10^{-10}$$

The Q_t for $(\text{BER})_{\text{sect}} = 6.2$ from Figure 4.9.

Step 2D.2: Revised Photonics Performance.   From Step 2A, the section layer added nothing; therefore, the revised photonics performance is the same as in Step 2D.1 for the single section.

$Q_t = 6.2$
$(\text{BR})_t = 700 \text{ Mb/s}$
$D_s = 30 \text{ km route distance}$
Code: NRZ scrambled

6.17.3 Stage 3: Refine Photonics-Layer Design (Section 6.10)

Step 3A: Link-bandwidth budget (Table 6.10).   For a digital transmission system, we are generally concerned with the band-limiting (ISI) effect that link components might have on receiver sensitivity (power penalty). Figure 6.13 helps determine the power penalty once we find $t_{d,\text{tot}}$ and/or $(\text{BW})_t$. See Table 6.19 for the link-bandwidth analysis. Although we are usually concerned with only the fiber band-limiting effect (with purchase transceivers), we include the effects of source and detector response time here for the analytical case. The bandwidth budget uses worst-case conditions and maximum fiber length D_{max}.

$$D_{\text{max}} = D_s(1 + \mu) + D_r = 30(1 + 0.1) + 2 = 35 \text{ km}$$

$$(\text{BW})_{t,\text{link}} \text{ (MHz)} = \frac{350}{t_r} = \frac{350}{0.56} = 625 \text{ MHz}$$

TABLE 6.19   LINK RISE-TIME BUDGET

Case: <u>Worst</u>; Conditions: <u>Nominal</u>

Solve For: <u>Optical Span Rise Time</u>

Required systems results:

       Transmission Bandwidth $(BW)_t = \underline{N.A.}$ Hz

       Bit Rate $(BR)_t = \underline{700 \times 10^6}$ BPS; Encoding $= \underline{NRZ}$

       Length $D_{max} = D_s(1 + \mu) +$

       $D_r = 30(1 + 0.1) + 2 = 35$ km

Equivalent rise time:

$$t_r = \frac{700}{BR(Mbps)} \text{ for NRZ} = \frac{700}{700} = 1.0 \text{ ns}$$

Component parameter	Total T_r	Total T_r^2
Source response: rise time =	0.3	0.09
Fiber multimode dispersion:		
$\quad t_r = 1.087 \times t_d(\text{ns/km}) \times D_t^\gamma (\text{km})$		
$\quad = 1.087 \times (0) \times (35)^1 =$	0.0	0.0
Fiber-material dispersion:		
$\quad t_r = 1.087 \times t_d(\text{ps/nm/km}) \times \Delta\lambda \text{ (nm)} \times D_t \text{ (km)} \times 10^{-3} \text{ ns}$		
$\quad = 1.087(3)(2.5)(35) \times 10^{-3} =$	0.285	0.08
Receiver response:		
$\quad$ Detector Rise Time =	0.3	0.09
$\quad$ Receiver Noise BW =	N.A.	N.A.
$\qquad\qquad$ Sum of the Squares =		0.26
Total Rise Time:		
$\quad t_{r,\text{tot}} = 1.1\sqrt{\text{sum of squares}} =$		0.56 ns
$\quad (BW)_{\text{tot}} = \dfrac{350}{t_{r,\text{tot}} \text{ (ns)}} = \dfrac{350}{0.56} = 625$ MHz		

Step 3B: Calculate power penalties (Table 6.11).

Bandlimiting.   Figure 6.13 is set up to determine power penalty as the ratio of optical link bandwidth to bit rate. Using this plot and the results from Step 3A:

$$\frac{(BW)_{e,\text{link}}}{(BR)_t} = \frac{625}{700} = 0.893$$

$$P_p = \text{negligible} = 0 \text{ dB}$$

Partition Noise.   This could more accurately be handled as an SNR penalty (Step 3C) since it is related in terms of Q, however, we use the approximation in this example for P_p. From Equations (6-90) and (6-89), or Table 6.11, we get

$$Q_n = \frac{\sqrt{2}}{[\pi (BR)D\sigma_\lambda t_{dm}]^2}$$

$$= \frac{1.414}{[\pi (700 \times 10^6)(35)(1)(3 \times 10^{-12})]^2}$$

$$= 26.5$$

where D is in km, σ_λ is in nm, and t_{dm} is in ps/nm/km.

$$P_p = -5 \log \left[1 - \left(\frac{Q_t}{Q_n} \right)^2 \right]$$

$$= -5 \log \left[1 - \left(\frac{6}{26.5} \right)^2 \right] = 0.12 \text{ dB}$$

Extinction Ratio.   Using the plot of Figure 6.12 with ER $= 0.1$ from Table 6.18: $P_p = 0.9$ dB.

Step 3C: Compute SNR penalty and Q_r (Table 6.13).   Since we treated partition noise as a power penalty above we do not need to include it here. In this example we will assume negligible source noise or reflections and no crosstalk therefore $Qr = Qt = 6.2$.

Step 3D: Using Q_r and (BR)$_t$, calculate MDP.   Tables 6.20 and 6.21 summarize the computation of MDP. If the receiver is a purchased product, this computation is not necessary; the value for MRP is found on the vendor's data sheet as a function of bit rate. If the receiver is to be designed analytically, then computation of MDP is necessary. It may also be necessary to compute receiver noise from the

TABLE 6.20 REQUIRED DETECTED OPTICAL POWER (MDP) RELATIONSHIPS (DIGITAL SIGNALING)—FROM TABLE 6.14

Receiver Sensitivity (MDP) for a PIN/FET:

$$\text{MDP} = \frac{Q}{r} \left[Qe(\text{BR})J_1 + \sqrt{\frac{\langle i_{nA}^2 \rangle}{1} + 2eI_{du}(\text{BR})J_2} \right]$$

where, for NRZ signaling, $J_1 = 0.5$, and $J_2 = 0.563$.

$$\text{MDP} = \frac{6.2}{0.6} \Bigg[(6.2)(1.6 \times 10^{-19})(700 \times 10^6)(0.5)$$

$$+ \sqrt{\frac{(5.34 \times 10^{-15})}{1} + (3.2 \times 10^{-19})(10^{-9})(700 \times 10^6)(0.563)} \Bigg]$$

$$= (10.3)[(3.47 \times 10^{-10}) + \sqrt{(5.34 \times 10^{-15}) + (1.26 \times 10^{-18})}]$$

$$= 7.56 \times 10^{-7}$$

or

$$\text{MDP (dBm)} = 10 \log [(7.56 \times 10^{-7})1000] = -31.21 \text{ dBm}$$

TABLE 6.21 AMPLIFIER NOISE FOR AN FET PREAMPLIFIER BASED ON DIGITAL TRANSMISSION—FROM TABLE 6.15

Conditions: Nominal
Noise power:

$$\langle i_{nA}^2 \rangle = \frac{4kT}{R_i}(BR)J_2 + \frac{16\pi^2 kT\tau C_t^2}{g_m}(BR)^3 J_x J_3$$

$$= \frac{(5.52 \times 10^{-23})(298)}{(1 \times 10^6)}(700 \times 10^6)(0.563)$$

$$+ \frac{(2.18 \times 10^{-21})(298)(1.8)(2 \times 10^{-12})^2}{(0.03)}(700 \times 10^6)^3(1.15)(0.0868)$$

$$= (6.48 \times 10^{-18}) + (5.33 \times 10^{-15}) = 5.34 \times 10^{-15}$$

where K = Boltzmann constant = 1.38×10^{-23} J/K

T = temperature in degrees Kelvin = 273 + °C = 298K at room temperature

R_i = amplifier input resistance = 1 megohm

C_t = total capacitance = $C_d + C_a$ = 1.0 + 1.0 = 2 pF

BR = bit rate

τ = channel noise factor = 1.8 for GaAs FETs

g_m = transconductance = 0.03

$$J_x = \left(1 + \frac{J_f f_c}{(BR)J_3}\right)$$

$$= 1 + \frac{0.184 \times 50}{700 \times 0.0868} = 1.15$$

f_c = $1/f$ corner frequency of the FET = 50 MHz

J_f, J_2, J_3 = noise bandwidth factors; for NRZ: J_f = 0.184, J_2 = 0.563, and J_3 = 0.0868

Note: FET noise dominates.

device parameters, as shown in Table 6.21. For a truly rigorous analysis, MDP should be computed at both temperature extremes. From Table 6.20:

$$MDP = -31.21 \text{ dBm}$$

Step 3E: Compute the link-power budget. Table 6.22 illustrates the computation of link-power budget. A 10% distance uncertainty is assumed for span length. A 2-km cable repair-length increase is also assumed, as might be required for a small reroute, as well as eight repair splices.

Step 3F: Test results for desired operation. Table 6.23 illustrates the performance tests performed to determine adequacy of design. In this case the design margin was adequate and therefore no changes are required unless cost optimization requires it.

TABLE 6.22 STEP 3E: LINK-POWER-BUDGET WORKSHEET

Operating conditions: Room-temperature electronics, outside plant
Cable length D: Span D_s: 30 km; uncert. μ: 10% margin D_r: 2 km
$D_s(1 + \mu) + D_r = 35$ km

Coupled Power P_t	$\bar{P}_s$	$-L_{\text{coup}}$ =	$\bar{P}_t$		$\bar{P}_t$	σ_{P_t}
Nominal (dBm):	ns	$-$ ns =	0.00	=	0.00 *	ns *
Minimum (dBm):	ns	$-$ ns =	-3.00 min			
Derate (dB):	-0.50 Temperature		-0.50 lifetime	=	-4.00	

Power loss L_o (dBm):	Type	N Units	Mean l_o	Standard Deviation	$N \times$ Mean	$N\sigma^2$
L_c connector	sm/sm	1.0	0.70	0.33	0.70*	0.109*
	sm/mm	1.0	0.50	0.33	0.50*	0.109*
	Derating	2.0	0.0	0.0	0.00	0.000
L_t coupler		0.0				
		0.0				
L_s splice	$33/2.7 + 1 =$	13.0	0.15	0.10	1.95*	0.130*
	N_r Repair	8.0	0.15	0.10	1.20	0.080
	Temperature	19.0	0.0	0.0	0.00	0.000
L_f fiber	$D_s(1 + \mu) =$	33.0	0.35	0.10	11.55*	0.330*
	D_r repair	2.0	0.35	0.10	0.70	0.020
	Temperature	35.0	0.05	ns	1.75	
	Wavelength	35.0	minimal	ns	0.00	
	Measure	35.0		0.10		0.350*

Loss Total (dB):		Mean	1 Standard Deviation
	$\bar{L}_{o,wc} 1\sigma_{wc} =$	18.35	1.062
	$\bar{L}_{o,bc} 1\sigma_{bc} =$	14.70*	1.014*

Received power (dBm):

		Mean	1 Standard Deviation
$P_{r,\min} = \bar{P}_{t,\min} - \bar{L}_{o,wc} - 1\sigma =$		-22.35	-1.062
$P_{r,\max} = \bar{P}_{t,\max} - \bar{L}_{o,bc} + 1\sigma =$		$-14.70*$	$+1.014*$

Receiver MRP (dBm):	$\overline{\text{MDP}}$	$+ L_{\text{coup}} =$	$\overline{\text{MRP}}$	$(\overline{\text{MRP}})$	σ_{MRP}
Nominal:	-31.21	$+$ 0.50 =	-30.71	-30.71	ns
Worst Case:	$\overline{\text{MRP}} + 1.00$ Temperature $+$		0.00 Wavelength =	-29.71	

Margin and Penalty:		M, P_p
	Operating Margin M_{op}	3.00
	Band-Limiting Penalty	0.00
	Modal-Noise Penalty	0.12
	Extinction Ratio and Other P_p	0.90
	Total $M + P_p =$	4.02

Excess Margin P_e		Mean $\bar{P}_{e,wc}$	Standard Deviation $\sigma_{Pe,wc}$
Worst-Case Mean $= \bar{P}_{t,\min} - \bar{L}_{o,wc} - (\overline{\text{MRP}})_{wc} - M - P_p =$		3.34	
Standard Deviation $= 1\sigma_{P_e} = (\sigma^2_{P_t} + \sigma^2_{L_{o,wc}} + \sigma^2_{\text{MRP}})^{1/2} =$			1.062

TABLE 6.23   SYSTEMS-PERFORMANCE TEST EXAMPLE

Excess power (P_e): Worst-Case Analysis:

$$P_{e,\text{wc}} = \overline{P}_{e,\text{wc}} - 2\sigma_{P_{e,\text{wc}}}$$

$$= (3.34) - 2(1.062) = 1.216 \text{ dB} \qquad \text{OK if the result} > \text{zero}$$

Saturation margin (M_{sat}): Best-Case Analysis Values:

$$M_{\text{sat}} = P_{\text{sat}} - P_{r,\max} = P_{\text{sat}} - (\overline{P}_{r,\max} + 2\sigma_{Pr,\text{bc}})$$

$$= (-10) - [-14.70 + 2(1.014)] = 2.672 \text{ dB; OK}$$

Dynamic range (DR): Use Best-Case Values:

Required Dynamic Range:

$$(\text{DR})_r = [\overline{P}_{r,\max} - \overline{P}_{r,\min}] + 4\sigma_{P_{r,\text{wc}}}$$

$$= [-14.10 - 22.35 + 4(1.062)$$

$$= 7.65 + 1.062 = 11.9 \text{ dB}$$

Specified Dynamic Range for the Receiver:

$$(\text{DR})_{\text{spec}} = P_{\text{sat}} - (\overline{\text{MRP}})_{wc} + 2\sigma_{\text{MRP}}$$

$$= (-10) - (-29.7) = 19.7 \text{ dB; OK if} > (\text{DR})_r$$

Operational probability (OP):

OP for Minimum Performance:

$$(\text{OP})_{\min} = \frac{\text{mean } P_{e,\text{wc}}}{1\sigma_{P_{e,\text{wc}}}} = \frac{3.34}{1.062} = 3.15 \text{ standard deviations}$$

$$(\text{OP})_{\min} = 3.15 \text{ standard deviations} = 99.9\% \text{ probability of success}$$

OP for Nonsaturation:

$$(\text{OP})_{\text{sat}} = \frac{P_{\text{sat}} - \overline{P}_{r,\max}}{1\sigma_{P_{r,\text{bc}}}} = \frac{-10 - (-14.70)}{1.014}$$

$$= 4.64 \; \sigma = 99.999\% \text{ probability of success}$$

REFERENCES

1. American National Standards Institute, *American National Standard for Telecommunications Digital Hierarchy Optical Interface Specifications: Single Mode*, ANSI/ECSA TIX1/ 87-126R1, New York, 1988.

2. Electronic Industries Association, *Single-Mode Fiber Optic Transmission Design*, EIA Document TSB 20, Washington, DC.

3. J. Millman and H. Taub, *Pulse Digital and Switching Waveforms*, McGraw-Hill, New York, 1965, pp. 138–139.

4. D. R. Smith, *Digital Transmission Systems*, van Nostrand Reinhold, New York, 1985, pp. 39–60, 330–335.

5. C. Byrne, B. Karofin, and D. Robinson, "Systematic Jitter in a Chain of Digital Regenerators," *Bell System Technical Journal* **42,** 6 (November 1963).

6. H. Rowe, "Timing in a Long Chain of Binary Regenerative Repeaters," *Bell Syst. Tech. J.* **37,** 6 (November 1958).

7. D. Dutweiler, "Waiting Time Jitter," *Bell System Technical Journal* **15,** 1 (January 1972).

8. American National Standards Institute, *Optical Fiber Digital Transmission Systems Considerations for Users and Suppliers,* ANSI/EIA TSB-19-1986, New York, 1986.

9. I. Garrett and J. Midwinter, "Optical Communication Systems," in M. J. Howes and D. U. Morgan (Eds.), *Optical Fibre Communications,* Wiley, New York, 1980, pp. 251–282.

10. R. C. Hooper and B. R. White, "Digital Optical Receiver Design for Non-Zero Extinction Ratio Using Simplified Approach," *Optical Quantum Electronics* **10** (1978): 279–282.

11. J. Gower, *Optical Communications Systems,* Prentice-Hall International, London, 1984, pp. 453–454.

12. S. D. Personick, "Receiver Design for Digital Fiber Optic Communications Systems, *Bell System Technical Journal* **52** (July–August, 1973): 843–874.

13. Rockwell International, *Product Design for LTS-3139 Lightware Transmission System,* Richardson, TX, 1985.

14. J. E. Midwinter, *Optical Fibers for Transmission,* Wiley, New York, pp. 372–373.

15. Western Electric Technical Publications, *Telecommunications Transmission Engineering, Vol. 1: Principles,* Winston-Salem, NC, 1980, pp. 240–241.

16. International Telephone and Telegraph Company, *Reference Data for Radio Engineers,* Howard W. Sams, Indianapolis, 1968, pp. 39–4 .

17. E. Rawson, J. Goodman, and R. Norton, "Dependence of Fiber Modal Noise on Source, Fiber and Splice Parameters," FOC 80 Proceedings, San Francisco, CA, September 16–19, 1980: pp. 118–122.

7

Systems Architecture

7.1 INTRODUCTION

With the ever-increasing integration of voice, video, and data over common (mostly digital) transmission medium, the classical division of transmission product into these three categories is diminishing. Product and technology today can be more appropriately differentiated by the scale and architecture of the transmission system. The division by wide-area, metropolitan-area, and local-area networks better differentiates technological product differences.

Wide-area networks.   Wide-area networks (WANs) imply long-haul trunking between channel-concentration points, telephone central offices or tandems, and switching centers or channel drop and insert points. Examples include communication satellites and common-carrier long-line terrestrial microwave and fiber-optic links.

Metropolitan-area networks.   Metropolitan-area networks (MANs) are those that transport signals within a traffic concentration area, such as within a city, between buildings, between LANs, or for local distribution of WAN traffic. These are generally considered self-contained networks, with dedicated and switched bandwidth, often with gateways to wide-area networks. Examples include telephone local loop, interexchange carrier alternate access DS-1 and DS-3 business networks, CATV subscriber and "institutional" networks, and private business networks linking buildings within a limited geographical area.

Local-area networks.   Local-area networks (LANs) are multiple-access transmission facilities within an office, a building, or a group of buildings within a small geographic area. They include private computer networks linking mainframes, storage devices, data terminals, and other peripherals. Examples include Ethernet, IBM Token Ring, and the fiber distributed-data interface (FDDI).

This chapter deals with the more common architectures and standards used in the application of fiber optics to the transmission of voice, video, and data in LAN, MAN, and WAN.

7.2 WIDE-AREA NETWORK ARCHITECTURE

Wide-area networks (WANs) are designed to trunk large quantities of data, voice, and video traffic between concentration centers or MANs. In public telecommunications networks, a WAN is typically a mesh-type network, as shown in Figure 7.1. The route redundancy that a mesh network offers is mandatory for fiber optics since physical media such as cable are very susceptible to cuts and physical damage on a regular basis. In the public switched telephone networks, traffic is routed from the user to an end office and then to the carriers' point of presence (POP). POP to POP communication occurs over one path through the network. In the event of an outage on this path, the traffic could be rerouted along an alternate path by the switch center within the POP.

Figure 7.2 illustrates the functional block diagram of a fiber-optic trunking system that might serve the public switched telephone network described in Figure 7.1. At the central office, the individual subscriber voice or data channels are received by a switch or digital crossconnect (DACS) that multiplexes them onto higher-speed trunk groups (generally, DS-1 or DS-3) and passes them, along with signaling information, to the interexchange carrier POP.

Within the POP, the channels enter the interexchange carriers switch or DACS to be routed to the appropriate destination area within the network. The channels bound for a particular long-haul trunk group are once again multiplexed to a DS-3 rate and connected through a protection switch to the fiber-optic terminal equipment. The fiber-optic terminal equipment contains high-speed multiplexers that combine the channels further (typically, $\times 12$, $\times 24$, $\times 48$) for transmission at a trunk transmission rate of typically 560 Mb/s to 2.4 Gb/s per fiber pair. Any overhead signaling, orderwire, alarm, or control data between POPs or regenerator sites is combined with the channel signals at the POP terminals. The composite digital data stream is encoded for transmission, with error encoding inserted, and converted to a light-wave signal in the optical transmitter. The signal is transmitted on a single fiber pair for each high-level multiplexer.

The protection switch is a $1 : N$ switch that protects against terminal and repeater outage along the terminal-to-terminal span or path. The degree of redundancy is designed to achieve the necessary equipment availability for each span and to allow on-line repair without span outage. The protection switch does not protect against cable cuts, which usually sever all fibers at once.

At a regenerator, the optical signal is detected and regenerated, that is, retimed and restored as a square-wave binary signal. This signal is then input to an optical transmitter and converted to an optical signal. Section orderwire and overhead information, including error and alarm monitoring, is dropped and inserted as well.

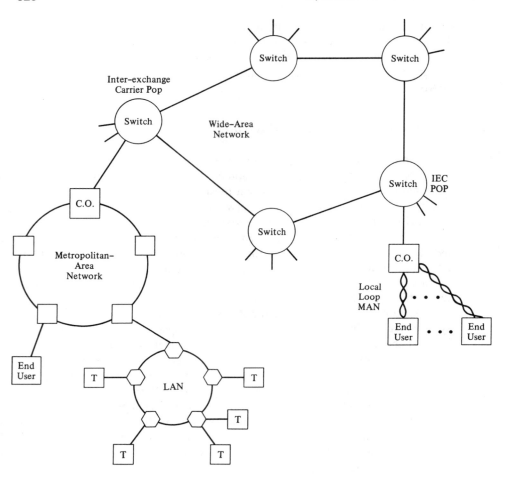

Figure 7.1 Wide-area network (mesh architecture) and its interconnection to the MAN and LAN

Figure 4.14, Chapter 4, shows the block diagram of an optical repeater at the photonics layer. If overhead functions such as orderwire and performance monitoring are required, then a section layer is inserted between the receiver regenerator and the transmitter.

In addition to the transmission equipment is an alarm, control, and orderwire system that supports network operation. This is generally carried in part as terminal-to-terminal overhead (section overhead) multiplexed in with the information channels and in part as alarm and alert information carried on separate overhead channels (line overhead) multiplexed into the high-level data stream.

Alarms and system status are generally sensed at all levels within the equipment including loss of signal, loss of frame synchronization, BER above threshold, "blue" signal (generated when no signal is present), individual module or channel failure, power-supply failure, backup power and protection status, and terminal configuration. Housekeeping alarms can also be inserted, such as door open, pri-

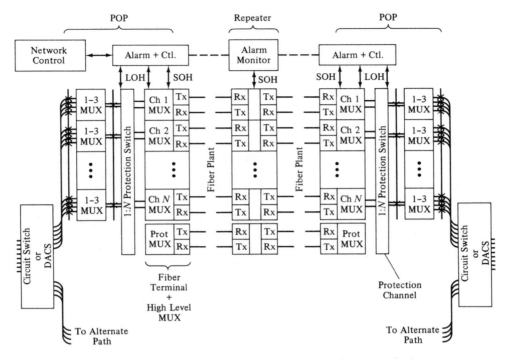

Figure 7.2 Fiber-optic trunking system for telecommunications application.

mary power failure, motor generator activated, temperature sensors, etc. There are generally three levels of status monitoring within a fiber-optics network of this type:

1. Individual equipment status displays that appear on the modules and front panels of the equipment shelves, generally in the form of LEDs. These are intended for on-site fault identification to the module level after failure notification by one of the higher-level status systems.

2. Section- and line-level alarms and loopback controls that are also intended for on-site or remote-site troubleshooting and fault isolation and restoration. These are generally found on a control console within the equipment bays at all terminal and regenerator sites.

3. Status, alarm, and alert information collected and transmitted to a network control center (NCC) for the purposes of network-level status, dispatch, and signal rerouting. The detailed alarm information is generally processed and filtered to present a high-level network status to the NCC operators. Most systems also provide the ability to "zoom in" on the status of individual sites and equipment.

The orderwire channels are also used for control signals. Control signals can include communications between protection switching equipment, repeater hut security controls, control of backup generators, etc. Voice orderwire is also provided from site to site along a section for troubleshooting.

7.3 WAN TRANSMISSION STANDARDS

7.3.1 Standards

To date there are few transmission standards specifically addressing fiber-optic trunking, with the possible exception of ANSI Draft Standard T1.105 for a synchronous optical network (SONET). This standard, developed by the Exchange Carriers Standards Association (ECSA) T1X1.2 working group, provides for a standard synchronous interface at both the optical and electrical multiplexing interfaces.

For WAN transmission quality standards, the designer generally refers to the transmission and trunking standards for the specific application to be supported on the network. Some of these are provided in Chapter 2. Interface standards for asynchronous fiber-optic transmission equipment generally follow the ANSI T1 standards for digital hierarchy electrical interfaces (DS-1 through DS-3), as listed in Chapter 2, and the alarm system standards follow the AT&T or ECSA standards for multiplexer and transmission equipment, since most of the fiber trunking product is designed to serve the telecommunications market. Some of the AT&T standards for digital networks include:

> AT&T Tech. Ref. Pub. 43801, *Digital Channel Bank Requirements and Objectives*
>
> AT&T Tech. Ref. Pub. 43802, *Digital Multiplexers—Requirements and Objectives*
>
> AT&T Compatibility Bulletin CB119, *Interconnect Specification for Digital Crossconnects*
>
> AT&T Tech. Ref. Pub. 41451, *High Capacity Terrestrial Digital Service*
>
> AT&T Compatibility Bulletin 147, *Engineering and Operations Plan for Synchronization of the Integrated Services Digital Network*
>
> AT&T Compatibility Bulletin 142, *Extended Framing Format Interface Specifications*
>
> AT&T Compatibility Bulletin 144, *Clear Channel Capability*

These AT&T standards all refer to the electrical interfaces to the system.

7.3.2 Asynchronous Transmission Characteristics

At the optical interface, the input to the fiber, there are no standards for asynchronous multiplexing fiber terminals. Each manufacturer uses its own signaling format, degree of overhead, error encoding, and synchronization; therefore, all the output data rates differ. Terminal equipment must be of the same manufacture and type end to end in any optical span. There does exist some degree of pseudostandardization as to the level of multiplexing (number of DS-3s per fiber). The various levels are shown in Figure 7.3.

Figure 7.4 shows the performance capability of fiber-optic digital trunking equipment based on typical asynchronous product. Actual performance varies, depending on manufacturer and outside plant components used.

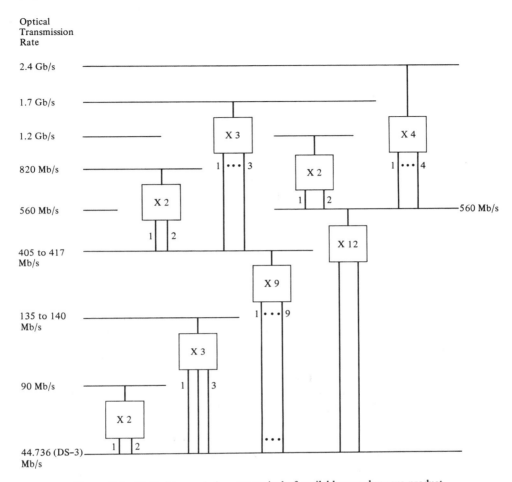

Figure 7.3   Optical transmission rate typical of available asynchronous product.

Performance will improve with time as fiber/cable design progresses, as splicing loss is reduced with new techniques and equipment, and as GaAs photodetector and receiver technology becomes more refined. Dispersion and partition noise will reduce as narrow-line lasers are used in the transmitters. The emphasis on these improvements will be in the 1550-nm wavelength range for the near future, where scattering loss in fiber is low.

7.3.3 Fiber-Optic Synchronous Transmission

ANSI T1.105 standard [1] and its companion standard, ECSA T1X1/87-128R1, for the single-mode fiber interface recommend a set of synchronous electrical and optical interfaces. The standard serves the North American heirarchy of DS-0, DS-1, DS-1C, DS-2, DS-3 standard interfaces, as well as the fourth heirarchical level, DS4NA, at its highest layer-termination point, the path layer. The path layer assem-

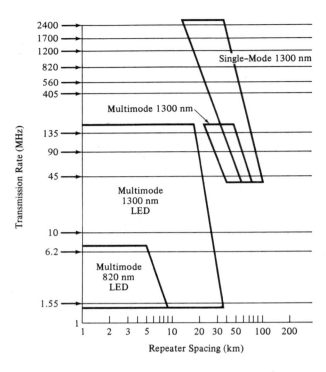

Figure 7.4 Capability of typical fiberoptic trunking products.

bles the services and path overhead (POH) into a synchronous payload envelope (SPE) for transport by the line layer as synchronous transport signals (STS-N).

The lowest-level STS is the STS-1 at a rate of 51.840 Mb/s. The STS-1 frame consists of 6480 bits (810 bytes). Operating at a rate of 51.840 Mb/s, the frame length is 125 microseconds (8000 frames per second). It consists of a transport overhead (27 bytes for section and line overhead) and a payload envelope (783 bytes, 9 for POH and the rest for data).

The line layer assembles multiple SPEs to an STS-N signal for transport by the section and photonics layers. The photonics layer converts the STS-N electrical signals into synchronous optical-carrier signals (OC-N) at the same bit rate. The basic OC-1 rate is 51.840 Mb/s.

Some equipment does not require all four layers to function. For example, in a repeater, only the photonics and section layers need be present. Terminals that do not drop and insert new SPEs do not require the path layer.

This process and the layers are shown in Figure 7.5. Although the standard indicates that the maximum STS-N rate multiplier N is 255, the STS-1 to OC-N multiplexing levels documented at this time range from X3 (155.520 Mb/s), to X48 (2488.320 Mb/s), as listed in Table 7.1. The SONET layers are defined as follows:

Path layer. A path is defined as a logical interconnection between a point at which a standard frame is assembled at a given rate and the point at which it is disassembled. The path layer is responsible for the transport of services between path-termination equipment and communicates within itself end to end via the path overhead

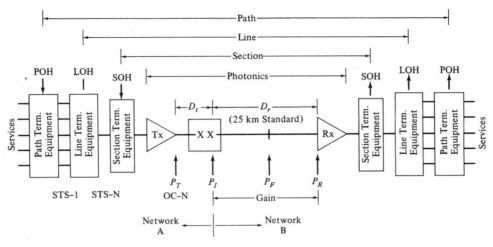

Figure 7.5   SONET model.

TABLE 7-1 SYNCHRONOUS MULTIPLEXING HEIRARCHY

STS level	OC level	Line rate (Mb/s)	Standard payload mapping
STS-1	OC-1	51.840	DS3 at 44.736 Mb/s
			SYNTRAN at 44.736 Mb/s
STS-3	OC-3	155.520	DS4NA at 139.264 Mb/s
STS-9	OC-9	466.560	
STS-12	OC-12	622.080	
STS-18	OC-18	933.120	
STS-24	OC-24	1244.160	
STS-36	OC-36	1866.240	
STS-48	OC-48	2488.320	
Virtual tributary rates			
VT1.5		1.728	DS1 at 1.544 Mb/s
VT2		2.304	2.048 Mb/s
VT3		3.456	DSIC at 3.152 Mb/s
VT6		6.912	DS2 at 6.312 Mb/s

Source: ANSI T1.105, March 9, 1988.

(POH). The function of the path layer is to map services and POH into the synchronous payload envelope (SPE) required by the line layer. Virtual tributaries are specified to transport sub-STS-1 payloads. Four sizes are specified at this time (V1.5, VT2, VT3, VT6), as described in Table 7.1. Although most of this mapping converts sub-STS-1 signals into STS-1 payload envelopes, a concatenated superframe STS-Nc format is specified to carry such multiple STS-1 services as a broadband ISDN H4 channel combined with POH.

Line layer.   The functions of the line layer are to provide synchronization and multiplexing for the path layer. It maps the path-layer SPE and line overhead

(LOH) into STS-N frame and transports them over the lower layers (the physical medium). LOH is accessed at all line-termination points where STS-N signals are created. The line layer also transports the STS-Nc channel within an STS-M line signal, where $M > N$. The line layer transmits SPEs and LOH to its peer entities. The LOH communications channel bandwidth is 192 kb/s and uses a packet protocol. In addition to providing a multiplexing function, the line layer provides line maintenance and protection. 1 : 1 and 1 : N protection switching are specified for this layer. Specified values of N are from 1 to 14 channels that can be switched to an optical protection channel.

Section layer.   The function of the section layer is to transport STS-N signals across the physical medium by mapping STS-N signals and section overhead (SOH) into pulses for the photonics layer. The section layer provides framing and scrambling of the signal in a manner compatible with optical transmission by the photonics. It also performs error monitoring. The section layer transmits STS-N signals and SOH to its peer entities.

The SOH data-communications channel bandwidth is 576 kb/s, uses LAPD protocol packet communications and ISO 8473 layer-3 interfacing. It is used for communicating operations, provisioning, and administration information, as well as local orderwire.

Photonics layer.   The photonics layer provides optical transmission over a physical medium. No overhead is associated with this layer. Its function is to convert STS signals to OC signals. The specifications for power levels, systems gain, wavelength, physical interconnection, and optical pulse shape are provided in T1X1/87-128R1 [2]. Some of the characteristics of the photonics layer as recommended by the standard include:

Fiber: Single-mode
    Operating window: 1310 nm for SMF, 1550 nm for DS-SMF
Laser center wavelength:
    1310-nm MLM laser = 1270–1340 nm and distance-dependent
    1310-nm SLM laser = 1280–1340 nm for sections <40 km
    1550-nm MLM laser = 1525–1575 nm and distance-dependent
Spectral width:
    1310-nm MLM laser = 3.5–30-nm, rate and distance-dependent
    1310-nm SLM laser = <1.0 nm
    1550-nm MLM laser = 3.5–30-nm, rate and distance-dependent
Section performance:
    10^{-9} BER for 40 km or less
    Standard section: 25 km
Connectors:
    Optical return loss: >20 dB
    Standards: EIA RS-455-XX and EIA 4750000-A
Digital rates and formats: See Table 7.1
Power levels: Determined by one of three methods

  1. Jointly Engineered per EIA TSB 20

  2. Dynamic: See Figure 7.5 and Table 7.2

  3. Fixed power point: See Figure 7.5 and Table 7.2

TABLE 7-2   SONET STANDARD INTERFACE-POINT POWER LEVELS

OC level	Line rate (Mb/s)	Systems gain (dB)	Minimum Fixed Power Point P_F (dBm)
OC-1	51.840	25.0	−21.0
OC-3	155.520	23.0	−20.0
OC-9	466.560	17.5	−17.5
OC-12	622.080	17.0	−17.0
OC-18	933.120	17.0	−17.0
OC-24	1244.160	17.0	−17.0
OC-36	1866.240	17.0	−17.0
OC-48	2488.320	17.0	−17.0

Source: ECSA Standard T1X1/87-128R1, March 9, 1988.

  Pulse shape:
   OC1 to OC18 $= t_r < T/3,\ t_f < T/3$
   OC24 to OC44 $= t_r < T/3,\ t_f < T/2$
   Overshoot (1 or 0): $<100\%$
   Undershoot: $< 30\%$ for "1," $< 20\%$ for "0"

These parameters are provided as general guidelines only. The reader should refer to the most recent standard before proceeding with a design where more specific values are given as referenced to specific line rates and transmission distances [2]. This is also necessary since standards change from time to time, particularly when in draft form.

  Three design methods are specified:

  1. For sections of 25 km or more, a "joint engineering" practice is specified per EIA TSB 20. This practice is much like the method described in Chapter 6.

  2. A dynamic power method is recommended for sections of 25 km or less. This practice defines a standard interface point and designs the power budget around this point. Figure 7.5 illustrates this practice and Table 7.2 lists the systems gain (G_s) requirements for different OC rates. The transmitting party must achieve a power level at the interface point of P_i given as follows:

$$-4 \text{ dBm} > P_i > -7 \text{ dBm} - (G_s - 3)\frac{D_t}{D_t + D_r}\,\text{dB} - 2l_c \qquad (7\text{-}1)$$

where G_s is the systems gain, D_t is the distance from the transmitter to the interface point, D_r is the distance from the receiver to the interface point, and l_c is the connector loss at the patch panel. 3 dB is specified for two connectors.

  3. A fixed power method specifies a fixed power level (P_f) at the midpoint of the optical section, where $D_t = D_r$. Table 7.2 also provides the value of P_f as specified in this draft.

7.4 METROPOLITAN-AREA NETWORKS

A metropolitan-area network (MAN) is one that transports signals within a city, between buildings, distribution nodes, or provides local distribution for WAN traffic.

From an applications standpoint, MANs generally involve:

(a) subscriber-loop communications;

(b) access from a customer premise to an interexchange carrier's point of presence (POP);

(c) private networks, interconnecting office buildings, business complexes, or LANs;

(d) cable television (CATV) networks involved in the transport of entertainment, educational, and business data traffic.

The transport medium is generally twisted-pair copper wire, coaxial cable, line-of-site microwave, or fiber. We focus on the most active area in MAN communications, the application of fiber for business communication, involving private networking and local-loop access.

7.4.1 Application of Fiber

Fiber optics has gained popularity as a transmission medium in the MAN environment primarily because it is low in cost (when its capacity is utilized) and very simple for commercial businesses to implement without a heavy technological or resource investment.

Fiber is a simple and high-performance transmission medium relative to other MAN transmission approaches. Once the fiber is selected and installed, concern for the transmission medium is minimal since there exists no outside plant electronics and the characteristics of the cable are stable with time, temperature, and electromagnetic interference. With microwave, noise interference, environmental fade, physical blockage, and FCC regulations change with time. Coax suffers from interference ingress, changes in characteristics with temperature, corrosion, and degradation (and thus constant maintenance) of plant electronics. Twisted-pair wire is constantly susceptible to changing amounts of interference and crosstalk.

Systems architecture is also very simple, generally consisting of fully modular end terminals with no repeaters. When following vendor guidelines, design is generally "cookbook." Installation is simplified by the small, light, rugged nature of the cable. Fiber is easier to install than other cable types, and nearly all major cable construction firms perform turnkey plant installations including splicing. The cable has properties that permit it to be overlashed in 2-to-3-km spans on an existing aerial plant, plowed directly into the ground in 5-to-10-km spans, or pulled continuously (using "figure-eight" techniques) through 2 to 3 km of duct around multiple bends.

Maintenance and operation is simplified because the equipment is generally modular, requires no setup or periodic adjustments, and generally contains microprocessor-based self-diagnostics. A mean time between maintenance action of 1 to 3 years is common. Other than for repair of physical damage to the cable, plant maintenance is nonexistent.

Fiber finds application in MANs when transmission capacities are large, when transmission distances are longer than achievable with twisted pair, and when environmental conditions (EMI, temperature, humidity) dictate. Fiber is generally applied to the transmission of the following:

(a) high-speed data or data packets (at the 1–10-Mb/s rate or above);

(b) PCM voice traffic where multiple DS-1s are required (N × 1.544 Mb/s);

(c) high-quality broadband video transport for broadcasters or CATV supertrunks; and

(d) systems requiring a high degree of availability and path redundancy between a few locations.

Fiber is less applicable where

(a) data rates are low (below 1 Mb/s);

(b) distance is within the range where twisted-pair or simple coaxial links can perform the function;

(c) single voice or data circuits are involved;

(d) the application requires distribution of low-capacity signals to a large number of users (such as CATV distribution, single-line telephone, etc.).

7.4.2 Network Architecture

In order to evaluate the application of fiber in MANs, some general network structures are assumed. For the most part, MAN transmission is aimed at the transport of voice, data, and video information between buildings or between clusters of buildings. Whether private or public, the MAN network contains two or more channel drop and insert nodes, with one being assigned as a network control center where network monitoring and provisioning take place.

Figure 7.6 shows four basic topologies for communication between locations in a cluster: the star, ring, bus, and tree architecture. Figure 7.7 shows the trunked-hub architecture employed in interconnecting clusters in a large-scale metropolitan area. Although these structures seem very similar to the star, ring, and bus of the LAN environment, there are usually basic differences in the MAN environment. With the LAN, the medium and channel capacity are generally shared equally among all users by virtue of some multiple-access protocol arrangement such as token passing or carrier-sense multiple access. Although some multiple-access approach can be accomplished on the MAN (such as packet switching), the applications at the MAN level generally require private channels or differentiation of channels for different services or user groups. The implementation of duplex MAN transmission is, therefore, generally on a point-to-point per-channel basis. The four topologies are described in what follows.

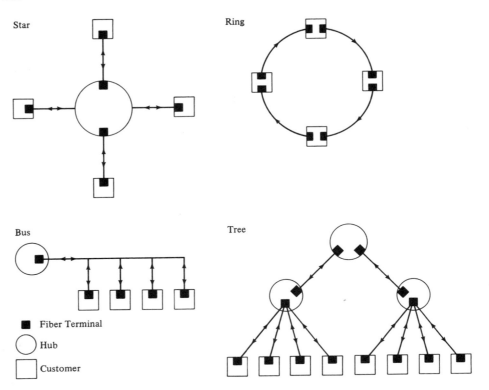

Figure 7.6 Metropolitan-area network architectures.

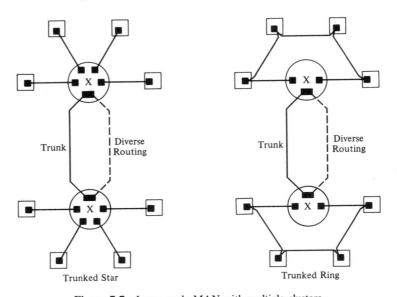

Figure 7.7 Large-scale MAN with multiple clusters.

(a) Star: This architecture provides direct point-to-point service to subscribers on dedicated lines emanating from a central master hub. Multiple subscribers can be physically connected to the cable, forming "spokes," but the channels are dedicated point to point by virtue of separate fiber pairs. Star architecture is common to the copper-pair telecommunications local loop in today's environment. In the event of a cable cut, all subscribers on that cable are out of service unless a backup cable along a different route is provided. In the event of a terminal outage, only the subscriber dedicated to that terminal is affected.

(b) Ring: This architecture provides a continuous loop of fiber interconnecting all nodes in the network. The bandwidth of the fiber ring is shared between all subscribers, who drop and insert their channels at the nodes. Although the ring can be configured as a chain of full duplex point-to-point sections, it is generally configured in a simplex fashion with the flow of information around the ring in one direction. In the duplex configuration, if a cable or terminal failure occurs, information still flows between connected operating nodes. In the simplex ring, if a cable is cut or a node fails, all subscribers are affected unless redundancy is provided. Redundancy in the simplex ring is generally configured as a counterrotating backup ring that loops back information in the event of a cable cut or node outage. Any node on the ring can act as a master node for status monitoring and reconfiguration. Circuits are logically connected in a point-to-point fashion, but physically connected in a TDM loop.

(c) Bus: The bus shares the bandwidth of a common cable by channelizing it and dropping/inserting channels at subscriber terminal points. It operates much like the ring except that it is not closed. There are many ways of configuring the multiple-access arrangement. The most common in a MAN is that used with CATV institutional networks, where subscribers are assigned different frequency "channels" on a single coaxial cable. Channels are crossconnected at the master node (head end) by frequency translation from the transmit frequency band to the receive band. If a cable is damaged with the bus architecture, all subscribers beyond that point are affected.

(d) Tree: Subscribers are connected in a heirarchical fashion to local nodes that can perform circuit concentration and switching functions. The concept of the local node permits high-quality supertrunking of channels between the master node and the local nodes and a lower-capacity, more economical distribution from the local nodes to subscribers. This architecture is common to the public switched telephone network, where local serving-office switches are connected in a heirarchical fashion to larger tandem switches at the Local Access and Transport Area (LATA) level that in turn feed channels from that LATA to interexchange carriers. This architecture is also common to the CATV industry for TV broadcast.

In a metropolitan area where there is much construction, traffic, and exposure of cable to damage, it is generally necessary to design route diversity into systems used for high-criticality business communications. Route diversity permits signal flow on a backup path in the event of primary-path failure. The ring structure with counterrotating redundancy is inherently diverse. The star configuration can also be made diverse by routing the cable to form a loop and by splicing fibers to form a redundant optical path in the opposite section of that loop. These techniques are discussed in the following sections.

7.4.3 Point-to-Point Transmission

Figure 7.8 illustrates the configuration of a single cable emanating from a master hub point. In the star configuration, each transmission channel is carried on a separate pair of fibers, one for transmit and one for receive. Crossconnection of channels from one node to another involves electrooptical conversion at the master hub. Termination to the fiber cable involves splicing into a pair of fibers within the cable at an appropriate splice point. This must be carefully planned since all fibers must be spliced at each splice point. For this reason, multiple terminals (and fiber pairs) are often dropped from each splice point. Terminals can be added as the fiber count permits.

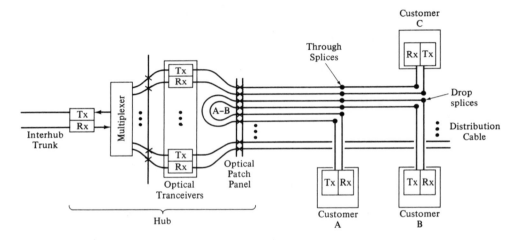

Figure 7.8   Fiber-optic metropolitan-area-network star distribution.

The problem with this configuration is that if a cable cut occurs, all terminals are effected. An alternative, which provides path redundancy, but operates as a set of private point-to-point nodes, is the optical-loop structure, shown in Figure 7.9. The interconnects are essentially point to point as with the star, but with diversely routed fibers bundled within the same cable. In this configuration, all signals are brought to the master node and crossconnected there. Since the terminals do not share fiber bandwidth, as is the case with the ring, additional terminal capacity can be added to those sections that have higher traffic loads without effecting the capacity limits of other terminals. A disadvantage is that this configuration uses two fibers in the entire cable (a pair in each direction) for each node on the network.

A functional block diagram of the terminal electronics that performs the redundant trunking function is shown in Figure 7.10. The terminal generally contains channel multiplexers to combine a number of lower-speed channels into a high-speed

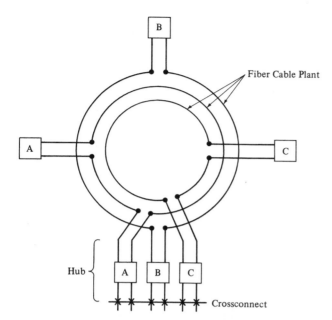

Figure 7.9   Optical-loop path redundancy structure.

channel for transmission on the fiber pair. Additional multiplexing at lower rates can be added to modularly customize the terminal to the application. For distribution trunks, fiber terminal equipment can be purchased with built-in multiplexers that are designed to combine from 4 to 28 DS-1s, or 1, 3, or 12 DS-3s.

In the protected configuration, the multiplexer equipment, fiber transmitters and receivers, and fiber pairs are redundant and switched by an automatic protection switch in the event of degradation or failure of the operating channel. Generally, the fiber terminals contain 1 : N low-speed (input-channel) protection and 1 : 1 high-speed (optical Tx/Rx) protection.

7.4.4 Active Fiber Optic Ring

The active ring, shown in Figure 7.11, is an ideal structure from the standpoint of redundancy and fiber economy. It requires only one fiber around the ring to transmit to all nodes and a second for counterrotating redundancy. The active (repeating) ring structure suffers, however, in that each terminal in the ring must handle all the signals passing around the ring. Traffic becomes additive at any cross section of the ring. For example, signals passed between locations A and D, must also be carried by terminals B and C. This situation can quickly reduce the overall capacity of the ring, requiring higher-speed terminals at each node. Another practical issue in a public ring network is the risk of placing terminals carrying shared public traffic within private premises. The ring is, therefore, generally reserved for private networking or as a backbone for interoffice trunking in a public network.

The nodes in the ring structure in Figure 7.11 consist of multiplexers (MUX) and demultiplexers (DUX) that drop and insert channels. Through channels are sim-

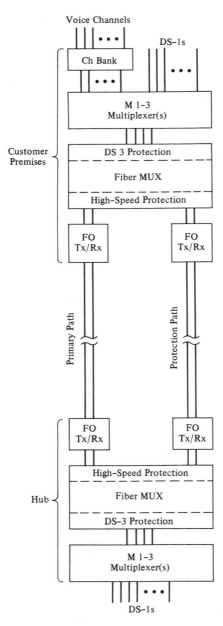

Figure 7.10 Fiber-optic distribution equipment.

ply crossconnected output to input at the low-speed interface to the multiplexer/de-multiplexer pair. In synchronous systems (such as SYNTRAN), the channels can be dropped and inserted without demultiplexing through channels.

The MUX/DUX pair is then served by a pair of primary optical transmitters and receivers that, in the figure, route all information flow clockwise around the ring. Dedicated node-to-node channels are dropped and inserted as they pass through their assigned terminating nodes.

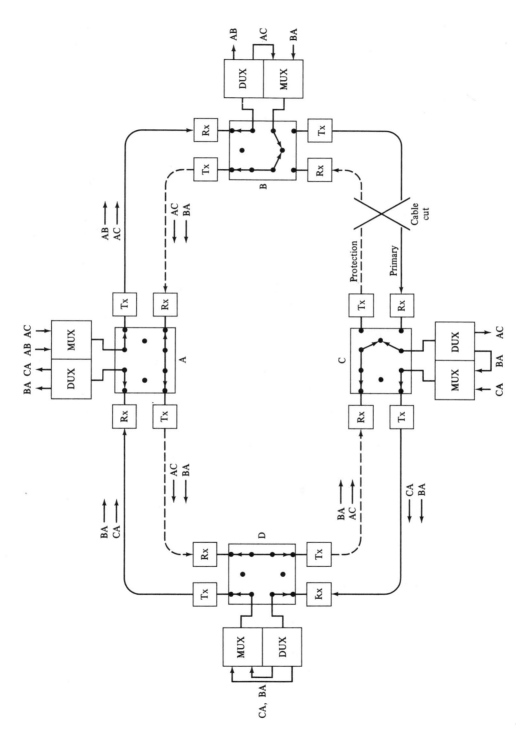

Figure 7.11 Counterrotating ring architecture.

In the event of cable or node failure, a second set of transmitter/receiver pairs is located at each node with an associated bypass switching function. The function is to loop the signal around at the node just downstream of the failure point and send it in a counterrotating direction around the loop, bypassing all other nodes, until it reaches the node just upstream of the failure point. The switching function is shown conceptually here as a double-throw switch. In reality, it is performed electronically and generally involves the signal multiplexing and SYNDES (Synchronise/Desynchronise) function of the line-termination equipment [3]. Each product implements it differently. The basic patent pending on the ring architecture belongs to FIBER-LAN, formerly a Seicor/Bell South company [4].

Loopback control is the function of the line-termination equipment that maintains status and configuration communication between all nodes. In the event of a fault condition, each node analyzes the condition (loss of signal and direction), takes appropriate local switching action, and notifies the systems-control function of the situation through the imbedded overhead communications channel. Whether central or distributed in each node's controller, the systems control reconfigures all nodes as appropriate to the fault condition and location.

Figure 7.11 shows failure condition with a cable cut between nodes B and C. If we follow channels A to C (AC) through the network, they illustrate the protection arrangement. Channel AC enters the MUX at node A, where it is transmitted on the primary fiber ring to node B, where it is looped through. Node B senses the upstream failure (absence of signal or fault condition from C) and switches all outbound traffic to the protection transmitter. Channel AC, therefore, is routed counterclockwise back to node A. Since there are no failures adjacent to node A, its protection path remains in the normal bypass mode, acting like a repeater, and passes channel AC through to node D. Node D treats the channel in the same manner and passes it on to the node C protection receiver. Node C recognizes the downstream cable cut (no signal from node B) and activates its downstream bypass switch, which routes signals from the protection receiver into the demultiplexer. The DUX at node C recovers the channel AC signal.

If designed with adequate dynamic range, the protection switch can be implemented with a passive optical switch rather than performing electooptical conversions at each node in the protection ring. This concept is part of the FDDI standard, which is discussed in Section 7.12.

7.5 MAN TRANSMISSION STANDARDS

Some of the key standards activity that apply to MANs include:

(a) IEEE 802.6, *Distributed Queue Dual Bus for Metropolitan Area Network* [5]
(b) ISDN Primary, Basic Rate, and Wideband [6]
(c) SYNTRAN [7]
(d) FDDI and FDDI-II [8]
(e) SONET [9]

With the exception of the SONET and FDDI standards, all others are media-independent. FDDI and SONET were particularly designed for fiber optics. SONET is discussed in Section 7.3. FDDI is covered in detail in Section 7.12.

7.5.1 IEEE 802.6

IEEE 802.6 is a part of a family of standards for local- and metropolitan-area networks, including 802.3 CSMA/CD, 802.4 token bus, 802.5 token ring, and 802.9 integrated voice and data LAN. It is interface compatible with these LAN standards using the 802.2 logical link-control protocol within the data-link layer. The standard defines a dual-bus subnetwork with two unidirectional contraflowing physical buses. The standard also describes bridging of bus networks to form a MAN. One purpose of the standard is to enable public-network providers to offer MAN services on the installed base of physical media and digital transmission equipment. Applications include high-speed packet communications service for LAN interconnection, office-systems interconnection (host and work station), ISDN, and video conferencing.

Each subnetwork uses a contraflowing unidirectional dual-bus pair, which can be configured in an open bus arrangement or closed loop. The looped-bus topology is capable of reconfiguration in the case of a break in the loop. Each node attaches to both buses in a fashion similar to a logical OR, acting in an essentially passive role in implementing the distributed queue dual-bus medium-access protocol (DQDB MAC), and, therefore, the reliability of the bus is not effected by a node failure. Timing and generation of the fixed-length slots and frames originate in the frame-generation or head-end node. Both buses operate at the same time, making the capacity of the network twice that of a single bus.

The DQDB access protocol can operate over a number of existing physical PMD (Physical Medium Dependent) sublayers, including the CCITT G.703 heirarchical interfaces (DS-3 or higher) as well as SONET. Initial realizations are targeted at rates of up to 150 Mb/s per bus, but future requirements are expected to be up to 560 Mb/s and beyond. One PMD under study uses the 139.264-Mb/s CCITT Rec. G.703 interface. An interface with B-ISDN protocols has also been recommended as a potential for the customer-premise network [10].

7.5.2 ISDN

ISDN is the acronym for integrated-services digital network. Table 7.3 indicates some of the ISDN interface channels that are emerging from the CCITT standards forum [6]. The primary reason for establishing ISDN is to integrate new services with existing services and on existing media to the degree possible. It expands the service capability of existing loop facilities by providing two voice channels (2B) and a data channel (D) on the single basic-rate subscriber pair. It also provides the ability for primary-rate subscribers to control channel allocation, bundle multiple 64-kb/s channels, and signal across the public ISDN network using the D channel.

TABLE 7-3 CCITT STANDARD ISDN CHANNEL STRUCTURES

Channel	Transmission rate	Potential application
B	64 kb/s	Digitized voice Wide-band 56- and 64-kb/s data Multiplexed lower-rate data Compressed 56- and 64-kb/s video Circuit and packet switched data
D	16 or 64 kb/s	Signaling Dynamic bandwidth allocation Packet switched data Transmission of ANI (Automated Number Identification)
HO	384 kb/s	Digitized video Circuit switched data
H1	1.544 Mb/s	DS-1 channels Circuit switched data Digitized video

Standard access configurations	
Rate class	Channel combinations
Basic Primary	2B + 1D (16 kb/s) 23B + 1D (64 kb/s) 4 HO channels 3 HO channels + 1D (64 kb/s) 1 H1 channel

ISDN will replace much of the data transmission that takes place today using modems over dedicated or switched analog lines.

Broadband ISDN (B-ISDN) was initially targeted for entertainment video, although it can be used for business needs as well [10]. It is envisioned as a vehicle for bringing existing and new services such as high-definition television into the home. In order to provide voice, video, and data on demand, future customer-premise networks may require the flexible multiplexing structures that the dynamic bandwidth allocation of ISDN provides. Initial focus is on the 150-Mb/s range, where the requirements of enhanced-quality television (90 to 138 Mb/s encoding) lie. High-definition TV, requiring 150 to 200 Mb/s of encoding, will likely drive higher-rate-standards activity as well.

7.5.3 SYNTRAN

The SYNTRAN standard provides for a synchronous DS-3 rate service with the capability to drop and insert subrate DS-1 channels without demultiplexing and remultiplexing all channels, including the through channels [7]. It is applicable to the bus or ring topology in a MAN. It has the capability of distributing the DS-1 crossconnect function with remote provisioning capability. Each SYNTRAN terminal can handle up to 28 DS-1s. It provides for a 64-kb/s imbedded data link for remote provisioning on demand and maintenance.

7.6 VIDEO ON FIBER IN THE MAN

7.6.1 Applications

Although data and voice transmissions have sustained the most near-term interest for fiber transmission in the MAN, the requirement for delivery of high-quality TV to the subscriber has stimulated much standardization activity, as was discussed in the previous section. Aside from the potential for high-quality video transmission to the suscriber, standard NTSC (National Television Systems Committee) quality video transmission on fiber has a definite application in areas where conventional methods (coax and microwave) are not totally satisfactory in performance or cost. Video transmission in the MAN has generally been applied to cable television (CATV), video conferencing, broadcast feeds, and closed-circuit television (CCTV) systems such as security or traffic-monitoring systems.

CCTV systems, such as video conferencing and security or traffic monitoring, generally require only a few channels of transmission, with color and black and white having weighted peak-to-peak-luminance-to-rms noise ratios, $(SNR)_{pp1/rms,w}$, in the 40-to-50-dB range. Channel capacity and transmission distance for closed-circuit systems vary widely, depending on application. Historically, most have been short coaxial systems within a building complex or longer coaxial traffic-monitoring systems for train systems or highways. With the maturing of video fiber systems, more municipalities are investigating total freeway or citywide traffic-control and monitoring systems using fiber optics.

Broadcast trunks generally require only a few (one, four, etc.) channels to be transmitted with $(SNR)_{ppwb/rms,w}$ of 62 to 67 dB over distances of 5 to 10 miles to reach a carrier, satellite, microwave site, or second studio (see Table 2.4). Historically, these systems have been coaxial or microwave, or, if fiber, single channel per fiber using frequency or digital modulation.

CATV requires about 52-to-55-dB $(SNR)_{ppwb/rms,w}$, and must carry between 50 to 100 channels over distances of 10 to 20 miles on supertrunks between head ends and distribution nodes. These signals are then distributed from each node to thousands of homes, arriving with from 40-to-45-dB $(SNR)_{ppwb/rms,w}$. Although the signal-to-noise requirements for CATV are not as stringent as for broadcast, the combination of SNR, channel capacity, and low cost makes the CATV requirement the most challenging for any new transmission technology.

This section describes techniques that apply to multichannel broadcast trunks. Although the analysis is performed for the CATV application, signal quality is primarily a function of the modulators selected and channel capacity per fiber, and, therefore, the basic approaches apply to broadcast-quality as well as low-quality CCTV.

A typical CATV network is shown in Figure 7.12. The potential application areas for fiber are (1) the feed trunks between signal source and head end, (2) the supertrunk between head end and hub, and (3) the distribution network. Fiber is limited in its ability to compete with coax in the distribution portion of the system. The fiber, connectors, and splitting components are too costly and power losses too great in comparison with coax with today's technology.

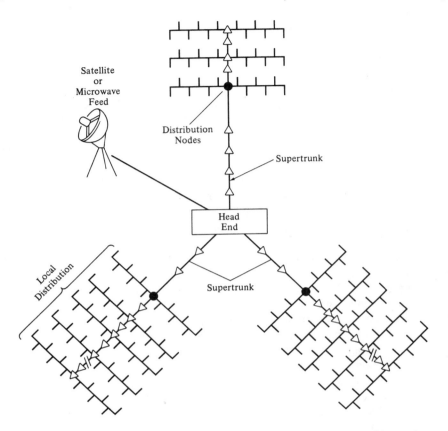

Figure 7.12 Large metropolitan-area CATV network using supertrunking and distribution nodes.

Historically, coax and microwave have been the only technologies that have been accepted into the CATV industry. Until single-mode fiber technology became practical, fiber proved too expensive for CATV. When single-mode fiber became available at production volumes and lower cost in the 1983–1984 time period, the situation changed. Development work and field trials performed by McDevitt and Hoss [11, 12] illustrate that multichannel single-mode fiber supertrunks are of lower cost and higher quality than FM coaxial supertrunks used in the larger CATV networks. Some successful installations followed by two of the largest multiple-systems operators, Warner Amex [13] and ATC [14]. Perceiving a new market in CATV supertrunks, video fiber-optic product development followed with a number of manufacturers. Fiber began to be the medium of choice for supertrunking on a cost/performance basis, where distance is great and signal quality high [15, 16]. A comparison of fiber against conventional coaxial and microwave trunking approaches is given for one installation case [12] in Figure 7.13. The equivalent AM or FM coaxial implementation for 100 channels required four $\frac{3}{4}$-inch cables and 56 amplifiers for a 10-mile span. In addition, the FM coax delivered only 53-dB SNR and

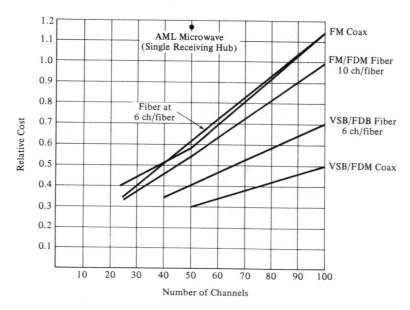

Figure 7.13   Cost versus channel capacity for a single 10-mile supertrunk.

the AM coax less than 50-dB SNR, whereas fiber delivered over 60-dB SNR end to end.

In all except very short systems or applications where few channels are required (one to six), some signal processing is required in order to trade bandwidth for SNR. This is because fiber is not an ideal medium for high-SNR signals. The nonlinearities of the source and small systems gain (30 to 40 dB) require a transmission CNR in the range of 20-to-30-dB rms signal to rms noise. Since single-mode fiber does have a lot of bandwidth, we can trade bandwidth for SNR with the appropriate modulation or encoding methods.

There are four popular methods for trunking video on fiber: (1) frequency multiplexing the AM-modulated video channels as vestigial sideband channels (VSB/FDM), (2) frequency-division multiplexing frequency-modulated channels (FM/FDM), (3) digitizing the video using pulse-code modulation and time-division multiplexing the channels onto the fiber (PCM/FDM), and (4) wavelength-division multiplexing individual channels (or previously multiplexed channel groups) onto the fiber (WDM). Four of these methods for trunking multichannel video on fiber are shown in Figures 7.15, 7.17, 7.19, and 7.20. The performance characteristics of these systems are discussed in the following sections.

The comparative performance of PCM/TDM and FM/FDM for high-quality video trunking (55-to-60-dB SNR weighted) is shown in Figure 7.14, which shows the practical channel-versus-distance performance achievable with conventional product technology. The plot was adapted from Muoi [11, 38].*

---

* The author wishes to acknowledge Dr. Tran Muoi for the information upon which this diagram was developed.

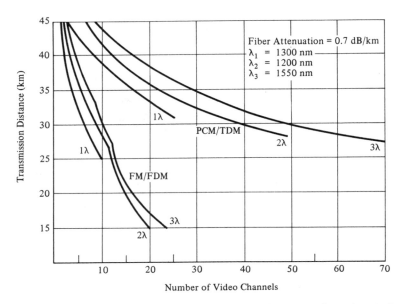

Figure 7.14 Relative comparison of video-transmission approaches using single and multiple wavelengths on single-mode fiber. The author wishes to thank Dr. Tran Muoi of PCO for submitting this previously unpublished figure to this publication.

The relative advantages will change over time as the technology changes. For instance, today, the most cost-effective approach for multichannel video is the FM/FDM approach. As fiber populates the subscriber loop for video, voice, and data, the volume production potential will drive the cost of digital encoding into the same range as analog approaches. The cost-versus-performance advantage will definitely move in the direction of PCM/TDM. By using coherent optics and integrated optics, WDM becomes a matter of optical frequency detection rather than bulk filtering, and, therefore, will become a technique of choice once low-cost product becomes available.

7.6.2 VSB/FDM

The standard broadcast approach in the CATV industry is the trunking and distribution of signals over coax using vestigial sideband channels of 6-MHz bandwidth each with center frequencies consistent with a particular CATV frequency plan. VSB is the simplest and least-expensive approach to broadcast, and, therefore, would be a desirable fiber transmission approach. Unfortunately, source linearity and power margins limit performance.

   Step 1A: Define end-to-end performance. We assume eight channels with an end-to-end video baseband $(SNR)_{ppwb/rms}$ of 60 dB measured in a 4.2-MHz bandwidth with an EIA RS250B weighting filter. For the VSB case, we assume a 6-MHz transmission band to accommodate the sound subcarrier.

Step 1B: Design systems architecture.   Figure 7.15 illustrates the approach with VSB modulators operating at eight separate center frequencies passively combined into an analog transmitter. Figure 7.16 provides frequency plans to illustrate the design considerations. These plans are not necessarily the optimum grouping for a practical design. In practice, the channel groups must match the CATV frequency plan. Through selective frequency planning, channel groups can be separately transmitted and filtered prior to recombination to reduce in-band harmonics, as shown in Figure 7.15.

Step 1C: Design path layer, determine intermediate-channel performance.   A series of VSB modulators is chosen for the path layer. With VSB, the relationship between $(CNR)_c$ and $(SNR)_b$ is from Table 5.2,

$$(CNR)_{c,\text{RMS/RMS}} = (SNR)_{b,\text{RMS/RMS}}$$

When transmitting video, there is a complex conversion between rms $(SNR)_b$ and the peak-to-peak luminance signal to rms weighted noise generally specified. In addition, the relationship between signal power and carrier power for VSB transmission is somewhat involved. Refer to Chapter 2, Section 2.6.4, for details. From Equations (2-10) through (2-13), we get

$$(CNR)_{b,\text{RMS/RMS}} = (SNR)_{b,\text{ppl/rms},w} - 20 \log (CF)_{\text{pp/rms,vsb}}$$

$$+ 20 \log (CF)_{\text{sw/lum}} + 10 \log \left(\frac{C_p}{SB_p}\right) \qquad (7\text{-}1)$$

$$+ 20 \log \left(\frac{1}{M_s}\right) - WF$$

where

$(SNR)_{b,\text{ppl/rms},w}$ is the baseband peak-to-peak luminance signal to rms weighted noise voltage

$(CF)_{\text{pp/rms,vsb}}$ is the peak-to-peak to rms baseband signal conversion for vistidual sideband

$(CF)_{\text{sw/lum}}$ is the sinc-to-white to luminance signal level ratio in IRE, typically 140/100 IRE

C_p is the relative carrier power

SB_p is the relative sideband power

M_s is the signal modulation index

WF is the noise filter weighting factor in dB from Table 2.3

If we assume that an EIA unified weighting filter is used to measure white noise in a 4.2-MHz bandwidth, then the weighting factor (WF) is 6.8 dB (per Table 2.3). For a 60-dB $(SNR)_{\text{ppl/rms}}$ weighted signal, the required received CNR is

$$(CNR)_{c,\text{RMS/RMS}} = 60 + 6 - 6.8 = 59.2 \text{ dB}$$

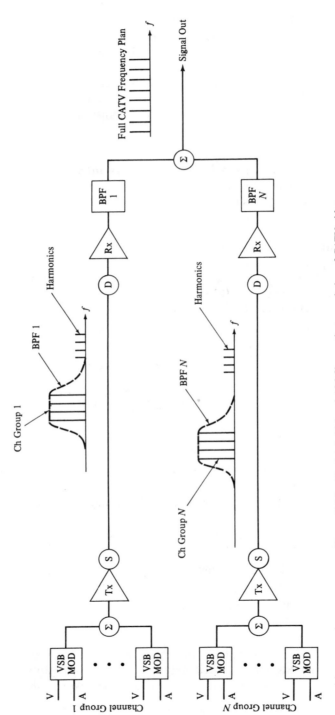

Figure 7.15 Vestigial sideband AM fiber-optic transmission of CATV video.

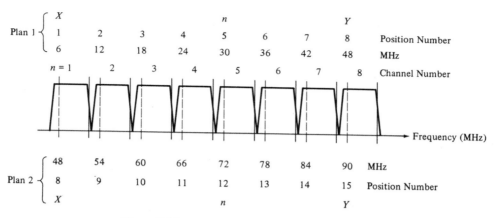

Figure 7.16   VSB frequency plans for video signals.

Step 1D: Design line multiplexing, determine optical section performance. With FDM in the presence of multiple signals, CNR is not only a function of received signal strength or received optical power, but also strongly dependent on harmonic products that may fall in band with the signals. In fact, the harmonic level can be considered a noise floor for the system. The level of this noise floor is analyzed using the relationships given in Section 5.4.3.

In our previous discussion of harmonics, we said that the effect is most pronounced within the center channels; we therefore concentrate our analysis there. With eight channels ($N = 8$), we perform the analysis on channel 5, i.e., $n = 5$. In frequency plan 1 of Figure 7.16, $X = 1$ and, therefore, $Y = 8$. If we assume a laser with a -50-dBc second harmonic and a -70-dBc third, we get the following results:

Second-Order Terms:

$$P_2 \text{ (dBc)} = 10 \log (T_1) + 6 + 20 \log \frac{k_2}{4} - 1.5$$

$$= 10 \log (T_1) + 6 - 50 - 1.5$$

$$= 10 \log (T_1) - 45.5$$

where $T_1 = \text{INT} \dfrac{N + 1}{2} - X + (Y - N - X + 1)$

$$= \text{INT} \frac{6}{2} - 1 + (8 - 5 - 1 + 1)$$

$$= 3 - 1 + 3 = 5$$

Therefore:

$$P_2 \text{ (dBc)} = 7 - 45.5 = -38.7 \text{ dBc}$$

Third-Order Terms:

$$P_3 \text{ (dBc)} = 10 \log \left(T_2 + \frac{T_3}{4} \right) + 15.6 + 20 \log \frac{k_3}{4} - 2.5$$

$$= 10 \log \left(T_2 + \frac{T_3}{4} \right) + 15.6 - 70 - 2.5$$

$$= 10 \log \left(T_2 + \frac{T_3}{4} \right) - 56.9$$

where
$$T_2 = \text{INT} \frac{8 - 5 - (2 \times 1) + 1}{2} \text{ INT} \frac{8 - 5 - (2 \times 1)}{2}$$

$$+ \text{INT} \frac{5 - 1}{2} \text{ INT} \frac{5 - 1 - 1}{2} + (5 - 1)(8 - 5)$$

$$+ \text{INT} \frac{8 - 5}{2} \text{ INT} \frac{8 - 5 - 1}{2}$$

$$= 0 + 2 + 12 + 1 = 15$$

$$T_3 = \text{INT} \frac{5 - 1}{2} - \text{INT} \frac{5}{3} + \text{INT} \frac{5 - 1}{3} - 1 + 1$$

$$+ \text{INT} \frac{8 - 5}{2} + \text{INT} \frac{5 - 1}{2} + \text{INT} \frac{8 - 5}{2} - 1 + 1$$

$$= 2 - 1 + 1 - 1 + 1 + 1 + 2 + 1 - 1 + 1 = 6$$

Therefore:

$$P_3 \text{ (dBc)} = 10 \log \left(15 + \frac{6}{4} \right) - 56.9$$

$$= 12.2 - 56.9 = -44.7 \text{ dBc}$$

The total harmonic distortion (THD) within the center bands, therefore, is the sum of the second- and third-order powers:

$$\text{THD}(5) = 10 \log \left(10^{-38.7/10} + 10^{-44.7/10} \right) = -37.8 \text{ dBc}$$

We have seen that to achieve the 60-dB weighted SNR at the output, we need a 59.2-dB $(\text{CNR})_c$ within the receiver in the absence of harmonic products. The level of harmonic products, however, limits the $(\text{CNR})_c$ to 37.8 dB. We are, therefore, unable to achieve the desired performance due to harmonic distortion. Possible solutions to the problem include the following:

(a) Choose a more linear laser. This may require the use of an advanced design such as an external cavity device. In order to achieve the 59.2-dB CNR, however, the THD of the laser would have to improve by more than 20 dB with this frequency plan.

(b) Use a lower modulation index in an attempt to gain greater linearity. For example, a 55-dB $(\text{CNR})_c$ has been observed with this quality laser by reducing the

M_o from 0.5 to about 0.1. This has the negative effect of reducing power per channel (7 dB for a fivefold M_o reduction) and thus shortening transmission distance.

   (c) Combine fewer channels. If we were to transmit only four channels, for example, a recalculation would yield a -53.4-dB THD for the initial frequency plan.

   (d) Design the per-fiber transmission frequency plan based on harmonic placement. For example, by simply moving the adjacent channel grouping out in frequency, the lower-order harmonics are moved up in frequency and out of band. An extension of this approach is to more selectively group channels on the basis of harmonic placement and recombine to achieve the full frequency plan at the receiver, after filtering, as shown in Figure 7.15.

   By following the approach in (d) for the adjacent-channel case, if $X > N/3$, for example, the $(f_a + f_b + f_c)$ and $(2f_a + f_b)$ terms do not appear in band, and the $(f_a - f_b - f_c)$ and $(f_a - 2f_b)$ terms do not appear if $X > (Y - N)/2$. If we were to move the carriers out to $f_1 = 48$ MHz ($X = 8$) and $f_8 = 90$ MHz ($Y = 15$), then the center channel (channel 5) would be $n = 12$. This is represented as plan 2 in Figure 7.16. Here all second-order terms fall out of band and only the $(f_a + f_b - f_c)$ and $(2f_a - f_b)$ terms dominate. The THD is now reduced:

$$T_2 = \text{INT}\,\frac{12 - 8}{2}\,\text{INT}\,\frac{12 - 8 - 1}{2} + (12 - 8)(15 - 12)$$

$$+ \text{INT}\,\frac{15 - 12}{2}\,\text{INT}\,\frac{15 - 12 - 1}{2}$$

$$= 2 + (4)(3) + 1 = 15$$

$$T_3 = \text{INT}\,\frac{15 - 12}{2} + \text{INT}\,\frac{12 - 8}{2} = 1 + 2 = 3$$

Therefore:

$$P_2 = \text{none in band}$$

$$P_3 = 10 \log\left(15 + \frac{3}{4}\right) - 56.9 = -44.9 \text{ dBc}$$

$$\text{THD(5)} = -44.9 \text{ dBc}$$

By adjusting the frequency plan, we achieved a 7.1 dB improvement in $(\text{CNR})_c$. Whereas the transmission bandwidth of the receiver doubled, the bandwidth requirements of the channel group remained the same. Although this plan still did not achieve the desired performance of a 59.2 dB CNR with the laser that was chosen, the example illustrates that the approach of selective frequency planning can relax the harmonic-distortion requirement on the laser to within a more practical range.

   Multichannel VSB is generally useful for the more moderate SNR requirements such as for CATV distribution or closed-circuit security systems where 45 to 50 dB $(\text{SNR})_{\text{ppwb/rms},\,w}$ is adequate.

Step 2A: Design optical section, determine photonics performance.
Even if we can improve SNR by choosing a more linear laser and, therefore, reducing THD, vestigial sideband AM still is limited by the amount of received power required. Assume for the eight-channel system, and a $(BW)_c$ of 6 MHz, that a THD greater than -59.2 dB is achievable with an M_o of 0.5 by proper laser selection and use of frequency plan 2.

Step 2B: Photonics design.   We assume a 1300-nm single-mode system with a PIN/FET receiver. The performance of the optical receiver and FET preamplifier is a direct function of the frequency spectrum choices made for FDM multiplexing. From Table 6.15, the relationship for preamplifier noise is

$$\langle i_{nA}^2 \rangle = \frac{4kT\,(BW)_c}{R_i} + \frac{16\pi^2 kT\tau C_t^2 (BW3)^3 (1 + fc/f)}{g_m} \tag{7-2}$$

$$(BW3)^3 = \frac{F_h^3 - F_l^3}{3} \tag{7-3}$$

Substituting $R_i = 1$ megohm, $T = 300$ K, $\tau = 1.8$, $C_t = 1.6$ pF, and $g_m = 0.03$ for the preamplifier, and $F_h = 95$ MHz, $F_l = 89$ MHz, $(BW)_c = 6 \times 10^6$, $fc = 10$ mHz, for the highest-frequency (worst-case) channel, we get

$$(BW3)^3 = \frac{(95 \times 10^6)^3 - (89 \times 10^6)^3}{3} = 5.08 \times 10^{22}$$

$$\langle i_{nA}^2 \rangle = 9.9 \times 10^{-20} + 5.1 \times 10^{-18} = 5.75 \times 10^{-18}$$

From Table 6.14, receiver sensitivity (MDP) for a PIN/FET is

$$\text{MDP} = \frac{2e\,(CNR)_c (BW)_c}{r} \frac{N^2}{M_o^2}\left(1 + \sqrt{1 + \frac{M_o^2}{N^2} \frac{\langle i_{na}^2 \rangle + \langle i_{nL}^2 \rangle}{2[e\,(BW)_c]^2 (CNR)_c}}\right) \tag{7-4}$$

Substituting $N = 8$ channels, $M_o = 0.5$, $(CNR)_c = 8.3 \times 10^5$ (59.2 dB), $\langle i_{nA}\rangle$ from before, $r = 0.7$, and $I_d = 10$ nA, we get

$$\text{MDP} = (6.8 \times 10^{-4})(1 + \sqrt{1 + 0.0147})$$

$$= 1.17 \times 10^{-3} \text{ watts} = 0.7 \text{ dBm}$$

With 0-dBm launch power, it can be seen that there is not enough power to provide eight channels per fiber at the 59.2-dB $(CNR)_c$. A more relaxed requirement is required for VSB/FDM on fiber. If we simply reduce the requirement to four channels per fiber, the MDP requirement becomes -5.31 dBM. If we, in turn, reduce the SNR requirement to 55 dB, $(CNR)_c = 54.2$ dB, then we get an MDP of -10.23 dBm, a more realistic value.

Continuing with the design on the basis of two 4-channel-per-fiber groups, we assume $P_t = 0$ dBm average, $l_c = 1$ dB, $l_s = 0.15$ dB, $l_f = 0.4$ dB/km, $t_{df} = 2$ ps/nm/km maximum, and source spectral width = 3.5 nm.

Step 2C: Preliminary performance estimate

Link-Bandwidth Limitation. In an FDM system, we measure performance on a per-channel basis within the channel bandwidth (BW); therefore, the link transmission bandwidth $(BW)_p$ does not effect the per-channel bandwidth. It does, however, reduce power of the channels at the band edges due to the attenuation slope of the passband of both the receiver filter and the fiber link. Since receiver bandwidth does not effect individual-channel noise performance, we set it such that the photonics bandwidth $(BW)_p$ is adequately greater than the highest transmission spectrum bandwidth $(BW)_t$ of 95 MHz by at least a factor of 1.33 in order not to have more than a 1-dB effect on the power of the highest channel (from Figure 6.9, use the thermal-noise limitation curve). The bandwidth of the fiber must be even greater than this. Computing distance based on the 1.33 $(BW)_t$ limit, we get

$$(BW)_{f, \text{elect}} = \frac{312}{t_d} \text{ MHz/ns}$$

$$T_{d, \text{tot}} = 2 \text{ ps/nm/km} \times 3.5 \text{ nm} \times D \text{ km}$$

Combining the two and using $(BW)_f \geqslant 1.33 \times 95 = 126$ MHz, we get

$$126 = \frac{312}{0.007D}$$

$$D_{\text{max}} = 354 \text{ km}$$

Bandwidth is, therefore, not a limiting factor. In order to equalize the response, assuming we do experience 1 dB of band limiting at the band edges, we must add a power penalty equal to $p_{m(-)}/2 = \frac{1}{2}$ dB (from Table 6.11).

With an MDP of -10.09 dBm, if we assume a 3-dB margin and one splice per 3 km ($l_s = 0.05$ dB/km), then distance is limited to

$$D = \frac{P_t - \text{MDP} - M - P_p - L_c}{l_f + l_s}$$

$$= \frac{-0 - (-10.23) - 3 - 0.5 - 2}{0.4 + 0.05}$$

$$= 10.5 \text{ km}$$

The VSB approach, therefore, is limited to moderate SNR channels, relatively short distances, and few channels per fiber.

7.6.3 FM/FDM Video Transmission

Frequency modulation (FM) using a wide deviation (8 MHz peak) provides SNR improvements of about 15 dB between baseband video $(SNR)_{b, \text{RMS/RMS}}$ and transmitted $(CNR)_{c, \text{RMS/RMS}}$ by trading a six-times bandwidth extension. Advantages of the FM/FDM approach are its simplicity, its use of mature product technology, and its low cost relative to digital encoding. The greatest disadvantage to FM/FDM is that, like

VSB, its end-to-end performance remains a direct function of the transmission link quality, which is, in turn, a direct function of distance and number of channels multiplexed. A further complication is, like VSB, the nonlinear increase in harmonic distortion products with an increase in the number of channels.

For a 20-km link, after about 10 channels per fiber, performance rapidly decreases as channels are added. Repeating the signal reduces SNR by 3 dB and reduces transmission bandwidth as well, which has further compounding effects on SNR. These effects are illustrated with an example.

Step 1A: Define end-to-end performance. The example is again an eight-channel-per-fiber video transmission with 60-dB $(SNR)_{ppwb/rms,\,w}$ per -EIA RS250B end to end.

Step 1B: Design systems architecture. The FM/FDM systems architecture is shown in Figure 7.17. Frequency-division multiplexing (FDM) is a natural extension of frequency modulation. The frequency-modulated carriers are multiplexed by simply making each local-oscillator frequency different for each video channel. The carriers are power combined onto the analog optical transmitter. At the receiver, they are separated by frequency-selective demodulators.

Step 1C: Design path layer

Step 1C.1: Path-Layer Processing. By using frequency modulation, the trade-off between SNR-to-CNR improvement and bandwidth extension is given in the following relationships from Table 5.2:

$$(CNR)_{c,\,RMS} = (SNR)_{b,\,RMS} - 20 \log \frac{\Delta f}{(BW)_b} - 10 \log \frac{(BW)_c}{(BW)_b} - 1.7 \text{ dB}$$

$$(BW)_c = 2[\Delta f + (BW)_b]$$

For transmission of video over fiber, we take advantage of the wide link bandwidth and use an FM modulator with one of the larger practical FM deviations available, a peak Δf of 8 MHz. We assume that the audio is transmitted on a separate subcarrier to the video and will leave spectrum accordingly.

Step 1C.2: Unit Conversion. Video SNR is given in terms of peak-to-peak values for luminance (picture) signal, whereas the FM relationship is in rms power terms. The conversion factor to rms, from Section 2.6.5, Equation (2-15) is

$$(SNR)_{b,\,RMS} = (SNR)_{b,\,ppwb/rms,\,w} - CF - WF - DF$$

If we assume a CCIR weighting filter and that preemphasis and deemphasis are used in the modulator/demodulator, the factor (WF + DF) for triangular noise in a 4.2-MHz bandwidth is 12.8 dB (from Table 2.3). For FM, we consider only the SNR of the white-to-blanking signal, therefore, the conversion factor from peak to peak to rms is 9 dB.

$$(SNR)_{b,\,RMS} = 60 - 9 - 12.8 = 38.2 \text{ dB rms}$$

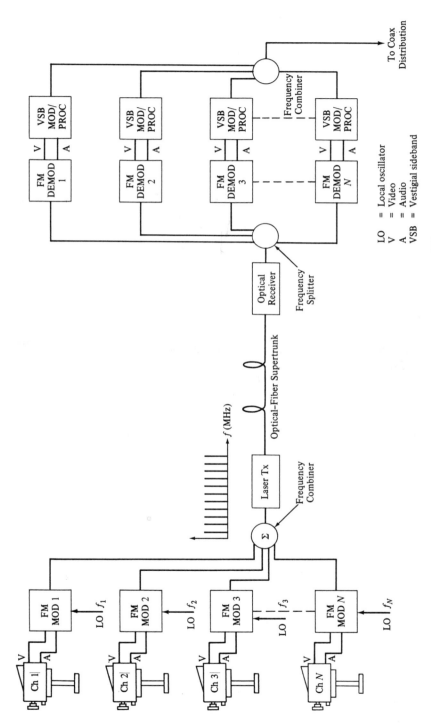

Figure 7.17 FM/FDM fiber-optic video trunk for CATV.

LO = Local oscillator
V = Video
A = Audio
VSB = Vestigial sideband

357

Step 1C.3: Intermediate-Channel Performance.   Based on the desired 60-dB $(SNR)_{b, ppl/rms, w}$ video signal, the per-channel $(CNR)_c$ is, therefore,

$$(BW)_c = 2(8 + 4.2) = 24.4 \text{ MHz}$$

$$(CNR)_{c,RMS} = 38.2 - 20 \log \left(\frac{8}{4.2}\right) - 10 \log \left(\frac{24.4}{4.2}\right) - 1.76 \text{ dB}$$

$$= 23.2 \text{ dB}$$

where the video portion bandwidth is 24.4 MHz.

Step 1D: Design line layer

Step 1D.1: Design Line Multiplexing.   Since the multiplexing approach is to be FDM, multiple carriers are transmitted simultaneously, and a harmonic interference situation, similar to VSB, exists. With the wide-channel bandwidth and multiple channels, we have little choice in frequency spacing for intermodulation other than adjacent spacing starting low in the band. The spectrum is shown in Figure 7.18. The 40-MHz bandwidth at the low end, $(BW)_z$, is reserved for the audio and a guard band of 15.6 MHz has been allocated to reduce adjacent-channel interference as well as position the channels such that unmodulated carriers fall between bands (40-MHz carrier spacing). Distortion products from the modulated signals fall within the bands since they extend over twice the channel bandwidth. With FM, we want the SNR improvement factor to overcome the harmonic-distortion level. Total transmission bandwidth is

$$(BW)_t = (BW)_z + N[(BW)_c + (BW)_g]$$

$$= 40 + 8(24.4 + 15.6) = 360 \text{ MHz} \tag{7-5}$$

Comparing Figures 7.18 and 7.16, we can see that the frequency plan is similar in

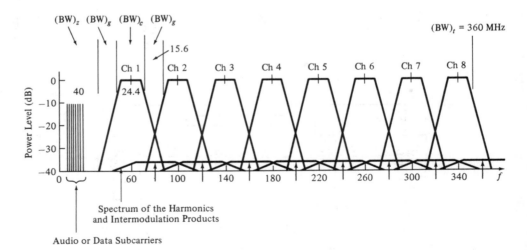

Figure 7.18 Frequency spectrum for the FM/FDM example.

relative channel positioning to plan 1 for the VSB case. Therefore, the number of in-band terms and THD are the same as in the previous example:

$$P_2 = -38.7 \text{ dBc}$$

$$P_3 = -45.0 \text{ dBc}$$

$$\text{THD}(5) = -37.8 \text{ dBc}$$

Step 1D.2: Determine Optical-Section Performance.
If a 23.2-dB $(\text{CNR})_c$ is required and the CNR limit due to THD is 37.8 dB, then the required optical-section $(\text{CNR})_{ct}$ to achieve 23.2 dB per channel at the output is

$$(\text{CNR})_{ct} = -10 \log (10^{-23.2/10} - 10^{-37.8/10})$$

$$= -10 \log (4.62 \times 10^{-3}) = 23.3 \text{ dB}$$

It is obvious that the FM SNR improvement permitted a negligible effect from harmonics.

Step 2A: Design optical section.
We assume for this example that 1300-nm single-mode optics is used with linear analog transmitters and receivers. A linear laser diode (-50 dB second harmonic, -70 dB third harmonic) was already assumed for the previous harmonics analysis. The receiver sensitivity must be determined.

Using the same component performance as in the VSB case, MDP can now be calculated using Equations (7-2) through (7-4). For the eight channels, we use $(\text{BW})_{ct} = 24.4$ MHz, $(\text{CNR})_{ct} = 213.8 = 23.3$ dB, $M_o = 0.5$, and $r = 0.7$. Note that for the FET preamplifier, noise varies with $(\text{BW})^3$. The results of the MDP analysis for each of the eight channels are shown in Table 7.4. Note that MDP requirements are 6.3 dB higher for channel 8 than for channel 1. If we equalize the channels by applying more optical power to the higher channels and less to the lower, per Equation (5-29), the approximate $(\text{MDP})_{\text{comp}}$ becomes

$$(\text{MDP})_{\text{comp}} = \frac{-(26.4 + 20.6)}{2} = -23.5 \text{ dBm}$$

TABLE 7-4   RECEIVER SENSITIVITY (MDP) FOR EIGHT-CHANNEL VIDEO
FM/FDM EXAMPLE

Channel number	Carrier Frequency (MHz)	f_l (MHz)	f_h (MHz)	Amplifier noise Power $\langle i_{nA}^2 \rangle \times 10^{-17}$	MDP (dBm)
1	60	47.8	72.2	1.1	-26.4
2	100	87.8	112.2	2.8	-25.0
3	140	127.8	152.2	5.2	-23.9
4	180	167.8	192.2	8.4	-23.0
5	220	207.8	232.2	12.4	-22.3
6	260	247.8	272.2	17.2	-21.6
7	300	287.8	312.2	22.8	-21.09
8	340	327.8	352.2	29.2	-20.6

Step 2A.2: Determine Photonics Performance.   Now that MDP has been established, a preliminary link-budget analysis can be completed. Power penalties for per-channel band limiting do not have to be established since $(BW)_t \gg (BW)_{ct}$ with FDM. Power penalties due to the attenuation of channels near the receiver band edges due to photonics band limiting, $(BW)_p$, must be considered, however (Section 6.12.3). Since we are solving for transmission distance in this case, we can only deal with these penalties once maximum distance is established by first solving the link-power budget. We then test for link bandwidth and readjust the power budget as necessary.

Using $P_t = -0$ dBm, $L_c = 2$ dB, $M = 6$ dB, $l_s = 0.05$ dB/km, and $l_f = 0.4$ dB/km in the link-power budget, we get

$$D = \frac{P_t - \text{MDP} - M - L_c}{l_f + l_s}$$

$$= \frac{-0 - (-23.45) - 6 - 2}{0.45}$$

$$= 34.3 \text{ km}$$

Testing for link bandwidth over 34.3 km with a $t_{df} = 2$ ps/nm/km and a source spectral width of 3.5 nm, and assuming source rise time at 0.32 ns and a maximum receiver bandwidth of $(BW)_r = 600$ MHz, from Table 6.10, we get a photonics bandwidth, $(BW)_p$, as follows:

(a) fiber $t_d = 2$ ps/nm/km $\times$ 3.5 nm $\times$ 34.3 km $= 0.24$ ns
(b) Receiver $t_r = 0.35/BW = 0.35/600$ MHz $= 0.58$ ns
(c) Total $t_{r,\text{link}} = 1.1/0.1 + 0.06 + 0.34 = 0.7$ ns
(d) $(BW)_{p,\text{elect}} = 312/t_d$ MHz/ns $= 312/0.7 = 445$ MHz

With a transmission spectrum of 360 MHz (from Figure 7.18), the link bandwidth appears equivalent to overall spectral occupancy. Since bandwidth is measured at the 3-dB point, however, this implies that the highest channel (at 340 MHz) is reduced in power. Approximated by the 1-pole passband power-penalty curve of Figure 6.9, with a $(BW)_p/(BW)_t$ ratio of $445/360 = 1.2$, we get a 2.2-dB reduction in power.

When band-limiting power reduction occurs, the options are (a) limit the transmission distance or (b) equalize the power by adding to the higher-frequency attenuated channels. Because we are within the passband, the attenuation slope is so gradual that length reduction is not helpful; in addition, the main limit on bandwidth is not fiber but the receiver bandwidth limit we imposed. The more appropriate approach is to equalize power. If we use the approximation from Equation (5-29), then the resultant overall power penalty is

$$P_p = \frac{2.2 \text{ dB}}{2} = 1.1 \text{ dB}$$

Another power penalty to consider with distance is the partition noise. From Equation (6-91):

$$\frac{i_{s,pp}}{\langle i_n \rangle} = \frac{0.5\sqrt{2}}{[\pi\,(\text{BW})D\sigma_\lambda t_{dm} \times 10^{-12}]^2}$$

$$= \frac{0.5\sqrt{2}}{(0.0184)^2} = 2088 = 66.4\ \text{dB} \tag{7-6}$$

Converting peak-to-peak to rms power (-9 dB), we get 57.4 dB. From Table 6.11, the power penalty is

$$P_p = -5\log\left(1 - \frac{(\text{SNR})_t}{(\text{SNR})_n}\right)$$

$$= -5\log\left\{1 - \frac{1.7 \times 10^2}{5.5 \times 10^5}\right\} = 0$$

Recalculating the link-power budget with the band-limiting power penalty added, we get

$$D = \frac{P_t - \text{MDP} - M - P_p - L_c}{l_p + l_s}$$

$$= \frac{-0 - (-23.45) - 6 - 1.1 - 2}{0.45}$$

$$= 31.9\ \text{km}$$

7.6.4 PCM/TDM Video Transmission

With pulse-code modulation, the same principle of trading bandwidth for SNR holds. With PCM, however, the transmitted signal is binary and the received baseband SNR is not a strong function of transmission SNR once a certain bit-error threshold is reached (about 10^{-4} to 10^{-5}). Fiber systems transmit at 10^{-9} minimum, well above this threshold, and, therefore, the received decoded baseband SNR is a function only of the encoding process.

PCM has the distinct advantage of placing control of signal quality in the design of the encoder and isolating it somewhat from the quality of the transmission channel. For this reason, repeating the signal many times without degradation is possible. Because of this and because SNR requirements are smaller than with FM, greater distances can be achieved with PCM. With PCM, decoded baseband video SNR is a function of the number of bits (n) used to represent a measured sample of the original signal, per the following expression from Equation (2-20):

$$n = \frac{(\text{SNR})_{b,\text{ppwb/rms},w} - 1.8 - \text{CF} - \text{WF}}{6}$$

and

$$\text{BR}_c = n \times f_s = 2.4n\,(\text{BW})_b$$

or

$$BR_c = n \times 10.74 \text{ Mb/s as multiple of colorburst}$$

where CF = peak to peak to RMS = 7.66 dB
   WF = CCIR white noise = 6.2 dB

For example, a 60-dB $(SNR)_{ppwb/rms}$ CCIR weighted color video signal of 4.2-MHz bandwidth can be transmitted with 7- to 8-bit encoding requiring a transmission bit rate of 70 to 86 Mb/s (bandwidth of just over 35 to 43 MHz with NRZ encoding).

The signal power and transmitted SNR requirement remain the same regardless of the number of channels, since the transmitted signal is binary. The transmission $(SNR)_t$ requirement is fixed at 15.6 dB (avg/rms) based on the BER of 10^{-9}.

With digital PCM encoding, time-division multiplexing (TDM) is a natural approach to combine multiple channels onto a single fiber. The functional block diagram of a PCM/TDM video transmission system for a CATV application is shown in Figure 7.19. The only effect that multiplexing has on signal quality is that of increasing bit rate (and thus receiver noise) in direct proportion to the number of channels multiplexed.

High performance can be achieved over long distances without repeaters (40 km or more) and over very long distances by repeating without degradation. The disadvantages of PCM/TDM are its cost and the fact that the multiplexing capacity is limited by available product. Linear encoded video (having the lowest cost) requires a data rate of 70 to 86 MHz, which is not near any standard interface rate. SONET and ISDN will eventually provide standard rates for video in the 139-Mb/s range. The cost of linear PCM/TDM has been between 3 : 1 to 5 : 1 greater than FM/FDM, depending on the implementation [11, 14].

The performance of an eight-channel video PCM/TDM system is determined in Chapter 6, Section 6.17. This example assumes a custom multiplexer at 700 and 80 Mb/s per channel with 8-bit encoding to achieve 65-dB $(SNR)_{ppwb/rms, w}$. By taking into account conservative margins and component tolerances, a 35-km maximum repeater spacing is determined practical.

7.6.5 Wavelength-Division Multiplexing Video

From Figure 7.20, WDM can be used as another level of multiplexing for FM/FDM or PCM/TDM channel groups, or can be used to multiplex individual channels directly onto a single fiber. WDM can combine individual channels onto a fiber with no preceding electronic multiplexing, which is useful when signal format or data rates are not compatible with available electrical multiplexer product. The greatest disadvantage of WDM today is the lack of source, filter, and WDM MUX/DEMUX coupler products that can supply enough multiple wavelengths to be practical. To be effective, WDM trunking systems must operate in the long-wavelength ranges (1300 to 1550 nm) and channel separations of 15 nm or better are required to fit even 6 channels within a 100-nm "window." With conventional optics, only from 2 to 4 channels per "window" are practical. The use of multiple fiber windows requires dispersion-compensated fiber at both windows. In addition, with single-mode fiber, the added WDM coupler attenuation becomes a significant percentage of overall link at-

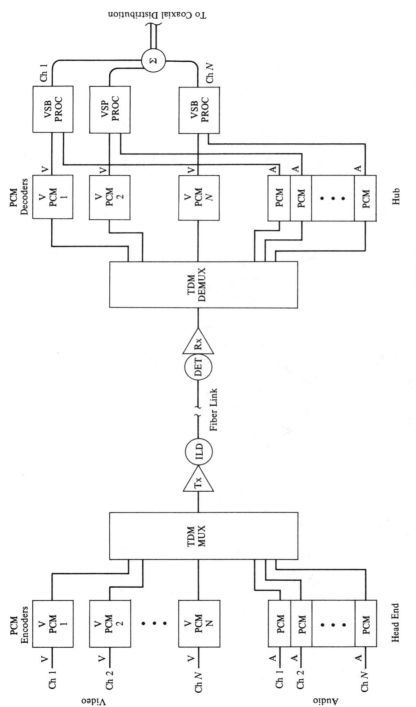

Figure 7.19 PCM/TDM transmission of CATV video.

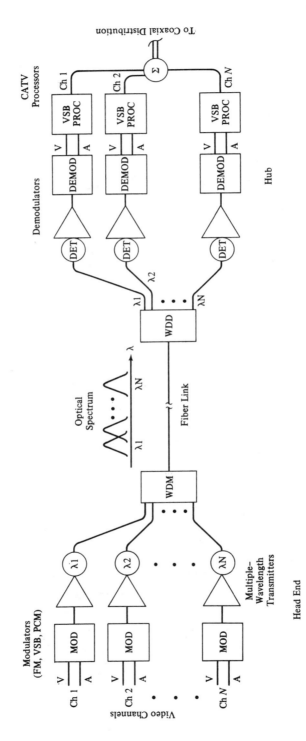

Figure 7.20 Wavelength-division multiplexed trunking of CATV video.

tenuation and, therefore, reduces the performance advantage. With conventional optics, WDM is not a conservative design approach. As wavelength-selectable single-line lasers and WDM grating couplers (3- to 5-nm separations) become more available or as coherent optics becomes a practical reality, WDM may then become the overall preferred approach for channel multiplexing.

In spite of the practical difficulties, many video transmission products employ WDM as a means of doubling or quadrupling channel capacity. The savings is in fiber or in the elimination of electronic multiplexing. Cost increases, however, for the electrooptics since multiple laser sources and receivers are required (one pair for each wavelength) and a WDM coupler is needed at both ends. Figure 7.14 illustrates the relative advantage of combining two and three wavelengths on a fiber, each carrying multiple FM/FDM or PCM/TDM channels. Note that in the case of FM/FDM, the curve has a steeper slope for the larger channel capacities. This is due to larger fiber material dispersions at 1.2 and 1.55 μm, causing bandwidth limitations.

7.7 FIBER IN THE DISTRIBUTION NETWORK

For the CATV application illustrated in Figure 7.12, one signal source broadcasts a common signal for distribution to multiple receiving terminals (converters) that tap onto a common distribution trunk. A second application of this tree distribution architecture is for signal monitoring, such as return signals from a subscriber loop or sensor monitoring, where multiple points along the trunk transmit to one common point.

7.7.1 Conventional Optics

Figure 7.21 shows a fiber implementation of a distribution trunk with passive Tee couplers. A common broadcast signal from a common source propagates down the cable and a portion of the energy is tapped off at each coupler and routed to the receiver. By using conventional optics, the number of terminals on such a system is limited by the loss associated with the couplers:

$$G_s = L_f + 2l_c + N(l_e + l_{th} + 2l_s) + l_t \qquad (7\text{-}7)$$

where G_s is the allowable operating power margin (systems gain), L_f is the total fiber loss, l_c is the connector loss at the source and receiving terminal, l_t is the power loss at the tap, l_e is the coupler excess loss, l_{th} is the power-splitting throughput loss, and l_s is the splice loss on either end of the couplers.

If, for example, we use 45 Mb/s for compressed video-service distribution to the home and a 26-channel system, we get a transmission rate of 1.2 Gb/s to the home. From Figure 6.18, we can assume a best-case G_s of about 30 dB using single-mode fiber and 1300-nm ILDs. At a 4-km mean distribution trunk distance and 0.4-dB/km fiber, $L_f = 1.6$ dB. If we use a 10 : 1 coupler, we get a power-splitting loss of 10 log $(1/11) = 10.4$ dB at the tap and a 0.4-dB throughput. Assuming a coupler excess loss of 0.2 dB, a connector loss of 1 dB, and a splice loss of 0.15 dB, we get a maximum number of terminals:

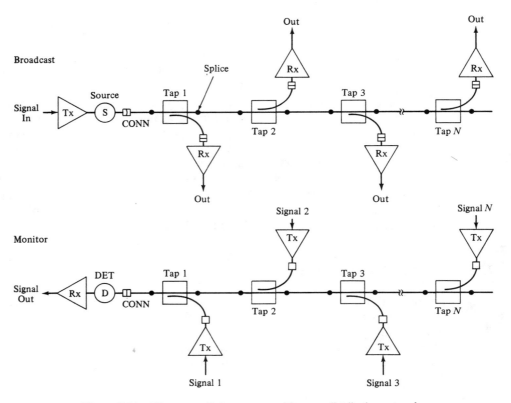

Figure 7.21 Fiber as applied to a tree-architecture distribution network.

$$30 = 1.6 + 2(1) + N[0.2 + 0.4 + 2(0.15)] + 10.4$$

$$N = 18 \text{ subscribers}$$

This relationship is plotted in Figure 7.23 for various coupler ratios and splice or connector losses.

In the return configuration, each terminal transmits into a common bus using optical couplers similar to those used for the broadcast case; therefore, the same power-splitting loss and terminal limitations result. Signals, in this case, are not common, so some means of separating them in frequency or time have to be implemented on the common trunk. The star architecture shown in Figure 7.22 is more suited to distribution using conventional optics. With this architecture, any terminal can transmit its signal into a common optical node (star) where its signal power is split equally to the fibers connecting all receivers. In the same way as with the bus, the power-splitting loss is the same whether it is being used as a splitter or as a combiner. With the star, however, the end-to-end power loss is not effected by the concatenation of multiple couplers and splices. The relationship is

$$G_s = L_f + 4l_c + [l_e + 10 \log (N)] \qquad (7-8)$$

where l_e is the excess loss of the coupler, and N is the number of terminals.

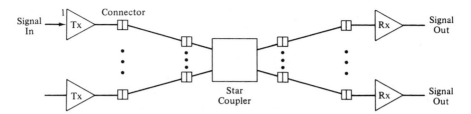

Figure 7.22 Fiber as applied to a star-architecture distribution network.

Using the same assumptions as before, but a coupler excess loss (plus uniformity tolerance) of 2 dB, we get

$$30 = 1.6 + 4(1) + [2 + 10 \log (N)]$$

$$N = 174 \text{ subscribers}$$

The number of subscribers is limited by the practical limits on coupler size (around 64 ports with conventional optics) rather than power. The relationship is plotted in Figure 7.23.

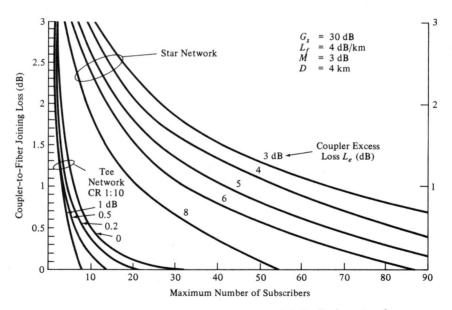

Figure 7.23 Comparison of star- vs. tee-coupled distribution network.

A practical implementation of the star system is a modified tree architecture. The broadcast source trunks the signals to star-coupler drop nodes that are located locally to serve a group of, say, 32 homes each. The drop cables from the node bundle each of the subscriber drop fibers in the same "breakout" cable and drop them as required along the route. The disadvantages of this system are the drop cable size and the planning of breakout points.

With conventional optics, fiber is hard pressed to compete with coax for widespread multiple distribution to residences using a tree architecture. Aside from the differential in component costs, the broadcast bandwidth becomes so large as to be impractical for more than a small group of channels. With FDM, the intermodulation as well as the bandwidth limits channel capacity in the same way.

A more appropriate architecture for the near term is the active star approach with active repeating or switching electronics serving a group of subscribers, as shown in Figure 7.24. In the CATV industry, this consists of "off-premise converters" serving a number of homes (generally, 12 to 24) with fiber drops from each converter to a home. This same video switching node could evolve to combine data and telephone as well. This approach has been put into practice with CATV off-premise converters in apartment complexes [17], as well as within the city of Alameda, CA [18]. The added cost of the fiber components (although small in these cases) over that of copper, and the more labor-intensive connector termination have presented a barrier to a full-scale introduction.

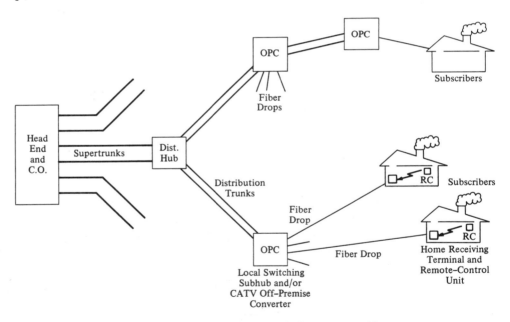

Figure 7.24   Fiber to the home: a practical near-term architecture.

7.7.2 Advanced Approaches

The next major area for fiber-optic technology advancement is coherent optics combined with integrated optic components and narrow-band wavelength-division multiplexing (WDM). The major advantages of coherent optics for the subscriber distribution network are the 10-to-19-dB improvement in receiver sensitivity [20, 21], the ability to transmit very broadband data rates, and the wavelength-selectable narrow-line lasers that are used. Line widths 10 to 30 MHz are achievable with distributed feedback lasers, 1 to 10 MHz with external coupled-cavity lasers and less than 1 MHz with external gratings [20]. With improvements in wavelength and line-width stabilization, WDM channel separations in the 10-GHz range will be possible. If we assume that a wavelength "window" is only 100 to 120 nm wide (say, 1250 to 1350 nm or 1480 to 1600 nm), using conventional optics with separations of 20 to 25 nm limits the number of channels to only 4 to 6 per window. As grating optics improves laser line-width stability and WDM coupler selectivity to permit 1-nm separation 50-channel capability will be possible. Coherent detection and such wavelength-detection technologies as interferometric filtering may permit over 100 channels per window [20].

With this in mind, Figures 7.25 and 7.26 show two possibilities for applying the advanced technology to the subscriber loop. The first, probably more near term than the second, consists of a narrow 1-nm separation WDM system whereby each subscriber is assigned a separate wavelength within a 100-nm spectrum of wavelengths. This permits 50 subscribers per trunk to be served. At the switching hub, the subscribers' requested service is switched to the narrow-line-laser modulator assigned to that subscriber. That laser's wavelength is then coupled with the 49 other wavelengths onto the distribution fiber with a narrowband WDM coupler. Along the distribution trunk are wavelength-selective couplers that couple off only the optical carrier at the specific wavelength assigned to that subscriber location. The carrier is then received and detected using a conventional noncoherent detector. Communications in the return direction, for service requests or interactive data, could be accommodated using packet or other multiple-access techniques on the same fiber using a common wavelength and directional couplers as shown.

With wavelength-selective couplers, there is no power loss to the optical carriers that are not selected by the coupler. Only the excess loss and coupler splicing loss are experienced by both the throughput signals as well as the signal tapped off to the subscriber. Also, since the service is selected and switched to the subscriber on demand, the transmission bandwidth to each subscriber is only that required for the services received at any one time.

As an example, let us assume that 139.264 Mb/s is chosen as the bandwidth per subscriber adequate to transmit compressed high-definition TV simultaneous with some other low-speed services. From Figure 6.18, the estimated systems gain in this case would be on the order of 35 to 40 dB. For the head-end WDM coupler, let us assume a 2-dB excess loss (L_{WDM}), and for each of the WDM trunk taps, a 0.3-dB excess and throughput loss combined (l_t). If we assume L_f is 1.6 dB, as before, a 1-dB connector loss, a 0.15-dB splice loss, and a 0.5-dB loss (l_{dc}) in the directional couplers, for the furthest subscriber, we get

$$G_s = L_f + 4l_c + L_{WDM} + 2l_{dc} + N(2l_s + l_t) + l_s$$

$$40 = 1.6 + 4(1) + 2 + 2(0.5) + N[2(0.15) + 0.3] + 0.15 \qquad (7\text{-}9)$$

$$N = 52 \text{ subscribers}$$

If these component-performance parameters can be realized reliably in the subscriber-loop environment, then the concept of a moderate-subscriber-count WDM trunk design is a potential.

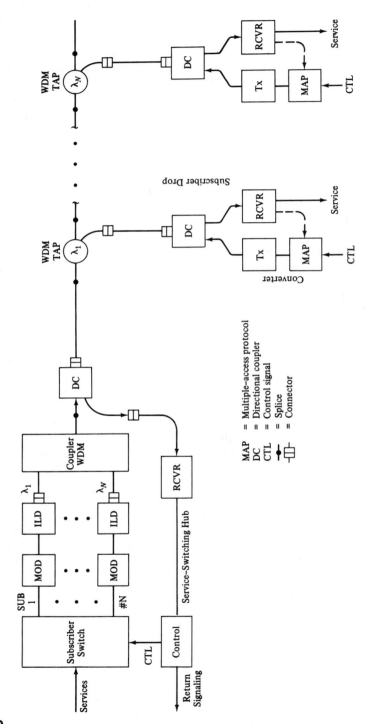

Figure 7.25 WDM subscriber distribution network.

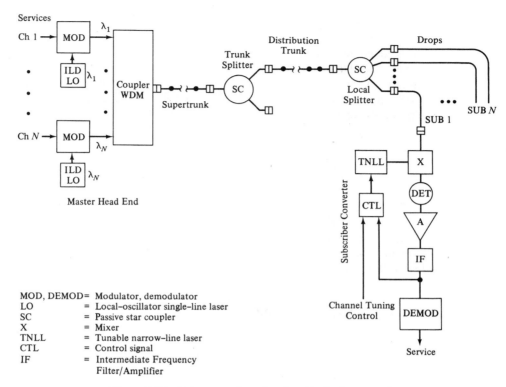

Figure 7.26   Coherent-optics subscriber distribution network.

Figure 7.26 illustrates a potential realization using coherent optics. The functional diagram is identical to that of a frequency-division multiplexing approach with the exception that the local oscillators in this case are ultra-narrow-line-width lasers (0.01 to 1 MHz). Each channel to be broadcast, amplitude, phase, or frequency modulates the optical local oscillator of a wavelength assignment corresponding to that channel. The modulation technique is technology-dependent, however, all approaches are conceptually possible just as with a broadcast radio carrier. The optical carriers are then combined in a WDM coupler onto a single fiber trunk. Following the tree architecture, we show the trunk being split at a hub point to feed multiple local-star distribution hubs. At the local star, the power is again split and all channels are dropped to the subscriber's "optical converter." There a tunable local-oscillator narrow-line laser selects the carrier wavelength corresponding to the channel of choice and downconverts it to a frequency that matches the IF frequency of the receiver (in the 0.1-to-1-GHz range). The signal then can be demodulated in a conventional fashion.

Although we achieve a 10-to-19-dB systems-gain improvement with coherent detection [21], the channel bandwidth at the receiver is that of the single channel, not of the entire channel spectrum. This approach, therefore, permits full broadcast capability with no high-speed demultiplexing required at the subscriber terminal.

For example, let us assume a 100-channel broadcast of digitized video at 139 Mb/s per channel and a WDM coupler that can combine 100 channels with a 3-dB throughput loss. If we assume only a 10-dB improvement in receiver sensitivity over that of noncoherent detection, the system gain becomes 45 to 50 dB. Assuming the same component losses as before, and a total of six in-line splices ($L_s = 0.9$ dB), a star excess loss (l_e) of 1 dB, and a fiber distance of 10 km ($L_f = 4$ dB), the power-budget relationship from Figure 7.26 becomes

$$G_s = L_f + 6l_c + L_s + L_{\text{WDM}} + 2l_e + 10 \log (N_{\text{TS}} + N_{\text{LS}})$$

$$50 = 4 + 6(1) + 0.9 + 3 + 2(2) + 10 \log (N) \qquad (7\text{-}10)$$

$$N = 1620 \text{ subscribers or splits per trunk}$$

The capability of coherent optics, therefore, provides the potential for simple distribution-network architectures employing passive fiber trunk systems, similar to that achieved today with CATV coaxial networks, but with the quality and bandwidth improvements that are required for high-definition digitized video and other interactive services.

7.8 BROAD-BAND FIBER NETWORKS

7.8.1 The Concept of Broad Band

A broad-band communications network provides multiple wide- and narrow-band communications services (voice, video, and data) to subscribers over a single cable. The network is generally interactive with broadcast, pay-per-view, one-way addressable, and two-way services accommodated on the same cable. Examples of broadband services and the bandwidth requirements estimated for each are shown in Table 7.5 [22, 23].

TABLE 7-5   BROAD-BAND SERVICES

Service classification	Estimated bandwidth requirements	
	Analog baseband	Digitized
Telephone	3 kHz	64 kb/s
Interactive data		1200–9600 kb/s
Wide-band data		64 kb/s
DS-1		1.544 Mb/s
Fax	4 KHz	9.6–64 kb/s
ISDN Basic $2B + D$		144 kb/s
FM radio	2 Mb/s	
TV video and audio	6 MHz	70–100 Mb/s
EQTV		90–138 Mb/s
HDTV	8.4–20 MHz	150 and 200 to 700 Mb/s
Metering and alarms		<1 kb/s
CD and stereo		2 Mb/s

Broad-band network trials have been underway for over 10 years. The first large-scale trial was the "Hi Ovis" trial in Japan, which started in the mid-1970s. Other trials followed in Canada, Germany, France, the United Kingdom, and the United States. The German BIGFON trials, with the first cut over in 1986, extended to seven cities. Some broad-band trials have become large-scale fully operational systems. The French system, installed initially in Biarritz, had 1500 subscribers by June 1986 and over 3 million projected for 1988 [19]. It is an all-optical system with services including telephone, videophone, TV, stereo, text, and data-base services.

In spite of these trials, with some minor exceptions, the demand for subscriber-supported broad-band networks has been disappointing. The CATV industry learned in the early 1980s that subscribers were unwilling to pay for all but the basic broadcast services. For CATV, fiber appeared to offer no advantage in the distribution network, except immunity to signal ingress, and was generally considered too expensive or too hard to install and maintain in the distribution network.

Broad-band networks will become a reality as the following circumstances alter the economics:

(a) broad-band networks are established in newly built or rebuilt areas as a part of an integrated telephone and CATV service, where both services share in the cost;

(b) private communities establish the network where the cost is absorbed in the construction of the community;

(c) the public becomes accustomed to interactive services and begins to demand them on a large-scale basis;

(d) technology and production volume drive the cost of integrated fiber-optic and digital termination components into an acceptable range.

7.8.2 Broad-Band Network Architecture

Figure 7.27 shows a general architecture common to most broad-band networks. The common elements are as follows.

Switched-hub architecture.   The network has a central point of signal origin and control, with remote switching and distribution hubs in areas where users are concentrated. All services are made available via supertrunks to the distribution hubs. From these local distribution hubs, only those services requested are switched to that requesting subscriber. In this way, the high-cost transmission trunks, and switching and terminating electronics are held to a minimum, and the cost is shared by the subscribers served by the distribution hub. By switching only a limited number of low-speed services at a time to each subscriber, the realization of the distribution network can be maintained at a reasonable cost per subscriber. The objective is to keep the subscriber drop and terminal at a minimum cost since these are the highest-quantity elements in the network.

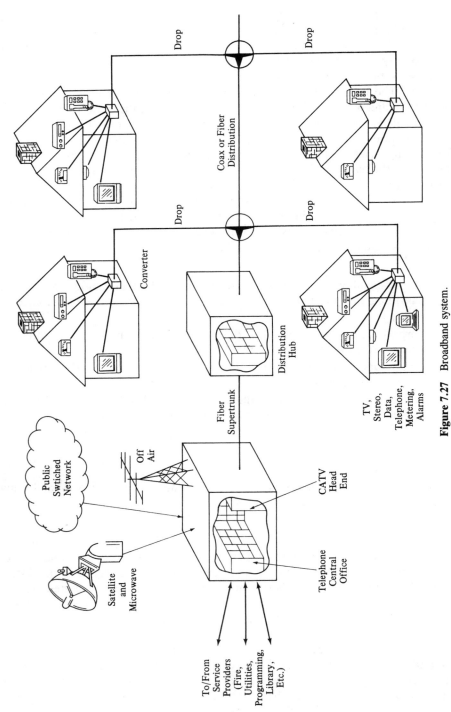

Figure 7.27 Broadband system.

Network center.   This location serves as the source and/or gateway for all services to the network and as a central control point. For CATV services, it houses the head end, and for Broadcast, it contains the satellite, off-air, and microwave receivers. For telephone, it is the central office. For data services, it is a digital cross-connect, switch, protocol converter, and gateway to other data networks and data services. For security and metering, it is a signal-concentration and distribution point for those businesses that provide the service. It must handle and switch both analog and digital information and encode and format the information for transmission over the supertrunks to the distribution hubs. The network center must also monitor the network equipment and traffic for maintenance, provisioning, and billing purposes.

Distribution hub.   This hub is a switch point for subscriber-requested services. It receives all available services from the network center and formats the signals for distribution to the users on demand. It therefore performs demultiplexing, demodulation, switching, remodulation, and multiplexing functions. Its primary functions are to switch services on demand and to modulate/multiplex these services into the distribution network.

Services received from the network center on the supertrunk are demultiplexed in order to separate services type and channel groups. Certain services can be remodulated and reformatted in order to facilitate retransmission on the distribution network. On request, or as assigned, specific services are routed or switched to each subscriber. Distribution-signal format and channel-multiplexing approaches are chosen so as to reduce the cost and complexity of channel extraction at the subscriber terminal.

Distribution network and drops.   This network serves each subscriber with all broadcast services and the on-demand services switched to that subscriber by the distribution hub. The network must be sufficiently broad band to supply the bandwidth required of simultaneous services to the subscriber, but low in cost. In today's environment, coax is the medium of choice due to its low cost and ability to distribute and drop in a tree topology. With proper design and service considerations, however, fiber will become the future choice. Initially, a point-to-point fiber drop from hub to subscriber is most feasible. This can be cabled to appear as a single cable with drop points. Eventually, technology will permit other configurations, as outlined in Section 7.7.2. It is likely that the fiber drops will be single mode operating at 1300 nm or greater with the possible exception of very short applications such as apartment complexes.

Subscriber terminal.   The functions of the subscriber terminal are to command the services and channels selected and to process the incoming and outgoing signals in a manner so that the attached appliances (televisions, radios, data terminals, telephones, etc.) can directly interface and utilize the services. The home terminal must interface with existing TV, radio, and data-terminal devices. Specifically, the home terminal will consist of a fiber-optic receiver and transmitter, with signal-processing (modulators, multiplexers, packet assemblers/disassemblers, and such) and port-interfacing electronics. It will be generally commanded by a remote hand-held control for selection of services.

Most home terminals will be required to serve multiple telephone lines, three to four channels of video, the FM broadcast band, high-fidelity subscription stereo, high-definition video, interactive data, and low-data-rate channels for control and metering (security, fire, etc.). More sophisticated terminals may return video for conferencing or monitoring purposes.

7.8.3 Transmission Technology Trade-Offs

The objectives in selecting a network architecture and the technologies to implement it are generally to limit the cost per subscriber to an affordable amount while providing the bandwidth and service delivery capabilities necessary for long-term operation. Most public entertainment networks must pass tens to hundreds of thousands of subscribers, and survive on the revenue provided by serving about 50% of these. Adding telephone service ensures nearly 100% penetration, but does not ensure the entertainment revenues. Analysts generally estimate that the network must be built for a cost per subscriber in the $1500 to $2000 range to be self-supporting on revenues [23]. Because the subscriber element is so important, we focus on the technology trade-offs surrounding the distribution plant and the home terminal. The time scale for implementation versus the time scale for technology evolution perhaps has the greatest impact on architecture.

Near term. The near-term single-mode technologies using bulk optical components, multimode LEDs and lasers, and GaAs PIN/FET or APD detection forces an architecture aimed at reducing the bandwidth of the drop and subscriber terminal so that it can be served with low-cost components. Cost is placed in the local switching hubs where services are selected and distributed. With installed systems to date, the cost constraints have dictated (a) low-cost fiber technology such as 850-nm laser or LED sources, PIN detectors, multimode graded or large-core fiber; (b) simple modulation and multiplexing techniques such as FDM VSB, which can be inexpensively processed on the receiving end; some WDM (usually, two to four wavelengths) for two-way fiber transmission or video channel separation; and (c) creative solutions to couplers or fiber/cabling technology that can reduce the cost of the glass used by each subscriber.

Since off-premise converter technology for broadcast CATV had matured to some degree, it has provided a viable option. By centrally locating the converters for a group of homes in the distribution hub and dropping fiber to the subscribers, a limited broad-band network is formed. Installations have been done on this basis [17, 18], but have not proven more economical than coax.

Regarding the drop fiber, the trade-off is between single-mode and large-core multimode. More light can be launched from an LED into the large-core multimode, and assuming the loss characteristics are comparable within 1 dB/km or less, the large core would intuitively seem to be the choice. Unfortunately, large-core fiber requires more glass and is not produced in large volumes as is single mode; therefore, it is more expensive. On analysis, the engineer may find that single-mode-fiber, even when interfaced with a multimode LED, may perform adequately and that the combination may be the most economical. For example, given a 40-dB optical margin, if a 1300-nm LED experiences a 30-dB coupling loss into single mode, the 10-dB remaining margin at 0.4 dB/km equates to 25 km.

A necessary approach to reducing the fiber cost is to transmit both ways over a single fiber. The trade-off is the cost of the fiber and duplex connectors versus the cost of directional couplers and a simplex connector. For example, assuming a cost of $0.25 per cable-fiber meter for drop cable, a 3-km two-fiber drop would cost $1500, whereas a single fiber drop would cost $750. If the cost of duplex couplers is less than $375 apiece (much more than should be feasible with volume production), the single-fiber approach is superior. Using single-mode fiber permits the couplers to be directional, thus permitting single-wavelength operation.

In the near term, careful selection of modulation and multiplexing at the hub is also necessary in order to keep the cost of demodulation and demultiplexing at the home terminal low. Digital multiplexing has been too expensive for video and voice. Frequency multiplexing of analog and digital has been the most economical choice. By using frequency bands and formats that can be inexpensively converted to standard "channel" assignments and interfaces compatible with telephone, TV, stereo, and data terminals, the home terminal can simply be a low-cost RF converter with simple modems for baseband signals.

Near future.   In the near future, wavelength-selectable narrow-line lasers and grating optics will make WDM a more viable technology, permitting the multiplexing of larger groups of channels than has been possible in the past. WDM can permit more channels to be transmitted simultaneously to the subscriber while maintaining a low data rate for the subscriber terminal to process. As discussed in Section 7.7.2, it also will permit a more conventional distribution trunk structure by using wavelength-selective couplers at the subscriber drop. This will reduce the cable cost per subscriber as well as the labor involved in handling and terminating large drop cable bundles.

With digital TV, digital telephone handsets, lower-cost LSI, and emerging digital subscriber-loop standards such as ISDN, it is likely that an all-digital approach will be universally adopted. Transmission rates of 139 Mb/s to 2.2 GHz to the home have been discussed. Combining digital technology with WDM should permit a multitude of services to the home while maintaining bandwidths to the subscriber in the 139-to-600-Mb/s range.

Architecturally, the switching distribution hub will still be required to limit service bandwidth on the subscriber drop. Although bandwidth capability is increasing, so are the bandwidth requirements for the services.

Longer term.   The maturing of coherent optics could bring about an entirely different architectural concept to broad band. With coherent optics, we treat the optical carrier much like one would an RF carrier. Conceptually, the same techniques for frequency multiplexing of channels within a narrow spectrum, as is done today for TV or CATV broadcast, can be used with coherent optics. One difference is that the frequency band is in the optical region of the RF spectrum. The other key difference is that because the carrier is in the terahertz range, line widths are 1 MHz or less, and separations are in the gigahertz range, there exists a capability to transmit a gigahertz or more of information per carrier.

The example given in Section 7.7.2 illustrates the capability of broadcasting over 100 channels of wide-band information to a large base of subscribers over a wide area using a totally passive optical network configured in a treelike heirarchy much like in today's CATV environment. The requirement for switching distribution nodes to limit bandwidth on the subscriber drops disappears. All broadcast services can be received by all subscribers and selected with a tunable optical converter much like the RF CATV converter used today. Obviously, the requirement for switched services exists, however, the location and architecture of the switches need not be dictated by the bandwidth limits of the subscriber drop.

7.9 LOCAL-AREA NETWORKS*

7.9.1 Introduction to LANs

A LAN is a multiterminal multiple-access communications system operating within a somewhat closed or private environment and within a limited-distance and fixed transport data rate. A LAN is generally a private computer-based data-transmission network operating within the office environment between hosts and terminals or between PCs, spanning a single floor, a single building, or multiple buildings within a complex.

A LAN is characterized by the fact that, unlike trunking systems, the transmission medium and the bandwidth are shared among the stations, through some multiple-access means. Multiple-access mechanisms can be classified as dedicated or contention. Dedicated implies that a portion of the communications channel is dedicated to each station, whereas contention implies that each station must bid for the use of the channel. Common approaches include the following:

1. Dedicated access (shown in Figure 7.28):
   (a) Time-division multiple access (TDMA)
   (b) Frequency-division multiple access (FDMA)
2. Contention access (shown in Figure 7.29):
   (a) Carrier-sense multiple access/collision detection (CSMA/CD)
   (b) Token passing

With TDMA, a controller on the LAN sets up a time dependent transmission frame that contains time slots, with one or more time slots dedicated to each station desiring access. Each station transmits only during its allotted time slot. A disadvantage of TDMA is that the bandwidth is used whether a station is transmitting or not. It is also constrained by having a master controller as a single point of failure, although some systems are designed to pass this function on in that event.

---

*The author would like to acknowledge and thank J. Richard Jones [39] for his contribution in supplying and organizing much of the material for this and the next three sections.

Time–Division
Multiple Access

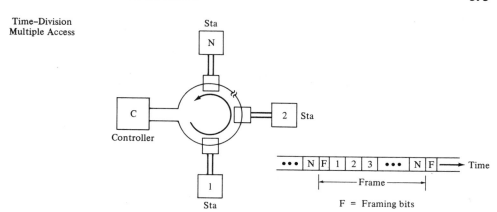

Frequency–Division
Multiple Access

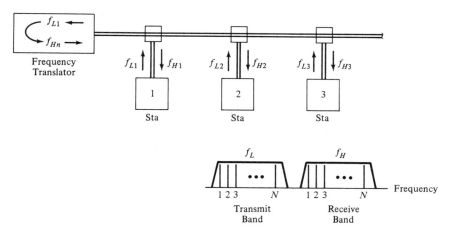

Figure 7.28   Dedicated multiple-access LAN approaches.

With FDMA, the medium is divided into frequency bands, one assigned to each station wanting access. A station transmits at its assigned frequency and receives on another. Translation from transmission frequencies to receiving frequencies occurs at a master translator point that is similar to the controller of a TDMA system. This is a common approach used with CATV systems for providing data services. Like the TDMA system, it has the disadvantage of using the assigned frequency slot even when the station is not transmitting. It also requires a translator, which becomes a single point of failure.

Contention methods are the most favored for general-purpose LANs. In CSMA/CD, each station listens (senses the carrier) before it attempts access and waits until the channel is clear to transmit. If two stations accidentally transmit at the same time, due to propagation delays causing the carrier from another transmitting station to be received after the listening station begins to transmit, then a collision

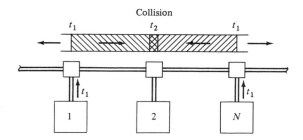

t_1 = transmission begins at terminals 1 and N simultaneously
t_2 = collision of leading edge takes place
t_3 = collision detected
t_4 = transmission ceases at all terminals
t_5 = terminals wait an indeterminate amount of time to
       transmit after sensing that the bus is clear

(a)

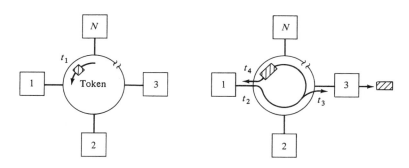

t_1 = terminal N gives up token and passes it to terminal 1
t_2 = terminal 1 takes token and transmits information intended
       for terminal 3
t_3 = terminal 3 accepts the information and passes on the total
       data stream
t_4 = terminal 1 senses its own transmitted data returning,
       deletes the data, and frees the token, passing it on to 2

(b)

Figure 7.29 Contention-based multiple-access LAN approaches. (a) CSMA/CD.
(b) Token passing.

occurs, as shown in Figure 7.29(a). The result is a distorted "collision" signal that is sensed by all stations. All stations then cease transmission for a probabilistically determined time period and then try again.

Token passing is somewhat like polling in that each station is sequentially given an opportunity to transmit by being offered a token. As shown in Figure 7.29(b), a control message called a token, which gives the right to access the medium, is passed around the network from station to station. Only the station receiving the token can transmit. If that station has nothing to transmit, then it regenerates the token and passes it to the next. If, for example, station 1 in Figure 7.29(b) receives the token and wants to transmit to 3, it transmits, and 3 receives and repeats the message back to 1. Station 1 then releases the token to the next station 2.

LAN topologies are shown in Figure 7.30. They consist of the bus, star, ring, and tree [39]. LANs are often interconnected to form hybrids of these topologies, as is shown in some practical configurations later in the chapter. As is demonstrated in the next section, the star and active ring (ring with active access couplers) topologies are the most applicable to fiber-optics implementation.

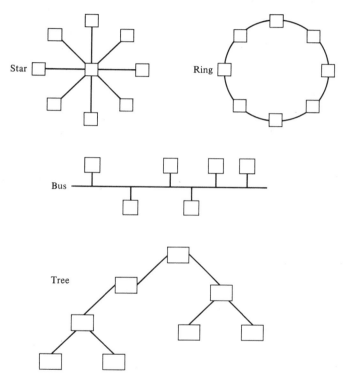

Figure 7.30   LAN topologies.

The role of fiber in the LAN environment is to replace the more conventional mediums of coax and twisted pair to reduce electromagnetic interference and to increase data rate and distance. Most LAN standards activities have concentrated on coax and twisted pair for transmission rates below 16 Mb/s, although fiber-replacement standards are evolving for those lower-speed LANs. The major impact that fiber has had on LANs is to provide the capability for extremely high-speed and large LANs. The FDDI standard, discussed further in Section 7.12, was developed around the unique capability of fiber. It provides for a 100-Mb/s token ring LAN that can extend to 100 km.

7.9.2 LAN Standards

The bulk of the LAN standards activity in the United States is being carried out by the IEEE 802 committee and the ANSI X3T9 committee. The standards have been written for compatibility with the ISO/OSI model. In order to accomplish this, the IEEE 802 committee has partitioned the data-link layer into two sublayers, the logical link-control (LLC) layer and the medium access-control (MAC) layer. The LLC is independent of access method and the MAC is dependent of access method. The MAC and physical (PHY) layers define the LAN approach and the physical medium. ANSI, in its development of the FDDI standard, has also made it compatible at the IEEE LLC layer. A comparison of the major IEEE and ANSI LAN standards to the OSI model is shown in Figure 7.31.

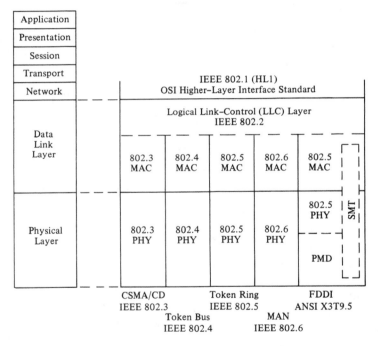

Figure 7.31 Relationship between IEEE, ANSI, and OSI LAN standards.

The IEEE has defined the 802.1 higher-layer interface (HLI) as the relational standard between the IEEE 802 standards and the ISO model [24]. Standard 802.2 LLC is responsible for the layer-2 functions of the OSI model [25]. The IEEE has formed 802.3 through 802.6 as committees responsible for MAC and PHY standards relating to specific LANs:

(a) Committee 802.3 is responsible for the CSMA/CD LANs [26], Ethernet being one example. Although this standard is a 10-Mb/s coaxial standard, the committee is also developing a twisted-pair version, Type 10, Base T [27]; a fiber version [28]; and a broadband version known as 10BROAD36.

(b) Committee 802.4 is responsible for the token-bus (TB) standards [29].

(c) Committee 802.5 is responsible for the token-ring (TR) standards [30]. Although the standard has been based on twisted pair, 802.5J has been drafted on fiber optics.

(d) Committee 802.6 is responsible for MAN standards [5, 31]. This standard, discussed in some detail in the previous sections, is media-independent.

Besides the specific LAN or layer-related groups, the IEEE has broad-band and fiber support committees. IEEE 802.7, the Broad-band Technical Advisory Group (BB TAG), supports MAC sublayer groups in problems related to broad-band design. IEEE 802.8, the Fiber-Optic Technical Advisory Group (F/O TAG), provides information on fiber-optic network design techniques and studies problems related to its implementation.

ANSI has set up under its X3 Information Processing group, and within its X3T9 I/O Interfaces committee, an X3T9.5 committee to study future interfaces. One standard evolving from this committee is the fiber distributed-data interface (FDDI) standard for a 100-Mb/s token-passing LAN [8]. This is discussed in detail in Section 7.12. As Figure 7.31 illustrates, the FDDI standard is compatible at the IEEE 802.2 LLC layer and contains a MAC and PHY layer specific to the FDDI LAN. The PHY layer in this case contains a physical medium-dependent (PMD) layer that is specific to the functioning of the fiber-optic ring medium. Also contained in the FDDI standard is a station-management (SMT) layer that bridges all three sublayers and provides the control necessary at the station level to manage the processes of all three sublayers. It provides such services as configuration management and fault isolation and recovery.

Table 7.6 lists the key characteristics of some of the more common standards that have emerged from IEEE and ANSI efforts.

7.9.3 LAN Application

Figure 7.32 shows how these various LANs might be used and interconnected to serve various applications within a large interoffice data network. The 100-Mb/s FDDI LAN is a natural to form a network backbone or a backbone within a building or office complex. Since it interfaces at the IEEE 802.2 LLC layer, it is compatible with all other 802 LANs and can, therefore, gateway into them as well as other OSI networks. FDDI II adds voice and video capabilities to the LAN and, by creating a 125-μs frame (up to 16 6.144-Mb/s wide-band channels), it is compatible with the public switched telephone network. Extension of the LAN across a WAN is therefore possible using the FDDI-II hybrid-ring control (HRC) [32]. Although the specifications for FDDI apply it in the MAN in terms of physical size (100 km), the LAN can also be extended to the 802.6 MAN through the LLC gateway and therefore operate with systems that might be adopted for fiber to the home and ISDN, for example.

TABLE 7.6 CHARACTERISTICS OF STANDARD LANS*

	IEEE				ANSI X3T9.5
Standard	802.3	802.4	802.5	802.6	
Designation	CSMA/CD	Token bus	Token ring	MAN	FDDI
Common product names	Ethernet StarLAN	MAP	IBM Token Ring		
Physical medium	Coax twisted pair Fiber	Coax Fiber	twisted pair Fiber	Any	Fiber
TOPOLOGY	Bus Star Tree	Bus	Ring Star	Ring Bus	Ring
Access method	CSMA/CD	Token passing	Token passing	Slotted	Token passing
Signaling	Baseband	Broad band Carrier band	Baseband	Baseband	Baseband
Data rate (Mb/s)	1 and 10	5 and 10	4 and 16	> 150	100
Configuration	1BASE5 1 Mb/s 500 ft twisted pair 10BASE2 10 Mb/s 200 ft RG-58 10BASE5 10 Mb/s 500 ft Coax 10BASET 10 Mb/s twisted pair Active Star 1024 terminals 4 km Fiber Passive Star 33 terminals 1 km Fiber		4 Mb/s twisted pair 260 devices 16 Mb/s Fiber twisted pair		100 km 500 terminals

*The author wishes to acknowledge and thank J. Richard Jones of Broadband Technologies for some of the materials for this table.

The 802.3 through 802.5 LAN standards are structured more for the office-size LAN. They are ideal for interconnects between PCs or between host Processors and peripherals. The 802.5 token-ring standard uses twisted pair and the 802.3 CSMA/CD standard has been adopted by some suppliers to twisted pair, which makes both standards compatible with existing building telephone-cabling standards. Data rates of up to 16 Mb/s have also been established. The argument of fiber

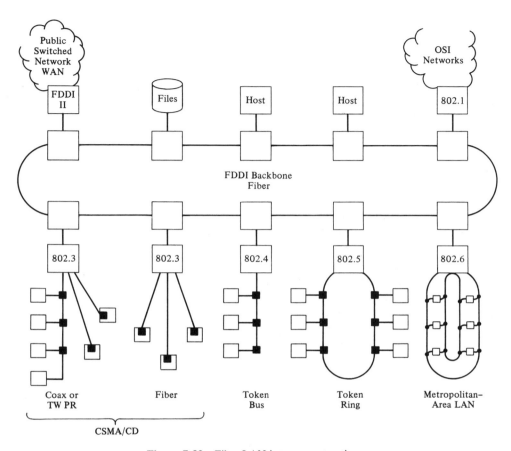

Figure 7.32   Fiber LAN interconnect options.

versus shielded twisted pair versus unshielded twisted pair for these LANs is somewhat application specific, being a tradeoff of cabling/connector cost, EMI constraints, distance and data rate.

7.9.4 CSMA/CD Versus Token-Passing-Capacity Limitations*

As fiber creates an atmosphere of higher data rates and longer-distance transmission, the relative capabilities of the multiple-access approach become an important consideration in the design of future fiber LANs. With contention access approaches, increasing transmission rate to take advantage of fiber bandwidth does not necessarily increase the data throughout accordingly. With contention schemes, time on the network is divided into (1) time contending for the channel and (2) time transferring data. Using this model, throughput can be thought of as the percentage of total time

*The author wishes to acknowledge and thank J. Richard Jones of Broadband Technologies for the materials for this section.

that is spent transferring data. Jones [39] has provided the following analysis of the two contention approaches.

In a CSMA/CD system, if the size of the information packet remains fixed while the data rate is increased, a smaller percentage of the total network time is spent transferring data. Stallings [34] formalizes this phenomenon with a study of the probabilities of collision and develops the following relationship for maximum throughput, $(BR)_{thru}$, in bits/second:

$$(BR)_{thru} = \frac{(BR)_{load}}{1 + \dfrac{3.4(BR)T_d}{PS}} \tag{7-11}$$

where $(BR)_{load}$ = offered load in bps
 BR = data rate in bps
 T_d = maximum propagation delay between stations (network size)
 PS = packet size in bits

Figure 7.33 [39] plots (dashed lines) the maximum throughput on a 2-km bus as a function of bit rate for various packet sizes for CSMA. There is a diminishing return in the maximum achievable throughput as data rate is increased. By increasing packet size, throughput increases.

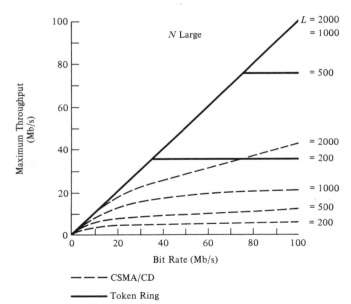

—— — — CSMA/CD

————— Token Ring

Figure 7.33 Throughput versus bit rate for CSMA/CD and token-passing LANs.

Unlike CSMA, with token-ring systems, maximum throughput improves linearly with increasing data rate as long as the transmission of a packet occupies the entire ring. This is because when many stations want to transmit, most of the network's bandwidth is spent transferring data not contending for the medium. In the IEEE 802.5 standard, a station cannot release the token until it sees the leading edge of the packet it has transmitted coming back to it. When a ring becomes so large or the data rate so fast that the packet transmission time does not occupy the entire ring, a portion of the bandwidth is wasted. Stallings [34] gives the following expression for maximum throughput of a token ring operating in this fashion:

$$(DR)_{thru} = (DR)_{load} \qquad for \frac{(BR)T_d}{PS} < 1$$

$$(DR)_{thru} = \frac{(BR)_{load} PS}{(BR)T_d} \qquad for \frac{(BR)T_d}{PS} > 1 \tag{7-12}$$

The expression is plotted as the solid line in Figure 7.33 for a 2-km ring using various packet sizes. By comparing this to the dashed-line plot for CSMA, it appears that token-ring LANs benefit more from increasing data rate than CSMA LANs and are less sensitive to physical network size (propagation delay T_d). These expressions were for heavily loaded systems. Under light loads, the opposite is true. CSMA outperforms token passing in terms of delay, and throughput improves more linearly with bit rate. The choice is, therefore, application-dependent. It is interesting to note that the ANSI committee has chosen the token-passing method for the fiber-optics FDDI standard.

7.10 FIBER-OPTIC-LAN TOPOLOGY CONSIDERATIONS

Figure 7.30 shows the possible topologies for a LAN. The three most common are the star, bus, and ring. The tree is in essence a hybrid of the star and bus in most implementations, whereby the node at the branch acts as a star feeding multiple-bus LANs or other stars. Applying fiber to these topologies raises issues peculiar to the nature of fiber optics, including (a) link-power-budget limitations caused by practical component power losses, (b) limited ability to detect line conditions and collisions in CSMA/CD LANs due to the unipolar nature of optical signaling and the dynamic signal range involved, and (c) broad-band capability permitting longer distances and higher data rates. This section concentrates on the practical limitations caused by component power losses that restrict fiber to the star and active-ring topologies. The following sections concentrate on signaling issues and the unique nature of fiber in such systems as the FDDI.

7.10.1 Passive Fiber-Optics Bus and Ring Topology

Figure 7.34 illustrates the physical components that make up a passive fiber-optic bus or ring. The bus can be unidirectional if connected end to end as a ring or bidirectional if configured as a linear bus. Passive two-port optical couplers are used to couple power from each station's transmitter onto the bus fiber and to couple common signal power from the bus to the receivers at each station. The station access electronics is called a medium access unit (MAU) and contains, at the PHY layer, the electrooptical transmitter and receiver electronics. All interconnections of the MAUs and the couplers are through optical connectors. The link-power budget calls

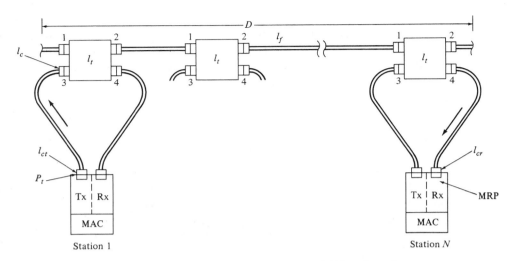

Figure 7.34 Passive fiber-optic ring or bus LAN configuration.

for systems gain to be greater or equal to loss. The relationships for the furthest separated stations on the bus, by tracing the signal from the transmitter at station 1 to the receiver at station N, is

$$P_t - \text{MRP} - M \geq [l_{ct} + N(2l_c) + l_{cr}] + [l_{32} + (N - 2)l_{12} + l_{14}] + Dl_f$$

$$(7\text{-}13)$$

where P_t = transmitted power before the connector, in dBm
MRP = minimum power required at the receiver after the connector; assumes all power penalties P_p as well, in dBm
M = assumed operating margin, in dB
l_{ct} = connector loss at the transmitter, in dB
l_{cr} = connector loss at the receiver; generally low, in dB
l_c = connector loss at the couplers, in dB
l_{nm} = loss from port n to port m of a coupler, where l_{nm} = excess loss (l_e) + power-splitting lossing (l_{sp}), in dB
D = longest fiber distance between stations, in km
l_f = fiber loss, in dB/km
N = number of stations

The simplified coupler loss relationship from Equation (3-74) is

$$l_{nm} = l_e + 10 \log \frac{1}{\text{CR}} \qquad (7\text{-}14)$$

$$\text{CR} = \frac{P_{oc}}{P_{o,\,\text{tot}}} = \frac{P_{ic}}{P_{i,\,\text{tot}}}$$

where P_{ic} and P_{oc} are the powers coupled at the input and output ports, respectively, and $P_{i,\,\text{tot}}$ and $P_{o,\,\text{tot}}$ are the total input and output powers, respectively. For a two-port coupler, if we assume that $l_{sp32} = l_{sp14} = 10 \log (P_{tot}/P_{oc})$, solving for number of stations N, we get:

$$N = \frac{P_t - \text{MRP} - M - l_{ct} - l_{cr} - 20 \log (P_{tot}/P_{oc}) + 20 \log (P_{o,tot}/P_{o2}) - Dl_f}{2l_c + l_e + 10 \log (P_{o,tot}/P_{o2})}$$

$$(7\text{-}15)$$

As an example, let us assume a 500-m 10-Mb/s multimode 820-nm bus with 4 : 1 couplers and the following parameters: $P_t = 0$ dBm, MRP $= -30$ dBm, $M = 3$ dB, $l_{ct} = l_c = 1$ dB, $l_{cr} = 0.5$ dB, $l_f = 3$ dB/km at 820 nm, and $l_e = 0.3$ dB. For the couplers: $P_{o4} = P_{i3} = P_c = 1$, $P_{o2} = P_{i1} = 4$ and $P_{o, tot} = P_{i, tot} = P_{tot} = 4 + 1 = 5$. The maximum number of stations is, therefore,

$$N = \frac{0 - (-40) - 3 - 1 - 0.5 - 20 \log (5/1) + 20 \log (5/4) - 0.5(3)}{2(1) + 0.3 + 10 \log (5/4)}$$

$$= 3.6$$

The maximum number of stations, therefore, is between 3 and 4 based on power limits alone. This relationship is plotted in Figure 7.35 for various component-parameter variations. This example and Figure 7.35 demonstrate the limitations of passive fiber bus and ring topology.

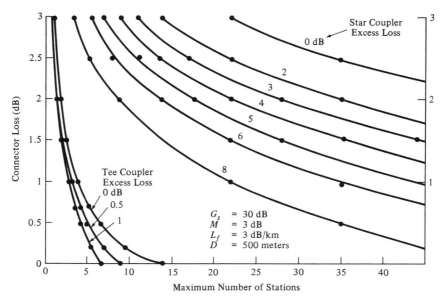

Figure 7.35 Comparison of star- versus tee-coupled LAN performance.

7.10.2 Passive Fiber-Optics Star Topology

Figure 7.36 shows the physical components that make up a passive fiber-optic STAR. Each of N stations access a common $N \times N$-port optical star coupler that acts as a power-sharing device. Each station's transmitter couples its power into the common star mixer and each station's receiver accesses that star mixer as well. The star coupler mixes all transmitted signal power and divides it equally among all output (receiver) ports. Each station, therefore, receives all transmitted signals at $1/N$th power less other in-line losses. As with the bus, all interconnections of the MAUs and the coupler are through optical connectors. The link-power Budget calls for systems gain to be greater than systems loss. The relationships for the furthest separated stations on the bus, found by tracing the signal from the transmitter at station 1 to the receiver at station N, is

$$P_t - \text{MRP} - M > (l_{ct} + 2l_c + l_{cr}) + L_{\text{star}} + Dl_f \qquad (7\text{-}16)$$

where P_t = transmitted power before the connector, in dBm
    MRP = minimum power required at the receiver after the connector, in dBm
        M = assumed operating margin (including power penalties), in dB
      l_{ct} = connector loss at the transmitter, in dB
      l_{cr} = connector loss at the receiver; generally low, in dB
       l_c = connector loss at the coupler, in dB
    L_{star} = loss from any of N input ports to any of N output ports, in dB
       D = longest fiber distance between stations, in km
       l_f = fiber loss, in dB/km
      N = number of stations

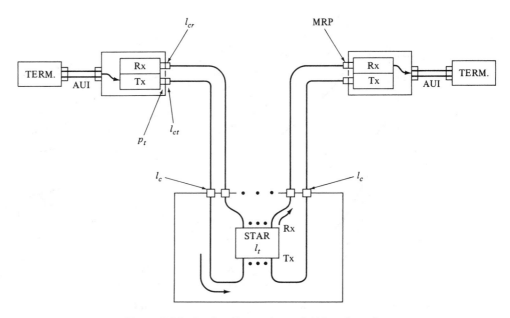

Figure 7.36   Passive fiber-optic star LAN configurations.

For the star coupler, $CR = 1/N$; therefore, from Equation (3-74), the coupler-loss relationship is

$$L_{star} = l_e + 10 \log \frac{1}{CR} = l_e + 10 \log (N) \qquad (7\text{-}17)$$

Solving Equation (7-16) for number of stations N, we get

$$\log N = \frac{P_t - MRP - M - l_{ct} - 2l_c - l_{cr} - l_e - Dl_f}{10} \qquad (7\text{-}18)$$

As an example, let us assume the same 500-meter 10-Mb/s multimode 820-nm bus with a large star coupler (maximum practical size of 64 ports) and the following parameters: $P_t = 0$ dBm, $MRP = -30$ dBm, $M = 3$ dB, $l_{ct} = l_c = 1$ dB, $l_{cr} = 0.5$ dB, $l_f = 3$ dB/km at 820 nm, and $l_e = 5$ dB to account for a large coupler with some nonuniformity. The maximum number of stations is, therefore,

$$\log N = \frac{0 - (-30) - 3 - 1 - 2(1) - 0.5 - 5 - 0.5(3)}{10} = 1.7$$

$$N = 1 \times 10^{1.7} = 50$$

The maximum number of stations, therefore, is 50 based on power limits alone. Since a practical coupler can only be built with about 64 ports, this example demonstrates that power margin is not a limiting factor for most passive star topologies. This relationship is plotted in Figure 7.35 for various component-parameter variations. This example demonstrates the advantage of the star topology over the passive fiber bus and ring.

7.10.3 Active Fiber-Optics Ring and Star Topology

Figure 7.37 shows the physical components that make up an active fiber-optic ring and star. In the case of the ring, each station is connected, transmitter to receiver, through a length of duplex optical cable. A ring topology of this sort works best with the token-passing access method. The function of each station, in addition to signal regeneration, is to implement the token-passing protocol per IEEE 802.5 or ANSI X3T9.5.

For the star topology, each of N stations is connected to an active port of an $N \times N$-port active star. The active star contains an optical transmitter and receiver that mates to the station transmitter and receiver connected to it. Optically, it is simply a point-to-point link between station MAU and the active star port. The function of the active star, in addition to signal regeneration, is defined by the LAN standard being implemented. In a CSMA/CD LAN, its functions are like that of a mini-Ethernet, detecting collisions and controlling the transmission as per the IEEE 802.3 standard.

In either topology, the fiber link is simply point to point between stations in the ring or between the station and the active star in the star topology. The link-power budget calls for systems gain to be greater than systems loss. The relationships for the furthest separation of stations on the ring, or the furthest separation from the active star in the case of the star, is

Active Ring

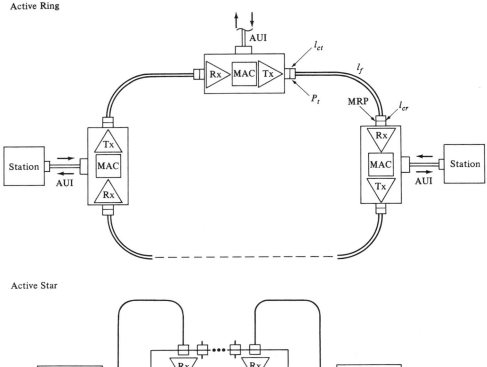

Active Star

Figure 7.37 Active fiber-optic LAN configurations.

$$P_t - \text{MRP} - M > l_{ct} + l_{cr} + Dl_f \qquad (7\text{-}19)$$

where P_t = transmitted power before the connector, in dBm

MRP = minimum power required at the receiver after the connector, in dBm

M = assumed operating margin (plus power penalties), in dB

l_{ct} = connector loss at the transmitter, in dB

l_{cr} = connector loss at the receiver, in dB

D = longest point-to-point fiber distance, in km

l_f = fiber loss, in dB/km

Since the fiber link is a series of point-to-point links, there is no splitting loss or direct power-related limit on the number of stations.

7.11 CSMA/CD FIBER-OPTIC LAN*

The standards activity related to CSMA/CD LANs by the IEEE 802.3 committee initially supported only coaxial- and twisted-pair based LANs. The IEEE FOSTAR Group, however, is completing the fiber version [28]. The main issue with the fiber version of the standard is collision detection.

7.11.1 Collision Detection

The 802.3 standard calls for 100% collision detection. Without collision detection, retransmission communications between stations would hinder the throughput of a LAN when heavily loaded. With the coaxial or twisted-pair version, the MAUs transmit a bipolar Manchester code with a constant duty cycle, thus providing a constant DC current or voltage component. If a collision occurs, the average DC component changes and is quickly detectable with simple circuitry. Each station monitors the DC voltage of the line, and if it changes, detects that a collision has occurred. The transmitting terminal upon detection ceases transmission, waiting for a clear line. This works well with copper, where the signal-level differences are small and the DC attenuation over the 500-m or less distance is negligible.

In a fiber implementation, the optical signal is unipolar and, although the Manchester code still produces a constant DC component, the wide variations in average signal strength from the various stations on the LAN make DC-signal sensing extremely difficult. Variations of 10 dB or more in signal strength can occur on a passive star system, for example, due to differences in cable length, coupler port uniformity, LED output, and connecter loss variances from coupling to coupling. The effect is that one signal can overcome the other at the collision point, creating a DC variance so small as to be difficult to detect reliably. Various approaches have been proposed for collision detection in fiber-optic LANs:

(a) average-power sensing approaches;
(b) detection of code violations as pulse-width violations after decoding of the data;
(c) time-delay violations, the detection of another packet before the echo of a station transmission is received; and
(d) use of active star couplers with collision detection centralized in the coupler.

Collision enforcement has also been considered, whereby if a MAU that was involved in the collision did not detect the event, any MAU detecting the collision would broadcast a signal indicating the event to all MAUs.

7.11.2 IEEE 802.3 Fiber-Optics CSMA/CD Standards

The IEEE 802.3 committee is studying two possibilities for a fiber-optics CSMA/CD LAN: first, a passive star approach, and, second, an active star. Standards are proposed for both implementations [28].

*The author wishes to acknowledge and thank J. Richard Jones of Broadband Technologies for the materials upon which this section was based.

The passive-star standard is being drafted for a 33-port coupler and a 1-km diameter LAN. The main issue with the passive star is that the coupler has no active components and, therefore, cannot detect or report a condition where more than one station is transmitting. This places the burden of collision detection totally on the individual station MAUs. Each MAU would have to detect average power variations, coding violations, or time-delay variations (depending on implementation) to determine whether a collision occurred, thereby increasing the complexity of the individual MAU. In addition, the MAUs may have to enforce the collision if detected by informing all other MAUs with a "jamming" signal. The problem is similar to the broad-band LAN case (802.3 Standard 10 BROAD 36), and, therefore, there should be a lot of similarity in MAU circuitry between the two standards.

The active-star draft standard permits up to 1024 ports and diameters up to 4 km. The active star receives the optical signal from the MAU attached to each port, converts it to an electrical digital signal, and routes it to all MAU receiving output ports, except the one that transmitted, where it is reconverted to an optical signal. In this implementation, collision detection still resides in the MAUs, only it is made simple because if only one station is transmitting (no collision), then that MAU receives no echo of its signal on the receive side. If more than one MAU is transmitting (a collision), that MAU sees activity on its receive side and can interpret it as a collision. Another feature of the IEEE implementation is that an "idle signal" is sent during the time period between transmitted data packets for systems maintainability and diagnostics. There is an asynchronous proposal and two synchronous proposals for dealing with the idle signals. In the synchronous proposals, the idle signal is synchronous to the data bits and the star regenerates clock. This synchronism permits the cascading of a large number of hubs without suffering the effects of interpacket gap shrinkage that limits the existing baseband 802.3 to no more than four repeaters between any two nodes.

7.11.3 Network Size as Determined by Round-Trip Delay*

The physical size of a coax CSMA/CD LAN is limited by the requirement that any station transmitting a minimum-length packet of 512-bits duration (51.2 μs at 10 Mb/s) must be able to sense a collision with a packet from any other station on the network [26]. The analysis to calculate the maximum network size assumes that the collision of a minimum-size packet occurs between the two most widely separated stations on the network. This analysis was performed as a part of the standards writing activity and was provided courtesy of J. Richard Jones. The result dictates a coax system design consisting of three 500-m segments with a total of 1 km of point-to-point repeater links interconnecting the three segments. Also included are 50-m twisted-pair AUI cables that interconnect the station to the MAU. According to 802.3, the maximum transmission path consists of five segments, four repeater sets (including their AUI cables), two MAUs, and two AUI cables at the stations. In this configuration, three of the segments can be 500-m coax cables and the other

---

* The author wishes to acknowledge and thank J. Richard Jones of Broadband Technologies for the materials for this section.

two segments are point-to-point links totaling 1 km in length. The AIU cables are each 50 m in length. By adding this up, the maximum length is 2.6 km.

Table 7.7 summarizes the propagation budget for the case in which a collision occurs between the most widely separated stations on the network. The timing shown is for a collision between a packet transmitted by station 1 and one at station N at the last possible instant before arrival at station N. The transmit path delay is the propagation delay for the packet to travel from station 1 to station N. The return path delay is the propagation delay experienced by the collision signal (sum of the signals of the two transmitted packets) as it propagates back to station 1 to inform it of a collision. The collision rise time of the coax is defined as the time it takes the summed levels of the colliding packets to reach 94% of its final value as detected at the end of a 500-m segment. With coax and the previous constraints on implementation, the total round-trip delay (48.95 μs) is less than the size of a minimum-length packet (51.2 μs), so that all collisions can be sensed and stations informed before the stations cease transmission.

TABLE 7-7 ROUND-TRIP DELAY IN AN 802.3 COAXIAL LAN

Transmission-path delay (μs)	
Electronic circuits	6.40
Cable propagation	
Coax segments	6.50
Repeater cable:	7.46
AUI	1.54
	$\overline{21.90}$
Return-path delay (μs)	
Electronic circuits	5.55
Cable propagation	
Coax segments	6.50
Repeater cable:	7.46
AUI	1.54
Collision rise time	6.00
	$\overline{27.05}$
Total round-trip delay (μs)	48.95

If the same methodology is applied to a fiber-optic implementation of 802.3 using a passive star and optical MAUs interconnecting the stations instead of coaxial segments and repeater links, and the maximum round-trip delay is defined to be the same as the coaxial implementation (48.95 μs), then we can calculate the maximum station-to-station distance. Figure 7.36 shows the implementation. Table 7.8 summarizes the propagation-delay budget. The additional budgetable rise time for fiber comes from the fewer electronic components and the elimination of collision-signal rise time. It is idealistically assumed, however, that collision detection is instantaneous, requiring no time delay (or rise time), which may not be the case in a practical sense. Using this assumption, we have 21.88 μs available for propagation delay, which at 200 m/μs translates to 4376 m of fiber. The 802.3 activity is considering the passive star as a 1-km maximum length and the active star as 4 km. Rawson [33] has shown that a 4-km network diameter can be supported in such a system as well.

TABLE 7-8   ROUND-TRIP DELAY IN AN 802.3 FIBER-OPTICS STAR LAN

Transmission-path delay (μs)		
	Electronic circuits	2.00
	Cable propagation	
	Fiber segments	21.88
	AUI	0.52
		$\overline{24.40}$
Return-path delay (μs)		
	Electronic circuits	2.15
	Cable propagation	
	Fiber segments	21.88
	AUI	0.52
		$\overline{24.55}$
Total round-trip delay (μs)		48.95

7.12 FDDI FIBER TOKEN RING

Although fiber has been applied in various fashions to the token-ring standard IEEE 802.5 and some standards activity is underway [35], perhaps the most significant to-ken-passing LAN standards activity for fiber has been the ANSI X3T9.5 activity on a fiber distributed-data interface (FDDI) 100-Mb/s token-passing ring. This standard recommends a 100-Mb/s dual counterrotating ring structure. FDDI is intended to be compatible at the IEEE 802.2 LLC layer. FDDI can be configured to support a sustained transfer rate of about 80 Mb/s (10 megabytes/s). Design values are 1000 physical connections and a total fiber path length of 200 km (100-km dual ring). Theoretically, it can serve an unlimited number of stations. The information supplied in this section is derived directly from the draft standard as it existed in July, 1988 [8, 36].

7.12.1 FDDI-Layer Definitions

Figure 7.31 illustrates the FDDI layer structure as it compares with IEEE 802 standards and the OSI model. FDDI consists of three layers: a physical layer divided into two sublayers, a data-link layer, and a station-management layer.

(a) Physical layer (PL):
  1. Physical medium dependent (PMD): This layer performs the physical transmission function of the digital baseband signal over the optical fiber. It defines the fiber-optic tranceivers, cables, connectors, code requirements, optical power budgets, bypass switches, and other physical characteristics.
  2. Physical-layer protocol (PHY): This layer is responsible for the connection between the PMD and the data-link layer. It defines such functions as clock synchronization, and decoding and encoding of the incoming/outgoing bit stream into symbols used in the DLL.
(b) Data-link layer (DLL): This layer controls the access to the medium and generates and recognizes addressing information and peer associations within the

network. This layer also generates and verifies frame check sequences to ensure that valid data are delivered to higher layers. The lowest sublayer is called the medium access control (MAC). The highest uses the IEEE 802.2 LLC and, therefore, is 802 compatible.

(c) Station management (SMT): This layer provides services such as configuration control and management, alarm and status monitoring, fault isolation and recovery control, and scheduling. It manages the processes of the various layers so that the station can operate cooperatively on the ring.

7.12.2 FDDI-Ring Operation

Figure 7.38 shows the interconnection of stations on a dual counterrotating FDDI ring. The network consists of stations physically serially connected in a ring, forming a closed loop. Physical connections to the FDDI network consist of the PMD layers of each station connecting to both a primary ring and a secondary ring. Information is transferred from active station to active station in one direction around each physical ring. The primary ring consists of an output, primary out (PO), and

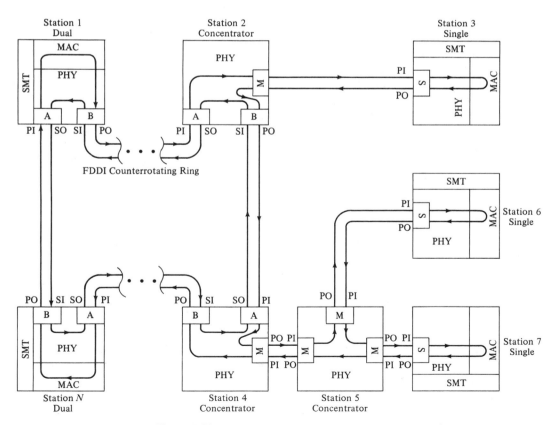

Figure 7.38 FDDI counterrotating ring interconnect example.

input, primary in (PI), and the secondary ring consists of an output, secondary out (SO), and input, secondary in (SI), all attached to the PHY of each station. The ring is subsequently logically closed by internal connections within the station's MAC layer. Connection to the physical medium through the PHY layer is controlled by the insertion and removal commands of the SMT layer. An optional optical bypass switching arrangement is specified for the PHY layer such that a failed or inactive station can be bypassed optically under the control of the SMT.

As shown, two classes of stations are defined: dual attachment and single attachment. A third class of station is called a concentrator, which functions to serve single attach stations. Only dual attach stations can form the physical FDDI ring because only they have dual tranceivers (A and B) to connect to both the primary and secondary fiber rings. The logical FDDI ring contains all classes of stations, as can be seen by following the primary ring as it connects through all stations in Figure 7.38.

7.12.3 FDDI-Station Operation

Figure 7.39 shows the functional block diagram of the PL and its relationship with the DLL and SMT layers. The MAC layer provides such functions as token-passing control, code-bit sequence control and validation, packet interpretation, and packet framing. The MAC passes 4-bit hexadecimal symbols on to the PHY layer, which must encode them into 5-bit NRZ code bits. The PHY then performs a second level of encoding to create a 4B/5B NRZI code bits for transmission on the fiber. For the received data stream, NRZI code bits are decoded to form NRZ code bits and then decoded again to hexadecimal symbols for use by the MAC.

All information on FDDI is transmitted as a sequence of code groups. The MAC determines the transmitted code sequence. The symbols can convey three types of information: (1) line state, (2) control, and (3) data. The 4-bit symbols from the MAC are encoded into 5-bit NRZ code groups and delivered to the transmit function. A 125-MHz local oscillator clocks the symbols and code bits through to the transmitter. The transmit function then encodes the NRZ code bit stream into an equivalent NRZI pulse stream for presentation to the fiber-optic transmitter in the PMD layer. FDDI uses the NRZI code and the "dual embedded" coding structure to combine data and clock and to maintain a constant DC balance within a ±10% variance. The code structure ensures at least two transitions per transmitted symbol in order to maintain clock recovery circuitry synchronization. The optical transmitter converts the NRZI electrical bit stream to an optical pulse stream.

The optical receiver converts the optical pulses to an electrical NRZI bit stream. The receiving function then decodes the NRZI bit stream into the equivalent 5-bit NRZ pulse stream. From the incoming pulse stream, it also derives clock at the code bit frequency of 125 MHz. The receiver recovery clock (RCRCLK) synchronizes the incoming code bit cell and optionally provides a frequency and phase lock indication to the SMT. Since the RCRCLK is locked to the previous upstream station, an elastic buffer is required to accommodate differences in the 125-MHz local clock (LOCLK). The minimum elasticity of 4.5 code bits is based on a clock tolerance of 0.005%. The elastic buffer also reports errors to the SMT. The NRZ code

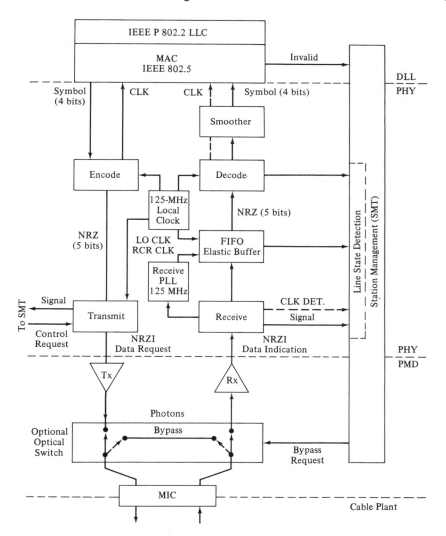

Figure 7.39 FDDI-station functional block diagram.

groups are sent to the decoder where they are converted to 4-bit symbols and sent to the MAC (possibly strobed once in every symbol or byte time). The smoother function inserts symbols into a preamble that might have been deleted by multiple elastic buffer functions so not to create preamble shrinkage.

In some instances, such as with a repeater, no MAC layer exists. In this case, a repeat filter functions prior to the encoder on the transmit side to change symbol states, preventing propagation of code violations and invalid line states, while permitting propagation of lost frames for correct detection by the next MAC in the ring.

The SMT layer serves all FDDI layers by supporting all systems-management applications, accumulating statistics, and initializing the station. It performs the functions of station-configuration management (SCM), physical-connection management (PCM), and media-interface management (MIM). The SMT commands three means for ring reconfiguration in the event of an outage or a station going inactive: (1) commanding the station bypass switch, (2) utilizing the secondary counter rotating ring, and (3) concentrator bypass. An optical bypass switch is specified for removal of a station under SMT control as well as the logical interconnection to the secondary ring.

FDDI uses a timed token protocol designed to guarantee a maximum token rotation time. The timing is decided by a bidding process upon initialization, which permits the station requiring the fastest time between token arrivals to dictate the token rotation time for the ring. This protocol offers both synchronous and asynchronous transmissions. With synchronous transmission, each station is given a predefined amount of bandwidth with each token. The amount of bandwidth for asynchronous transmission is dynamic and depends on the automatic allocation of bandwidth not used for synchronous transmission to those stations transmitting in the asynchronous mode. The FDDI token and frame structures are shown in Figure 7.40 [32, 37].

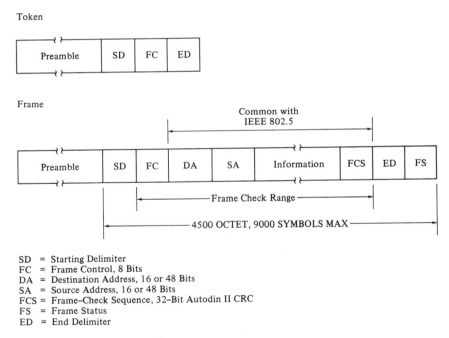

SD = Starting Delimiter
FC = Frame Control, 8 Bits
DA = Destination Address, 16 or 48 Bits
SA = Source Address, 16 or 48 Bits
FCS = Frame–Check Sequence, 32–Bit Autodin II CRC
FS = Frame Status
ED = End Delimiter

Figure 7.40   FDDI token and frame structure.

7.12.4 FDDI-Performance Characteristics

The standard specifies optical bypass-switch characteristics as well as a media-attachment connector, media-signal interfaces, cable plant, and test methods. Some of the key characteristics are summarized in what follows; however, the reader should always refer to the latest issue of the standard for current information.

(a) Media-interface connector:

> Duplex fiber-optic keyed connector, ferrule-in-sleeve design
> Insertion loss: Not specified; included in terminal power measurements or cable loss specifications
>
> Four types: A: Primary in/secondary out
>   B: Secondary in/primary out
>   M: Single attach to concentrator
>   S: On single-attach station

(b) Network performance:

> Data rate: 100 Mb/s, clock rate of 125 Mb/s
> Number of stations: 500 maximum
> Network size: 100-km duplex ring, 200-km fiber path length
> Maximum station spacing: 2 km maximum between stations
> Maximum station delay: 756 ns
> Maximum loop delay: 1.733 ms based on 5085-ns/km fiber delay and 200-km loop with 500 stations
> Addressing: 16- or 48-bit, individual and multicast
> Bit-error rate: $< 10^{-9}$ BER total ring
> BER through station: $< 2.5 \times 10^{-10}$ under all conditions at minimum power; $< 1 \times 10^{-12}$ when input power is 2 dB above minimum
> Tx coding: NRZI 4B/5B

(c) Station optical transmitter:

> Center wavelength: 1270 to 1380 nm
> Average output power: -20 to -14 dB at the output of the test connector
> Duty-cycle distortion: 1 ns maximum
> Data dependent jitter: 0.6 ns maximum
> Random jitter: 0.76 ns maximum
> Extinction ratio: 10% maximum
> Spectral width: Width plus chromatic dispersion plus source rise time must achieve an optical rise time less than 5 ns in a 2-km length of fiber; ranges from a width of 100 to 200 nm

Optical pulse: 80 ns $\pm$500 ppm time interval; 40 ns $\pm$0.7 ns
                               mean pulse width; 0.6 to 3.5 ns rise-time
                               window

(**d**) Station optical receiver:

Average received power: -31 to -14 dBm at the input to the test
                               connector
Rise/fall time: 0.6 to 5 ns
Duty-cycle distortion: 1 ns maximum
Data dependent jitter: 1.2 ns maximum
Random jitter: 0.76 ns peak-to-peak maximum

(**e**) Bypass switch:

Attenuation: 2.5 dB maximum
Isolation: 40 dB worst-case
Time to switch: 25 ms from command
Media interrupt: 15 ms maximum

(**f**) Fiber plant:

Fiber type: Multimode
Core diameter: 62.5 μm per EIA 455-58
Cladding diameter: 122.0 to 128.0 μm per EIA 455-27, -48
Numerical aperature: 0.275 per EIA 455-177
Attenuation at 1300 nm: <2.5 dB/km typical, 11.0 dB maximum
                               end to end per EIA 455-53
Modal bandwidth: 500 MHz $\cdot$ km minimum at 1300-nm opti-
                               cal 3-dB BW per EIA 455-30, -51, -54
Chromatic dispersion: See standard for profile, 0.11 ps/nm/km
                               between 1300 and 1348 nm
Optional fiber types not specified to date:
                               50/125 μm at 0.2 and 0.22 N.A.
                               85/125 μm at 0.26 N.A.
                               100/140 μm at 0.29 N.A.

7.12.5 Optical Bypass-Switch Operation

Figure 7.41 shows the operation of the optical bypass switch (only one ring shown).
In this case, station 2 is bypassed. The number of simultaneously bypassed stations
can be calculated by performing a link-bandwidth and a link-power-budget analysis
on the ring, assuming N number of bypass switches thrown over $N + 1$ segments of
cable of length d.

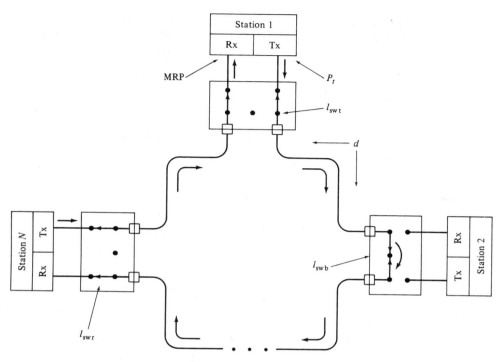

Figure 7.41 FDDI-ring optical bypass-switch configuration.

The worst-case link-power budget is derived by tracing the signal from station 1 to station N:

$$P_t - \text{MRP} - M = l_{swt} + l_{swr} + 2l_c + N(2l_c + l_{swb}) + (N + 1)(d \times l_f) \qquad (7\text{-}20)$$

Solving for the number of bypassed stations N, we get

$$N = \frac{P_t - \text{MRP} - M - l_{swt} - l_{swr} - 2l_c - dl_f}{2l_c + l_{swb} + dl_f} \qquad (7\text{-}21)$$

Assuming $P_t = -20$ dBm, MRP $= -31$ dBm, $l_{swb} = 2.5$ dB, $l_c = 0$ (since it was included), and $l_f = 2.5$ dB/km from the standard, and assuming l_{swt} and l_{swr} both $=$ 1 dB maximum and $M = 0$ dB, we get

$$N = \frac{9 - d(2.5)}{2.5 + d(2.5)}$$

If the distance between stations is less than 150 m, then $N = 3$ bypassed stations at the maximum. If d is between 150 and 500 m, then $N = 2$ bypassed stations maximum. From 500 m to 1.3 km, distance N goes to 1. Beyond 1.3 km, no bypass is possible using these parameters. Obviously, one must choose equipment performance specific to the size of the network and application in order to use the optical bypass option. The standard states its intended use in the data-center environment, where maximum length between stations is 400 m.

Cable bandwidth and signal rise time (5 ns maximum) must be considered in the bypass switch operation as well. If a maximum transmitter rise time of 3.5 ns is permitted, from Equation (6-39), this leaves a cable rise time of

$$t_{rcable} = \sqrt{\frac{t_{r,link}^2}{1.1} - t_{r,source}^2}$$

$$= \sqrt{22.7 - 12.25} = 3.23 \text{ ns}$$

If we assume a total cable modal bandwidth of 500 mHz·km and zero chromatic dispersion from Equations (4-40) and (4-41), we get a rise time of

$$t_r = 1.087 t_{d,FWHM} = 1.08 \frac{0.442}{(BW)_o}$$

$$= 1.087 \frac{442 Nd}{(BW)_f (MHz)} \text{ ns}$$

$$3.23 \text{ ns} = \frac{480 Nd}{500} = 0.96 Nd$$

$$Nd = 3.36$$

where d is in km.

If $N = 3$ stations bypassed in the maximum case from the preceding, then the bandwidth budget permits d to be 1.12 km between stations. In the case of 1 station bypassed, the bandwidth budget permits d to be 3.36 km between stations. The power budget is the limiting factor in this example.

REFERENCES

1. American National Standards Institute, *Digital Hierarchy Optical Interface Rates and Formats Specification*, ANSI T1.105-1988, ECSA document T1X1/87-129R1, New York, 1988.

2. American National Standards Institute, *American National Standard for Telecommunications Digital Hierarchy Optical Interface Specifications: Single Mode*, ANSI/ECSA T1X1/87-126R1, New York, 1988.

3. *Northern Telecom FD 565 Ring System Optical Transmission System Applications Guide* 1 (December 1987).

4. FIBERLAN Ring Architecture patent pending.

5. Institute of Electrical and Electronics Engineers, *Distributed Queue Dual Bus Metropolitan Area Network*, IEEE 802.6, Draft Standard, New York, 1988.

6. CCITT Subgroup XVIII.

7. American National Standards Institute, *Digital Hierarchy Synchronous DS3 Format Specification*, ANSI T1.103-1987, New York, 1987.

8. American National Standards Institute, *FDDI Physical Layer Medium Dependent (PMD)*, ANSI X3T9.5/84-48 REV 8, X3T9 Technical Committee Draft Proposal, New York, 1988.

9. American National Standards Institute, *Standards Title*, ANS T1X1/87-128R1 and -129R1, New York, 1987.

10. American Telephone & Telegraph, "Interworking the IEEE 802.6 and Broadband ISDN Protocols," presentation to the Working Group IEEE 802.6, Holmdel, NJ.

11. R. Hoss and F. McDevitt, "Fiber Optic Supertrunking, FM vs. Digital Transmission," presented at the NCTA 33rd Annual Convention and Exposition, June 3–6, 1984, Las Vegas: 168–72.

12. F. McDevitt and R. Hoss, "Repeaterless 16 Km Fiber Optic CATV Supertrunk Using FDM/WDM," paper presented at the NCTA 32nd Annual Convention and Exposition, Houston, June 12–15, 1983: 20–211.

13. Installation in Queens, New York, by Warner Amex Cable, a joint venture between American Express and Warner Communications.

14. J. A. Chiddix, "Optical Fiber Supertrunking, the Time Has Come, A Performance Report in a Real-World System," paper presented at the NCTA 35th Annual Convention and Exposition (1986).

15. R. Hoss, "Fiber Optic Options for Video Transmission," *Proceedings of the Newport Conference in Fiberoptics Markets,* Kessler Marketing Intelligence, Newport, RI, 1985.

16. J. A. Chiddix, "Fiber Optics Technology for CATV Supertrunk Applications," paper presented at the NCTA 34th Annual Convention and Exposition, Las Vegas (1985): 157–65.

17. R. McDevitt and R. Hoss, "Application of Fiber Optics Networking in the CATV Industry," paper presented at Fiber Optic Communications and Local Area Networks Exposition '83, Atlantic City, 1983.

18. Times Fiber, Inc., *Installation of the Times Fiber "Minihub,"* Wallingford, CN.

19. S. A. Esty, "Fiber in the Feeder," paper presented *Lightwave Magazine* Symposium, "Fiber in the Subscriber Loop," 1986.

20. H. Kobrinski and C. A. Brackett, "A Survey of Optical Frequency Multiplexing Techniques for Subscriber Loop Application," paper presented at the *Lightwave Journal* Symposium, "Fiber in the Subscriber Loop" Boston, 1986.

21. S. S. Cheng, "Optical Network Technologies for Subscriber Loops," paper presented at the *Lightwave Journal* Symposium, "Fiber in the Subscriber Loop," Boston, 1986.

22. American Telephone & Telegraph, "Interworking the IEEE 802.6 and Broadband ISDN Protocols," presentation to the IEEE 802.6 Working Group, Palm Desert, CA, 1988.

23. E. H. Hara, "An Approach to Broadband ISDN," paper presented at the *Lightwave Journal* Symposium, "Fiber in the Subscriber Loop," 1986.

24. Institute of Electrical and Electronics Engineers, *Tutorial Guide to 802.1B, Systems Management,* IEEE Standard 802.87*1.435, REV. M 2/87, Piscataway, NJ, 1987.

25. Institute of Electrical and Electronics Engineers, ANSI/IEEE Standard 802.2-1985, SHO 9712, Piscataway, NJ, 1985.

26. Institute of Electrical and Electronics Engineers, ANSI/IEEE Standard 802.3-1985, SHO 9738, Piscataway, NJ, 1985.

27. Institute of Electrical and Electronics Engineers, *Twisted Pair Medium, Type 10, Base T,* IEEE P802-31/D2-88-10, Piscataway, NJ, 1988.

28. M. E. Abraham, "Fiber Optic Ethernet," *LAN Magazine* (November 1988): 39–44.

29. Institute of Electrical and Electronics Engineers, ANSI/IEEE Standard 802.4-1985, SHO9720, Piscataway, NJ, 1985.

30. Institute of Electrical and Electronics Engineers, ANSI/IEEE Standard 802.5-1985, SHO9944, Piscataway, NJ, 1985.

31. Institute of Electrical and Electronics Engineers, IEEE 802.6-88/58 MAN Standard Draft, Piscataway, NJ, 1988.

32. American National Standards Institute, ANSI X3 Project 503D, New York.

33. E. G. Rawson, "The Fibertnet II Ethernet Compatible Fiber Optic LAN," *IEEE Journal Lightwave Technology* **LT-3** (June 1985): 496–501.

34. W. Stallings, "Local Network Performance," *IEEE Communications* **22** (February 1984): 27–36.

35. Institute of Electrical and Electronics Engineers, *Draft 9 on Fiber Optics,* IEEE802.5J-88/45, Piscataway, NJ, 1988.

36. American National Standards Institute, *FDDI Physical Layer Protocol (PHY),* Draft Standard ANSI X3T9.5/83-15 REV 15, New York, 1987.

37. J. F. McCool, "The Emerging FDDI Standard," paper presented at the Systems Design and Integration Conference, 1987.

38. T. Muoi, "CATV Supertrunking Study," unpublished paper, 1984.

39. J. R. Jones, unpublished work, 1988.

8

Network Availability and Cost Performance

8.1 INTRODUCTION

This final chapter discusses the last two stages in network design analysis, evaluating the availability and cost of the design. An in-depth treatment of both topics is beyond the scope of this book, however, the summary provided here should be adequate for most first-order analyses.

8.2 RELIABILITY AND NETWORK AVAILABILITY

The availability of a system and the reliability of the components that make it up are important performance parameters that effect architecture and component selection. Figure 8.1 shows the difference in concept between component or subsystem reliability and system or network availability. Reliability terminology is generally applied at the component or "black-box" level to describe the performance over time of a subelement of the system. Availability is the term generally applied to the performance over time of the entire network. By adding redundancy and protection switching to provide alternate signal paths, for example, high levels of network availability can be achieved in spite of a relative low reliability of the subsystem elements.

8.2.1 Standards

Some of the commonly used standards for defining systems reliability and availability are

Component Reliability

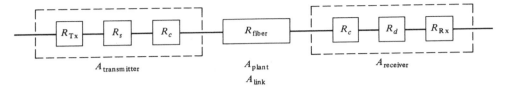

Network Availability

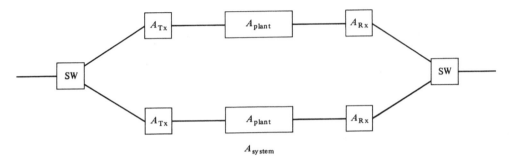

Figure 8.1   Reliability analysis methodology.

(a) Bell Communications Research Technical Advisory 10425, *Reliability Prediction Procedure for Electronic Equipment:* Provides prediction procedures and failure rates for generic components [1].

(b) MIL-HDBK-217: Defines component failure rates and gives relationships for calculation of component and subsystem reliability [2].

8.2.2 Terminology

(a) Failure: This is the departure of a component in the system from specified performance.

(b) Mean time between failure (MTBF): This is the statistical mean failure probability between failure occurrences for a component or system. MTBF is generally expressed in hours or years per failure, where 1 year = 8760 hours.

$$\text{MTBF} = \frac{\text{total operating time}}{\text{number failures}} \tag{8-1}$$

$$\text{MTBF (years)} = \frac{\text{MTBF (hours)}}{8760} \tag{8-2}$$

(c) Mean time between outages (MTBO): This is the statistical mean probability between service interruptions. This term takes into account redundancy and protection, such that a failure within the system does not always result in a failure to provide service.

(d) Failure rate (FR): This is the failures per unit time (failures per hour) or rate of expected failures over the useful life of the system.

$$\text{MTBF} = \int_0^\infty \exp \left[-(\text{FR})t\right] dt = \frac{1}{\text{FR}} \tag{8-3}$$

$$\text{FR (f/h)} = \frac{1}{\text{MTBF (h)}}$$
$$= \frac{1}{8760 \text{ MTBF (years)}} \tag{8-4}$$

(e) Reliability (R): This is the probability that the equipment will perform for a period of time t.

$$\text{Reliability over time, } R(t) = \exp \left[-(\text{FR})t\right] \tag{8-5}$$

$$\text{Reliability at a point in time, } R = 1 - \text{FR} \tag{8-6}$$

$$R = 1 - \frac{1}{\text{MTBF (years)}} = 1 - \frac{8760}{\text{MTBF (h)}} \tag{8-7}$$

$$\text{MTBF} = \frac{1}{1 - R} \tag{8-8}$$

(f) Failures in time (FITS): These are the failures per billion (1×10^9) operating hours.

$$\text{MTBF (h)} = \frac{(1 \times 10^9)}{\text{FITS(billion h)}} \tag{8-9}$$

(g) Mean time to repair (MTTR): This is the statistical mean time to repair an element in the system to operating condition, including fault isolation, repair, and test.

(h) Mean time to restore (MTR): This is the statistical mean time to restore the system to operational condition after the occurrence of a failure.

$$\text{MTR} = T_t + \text{MTTR} + (1 - P_s)T_s \tag{8-10}$$

where: T_t = failure notification, dispatch, and travel time
P_s = probability of having a spare on hand
T_s = time to obtain a spare

(i) Availability (A): This is the fraction of time the system is available for operation.

$$A = \frac{\text{MTBF}}{\text{MTR} + \text{MTBF}} \tag{8-11}$$

$$A = \frac{1}{1 + (\text{FR} \times \text{MTR})} \tag{8-12}$$

$$A = 1 - \frac{\text{MTR}}{\text{MTBF}} \tag{8-13}$$

$$A = 1 - (\text{MTTR})(1 - R) \tag{8-14}$$

(**j**) Unavailability (U): Unavailability, or fractional outage, is the fraction of time the system is unavailable for operation.

$$U = \text{MTR} \times \text{FR} = \frac{\text{MTR}}{\text{MTBF}} \tag{8-15}$$

$$U = 1 - A \tag{8-16}$$

(**j**) Outage time (OT): This is the time that a channel is not operating within specified performance.

$$\text{OT (hours/year)} = 8760 \times U_{\text{ch}} \tag{8-17}$$

Systems availability can be defined on the basis of unacceptable performance as well as on the basis of total channel failure. CCITT recommendation G.821 defines availability on the basis of five time-period classifications [3, 4]:

1. Total time over which availability is measured.
2. Unavailable time: 10 consecutive seconds or more of 10^{-3} or worse error rate (severely errored seconds).
3. Available but unacceptable performance: Periods of excessive (10^{-3} or worse) error rate of less than 10 seconds in duration.
4. Available but degraded time: Time in which the error rate is between 10^{-3} and 10^{-6}.
5. Acceptable performance time: Periods in which the error rate is less than 10^{-6}.

Another way to measure availability is to measure errored seconds within a certain test period. For example, acceptable performance might be defined as 95% of the total seconds in a 24-hour test period must be error-free.

8.3 SYSTEMS-AVAILABILITY RELATIONSHIPS

The calculation of systems availability depends on the measured or calculated reliability of each of the modules or subelements that make it up. In order to calculate element reliability, the individual component reliabilities must be combined. Likewise, in order to compute system or network availability, see the various basic configurations in Figures 8.2 and 8.3.

8.3.1 Series Elements

Series elements refer to those that have to perform in order for a channel within the system to be available. The relationships for channel availability and reliability for series-connected elements are as follows:

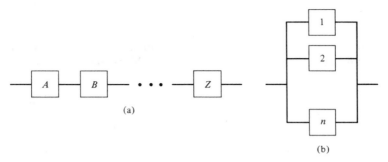

Figure 8.2   Availability relationships. (a) Series dependent elements and (b) parallel dependent elements.

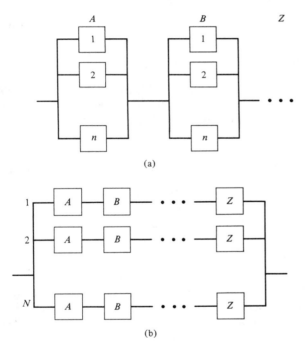

Figure 8.3   Hybrid relationships. (a) Series–parallel and (b) parallel–series.

$$R_{ch} = R_a \times R_b \times \cdots \times R_z \qquad (8\text{-}18)$$

$$FR_{ch} = FR_a + FR_b + \cdots + FR_z \qquad (8\text{-}19)$$

$$(MTBF)_{ch} = \frac{1}{1/(MTBF)_a + 1/(MTBF)_b + \cdots + 1/(MTBF)_z} \qquad (8\text{-}20)$$

$$A_{ch} = A_a \times A_b \times \cdots \times A_z \qquad (8\text{-}21)$$

$$U_{ch} = 1 - A_{ch} = 1 - A_a \times A_b \times \cdots \times A_z \qquad (8\text{-}22)$$

When U is very small, product $(U_a \times U_b \times \cdots \times U_z)$ is negligible and therefore:

$$U_{ch} = U_a + U_b + \cdots + U_z \qquad (8\text{-}23)$$

8.3.2 Parallel Elements

Parallel elements refer to elements that can fail but do not cause a channel outage unless all fail at the same time. Channel reliability and availability relationships are as follows:

$$R_{ch} = 1 - (1 - R_1)(1 - R_2) \cdots (1 - R_n) \tag{8-24}$$

$$FR_{ch} = FR_1 \times FR_2 \times \cdots \times FR_n \tag{8-25}$$

$$(MTBF)_{ch} = (MTBF)_1 \times (MTFBF)_2 \times \cdots \times (MTBF)_n \tag{8-26}$$

$$A_{ch} = 1 - (1 - A_1)(1 - A_2) \cdots (1 - A_n) \tag{8-27}$$

$$A_{ch} = 1 - U_1 \times U_2 \times \cdots \times U_n \tag{8-28}$$

$$U_{ch} = U_1 \times U_2 \times \cdots \times U_n \tag{8-30}$$

8.3.3 Hybrid Combinations

As series and parallel elements are further combined, forming "hybrids," the basic relationships combine directly. Because of its simplicity, we use unavailability computation in lieu of availability. For the series–parallel configuration in Figure 8.3(a), we have

$$U_{ch} = (U_1 \times U_2 \times \cdots \times U_n)_a + \cdots + (U_1 \times U_2 \times \cdots \times U_n)_z \tag{8-31}$$

For the parallel–series combination in Figure 8.3(b), we have

$$U_{ch} = (U_a + U_b + \cdots + U_z)_1 + \cdots + (U_a + U_b + \cdots + U_z)_n \tag{8-32}$$

8.3.4 Protection Switching

A method of greatly improving network availability is to provide p redundant protection channels, each of which is automatically switched in whenever one of n operating channels fail. The systems configuration is shown in Figure 8.4. Under this configuration, a channel outage can only occur when:

(a) more parallel channel paths (1 thru N) fail simultaneously than the number of protection channels provided; and

(b) failure of the protection circuitry to detect a failure and/or switch when a failure of an operating path occurs.

The unavailability of a given operating channel (U_{ch}) in a $p : n$ protected transmission span is [1, 3, 4]

$$U_{ch} = \frac{(n + p)! U_n^{p+1}}{n!(p + 1)!} \tag{8-33}$$

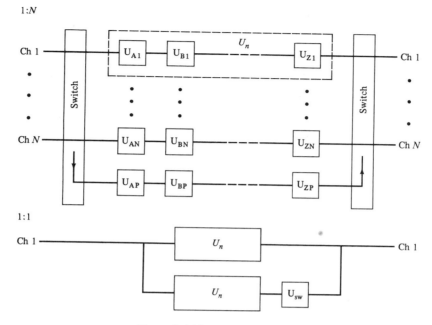

Figure 8.4   Protection switching.

where n = number of required operating channels

p = number of protection channels in the n group

U_n = fractional probability of any one of n operating channels being in a failed state (unavailable)

For example, a single redundantly protected channel (1 : 1) has an unavailability of

$$U_{ch} = \frac{(1 + 1)!U_n^{1+1}}{1(1 + 1)!} = U_n^2 \tag{8-34}$$

This expression assumes that the protection-channel unavailability (U_p) is the same as that of an operating channel (U_n). Protection-channel reliability (R_p) is generally lower because of

(a) added equipment for sensing, communicating and switching; and

(b) failures that are silent or unprotected.

The unavailability of the protection channel can be computed as the combined unavailabilities of the switch and the series components in the protection path:

$$U_p = U_{sw} + U_{path} \tag{8-35}$$

For 1 : n protection, when U_p is different than U_n, Equation (8-33) can be written as:

$$U_{ch} = \frac{n - 1}{2} U_n^2 + U_n \times U_p \tag{8-36}$$

For the case where the protection path is identical in composition to an operating channel, $U_{\text{path}} = U_n$, then

$$U_p = U_{\text{sw}} + U_n \tag{8-37}$$

and for 1 : 1 protection, we get

$$U_{\text{ch}} = U_n(U_n + U_{\text{sw}}) \tag{8-38}$$

From Figure 8.4, this can be seen as a combination of a signal path (U_n) in parallel with a series combination of a switch (U_{sw}) and a signal path (U_n).

8.4 SYSTEMS-AVAILABILITY EXAMPLE

8.4.1 Parameter Relationships

Figure 8.5 shows a typical functional block diagram for a protected fiber-optic transmission system as represented from a reliability standpoint. Channel availability is generally calculated in terms of two-way availability, i.e., that required for transmitting and receiving a full duplex signal.

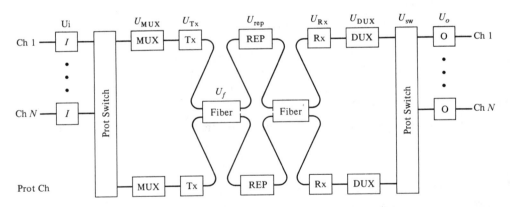

Figure 8.5 Digital fiber-optic trunking system availability model.

From Equation (8-23), one-way unavailability of the series-connected electronics for an individual channel signal path is

(a) on the protected side:

$$U_n = U_{\text{MUX}} + U_{\text{Tx}} + N_r U_{\text{rep}} + U_{\text{Rx}} + U_{\text{DUX}}$$

$$= U_{\text{term}} + N_r U_{\text{rep}} \tag{8-39}$$

(b) and on the unprotected side:

$$U_{nu} = U_{\text{in}} + U_{\text{out}} = U_{io} \tag{8-40}$$

One-way unavailability for the protection-channel equipment connected in series is

$$U_p = U_{sw} + U_{term} + N_r U_{rep} \qquad (8\text{-}41)$$

Power sources are treated as series elements in the analysis since the failure of power at one location eliminates all signal channels, including the protection channels if they are operating from the same supplies. Power supplies are generally made redundant for that reason. From Equations (8-23) and (8-36), the relationship is, therefore,

$$U_{pwr} = \frac{m + 1}{2} U_{ps}^2 \times (2 + N_r) \qquad (8\text{-}42)$$

where m = number of operating supplies in a $1 : m$ group
  $2 + N_r$ = number of terminals and repeaters

This formula should be modified for different configurations as appropriate.

Fiber plant availability should also be included in the analysis as well. The primary failure mechanism here is inadvertent cuts that usually damage all fibers, and, therefore, effect all channels simultaneously. Such outages are not protected unless the protection routes the signal around a separate diverse path. Fiber plant unavailability (U_f), like power supplies, is treated as a series element in the analysis, separate from the protection. For a nondiverse-route one-way unavailability, the unavailabilities of fiber, power supplies, and the I/O channel are added, per Equation (8-23), to the unavailabilities of the protected elements, per Equation (8-36):

$$U_{ch} = \frac{n - 1}{2} U_n^2 + U_n U_p + U_f + U_{pwr} + U_{io} \qquad (8\text{-}43)$$

For a diverse protection path, fiber is in series with the electronics in each path; therefore, fiber unavailability U_f adds to U_n and U_p per Equation (8-23):

$$U_{ch} = \frac{n - 1}{2} U_n^2 + (U_n + U_f)(U_p + U_f) + U_{pwr} + U_{io} \qquad (8\text{-}44)$$

The two-way channel unavailability for a $1 : n$ protected nondiverse route is a series combination of the unavailabilities of two one-way paths:

$$U_{ch} = 2U_{io} + (n - 1)U_n^2 + 2(U_n U_p) + U_{pwr} + U_f \qquad (8\text{-}45)$$

8.4.2 Network-Availability Example

Figure 8.5 illustrates a $1 : N$ protected point-to-point fiber-optics trunk containing a number of repeater spans. For this example, assume five repeaters (six spans) and $1 : 1$ power-supply redundancy. The system contains elements with the reliabilities shown in Table 8.1. Plant unavailabilities assume one complete cut per route every two years. We analyze two-way (full-duplex) systems availability for the following cases:

TABLE 8.1 FAILURE RATES FOR AVAILABILITY EXAMPLE

Component	MTBF (yr)	FR (10^{-6}/h)	R (yr)	MTR (h)	U (10^{-6})
F.O. Tx	16.3	7.0	0.9386	4	28
F.O. Rx	11.4	10.0	0.9123	4	40
Repeater	8.15	14.0	0.8773	4	56
MUX	9.5	12.0	0.8947	4	48
DEMUX	16.3	7.0	0.9386	4	28
Protection switch	16.3	7.0	0.9386	4	28
Channel I/O	570.0	0.2	0.9982	4	0.8
Power supply	11.4	10.0	0.9123	4	40
Fiber	50.0	2.28	0.9800	12	27
[a] Plant damage	2.0	57.1	0.5000	12	684

[a] Plant damage rate is dependent on construction method, route, and hostile conditions.

Case 1: A simple nonprotected path

Case 2: 1 : 1 electronics protection without path diversity

Case 3: 1 : 1 protection with protection-path diversity

Case 1.   For a simple nonprotected signal path, the channel becomes unavailable when any series element fails. In this case, all elements assume a series relationship, as in Figure 8.2a. Two-way channel unavailability (U_{ch}) becomes the sum of the unavailabilities of the duplex interface cards (U_{io}), the transmission electronics (U_n), the fiber plant (U_f), and the redundant power system (U_{pwr}):

$$U_{ch} = 2U_{io} + 2U_n + U_f + U_{pwr} \qquad (8\text{-}46)$$

The individual elements are computed from Table 8.1 and Equations (8-39) through (8-42):

$$U_{io} = 0.8 \times 10^{-6}$$

$$U_n = U_{MUX} + U_{Tx} + N_r U_r + U_{Rx} + U_{DUX}$$

$$= 48 \times 10^{-6} + 28 \times 10^{-6} + 5 \times 56 \times 10^{-6} + 40 \times 10^{-6} + 28 \times 10^{-6}$$

$$= 424 \times 10^{-6}$$

$$U_f = 684 \times 10^{-6}$$

$$U_{pwr} = \frac{1 + n}{2} U_{ps}^2 \times (2 + N_r)$$

$$= (2/2)(40 \times 10^{-6})^2 \times (2 + 5)$$

$$= 0.0056 \times 10^{-6}$$

By substituting these results in Equation (8-46), the two-way channel unavailability, availability, and outage time are

$$U_{ch} = 2(0.8 \times 10^{-6}) + 2(424 \times 10^{-6}) + (684 \times 10^{-6}) + (0.0056 \times 10^{-6})$$

$$= 1533.6 \times 10^{-6}$$

$$A_{ch} = (1 - U_{ch})100 = 99.8466\%$$

$$OT = 8760 \times U_{ch} = 13.43 \text{ h/yr}$$

Note that the availability is dominated by both transmission equipment reliability as well as cable failure.

Case 2.   From Equation (8-45), if equipment redundancy is added, the relationship for two-way channel unavailability is

$$U_{ch} = 2U_{io} + (n - 1)U_n^2 + 2(U_n U_p) + U_{pwr} + U_f$$

The individual elements are computed as:

$$U_{io} = 0.8 \times 10^{-6}$$

$$U_n = 424 \times 10^{-6} \text{ from Case 1}$$

$$U_p = U_{sw} + U_{term} + N_r U_{rep}$$

$$= 28 \times 10^{-6} + 424 \times 10^{-6} = 452 \times 10^{-6}$$

$$U_{pwr} = 0.0056 \times 10^{-6} \text{ from Case 1}$$

$$U_f = 684 \times 10^{-6}$$

By substituting these results into Equation (8-45), the two-way channel unavailability, availability, and outage time are

$$U_{ch} = 2(0.8 \times 10^{-6}) + (0) + 2[(424 \times 10^{-6})(452 \times 10^{-6})] + (0.0056 \times 10^{-6})$$

$$+ (684 \times 10^{-6})$$

$$= 2 \times 10^{-6} + 684 \times 10^{-6} = 686 \times 10^{-6}$$

$$A_{ch} = (1 - U_{ch})100 = 99.931\%$$

$$OT = 8760 \times U_{ch} = 6.01 \text{ h/yr}$$

Note that with equipment redundancy, the unavailability of the equipment is reduced by over two orders of magnitude to the point where availability is almost totally dominated by cable plant damage.

Case 3.   If the protection-signal path is routed over a diverse cable route, then the relation for two-way channel unavailability is from Equation (8-44),

$$U_{ch} = 2U_{io} + (n - 1)U_n^2 + 2(U_n + U_f)(U_p + U_f) + U_{pwr}$$

The individual elements are computed as:

$$U_{io} = 0.8 \times 10^{-6}$$

$$U_n = 424 \times 10^{-6} \text{ from Case 1}$$

$$U_p = 452 \times 10^{-6} \text{ from Case 2}$$
$$U_f = 684 \times 10^{-6}$$
$$U_{pwr} = 0.0056 \times 10^{-6} \text{ from Case 1}$$

The resultant two-way unavailability, availability, and outage time are

$$U_{ch} = 2(0.8 \times 10^{-6}) + 0 + 2(424 \times 10^{-6} + 684 \times 10^{-6})$$
$$\times (452 \times 10^{-6} + 684 \times 10^{-6}) + 0.0056 \times 10^{-6}$$
$$= 4.12 \times 10^{-6}$$
$$A_{ch} = (1 - U_{ch})100 = 99.99958\%$$
$$OT = 8760 \times U_{ch} = 0.03609 \text{ h/yr} = 2 \text{ min/yr}$$

With a fully redundant and diversely routed system, therefore, the availability is increased significantly. Diverse signal routing is a necessity in areas where the cable plant can be damaged.

8.5 COST ANALYSIS

In general, it is difficult to isolate the architecture and performance analysis of a system from the cost analysis. This is because the objective of any design analysis is usually to determine the approach that will provide the best performance for the lowest cost.

A cost model is generally developed with one or two objectives:

1. First-cost analysis: To study the effects of systems design on the per-channel unit cost or total installed cost. This often takes the form of a sensitivity analysis, with channel capacity, repeater spacing, and product selection as the variables.
2. Time-based analysis: To determine the effect over time of the systems-design variables on return on investment or life-cycle cost.

8.5.1 First-Cost Analysis

The first-cost analysis involves only the amount of money and resource it takes to design, build, and install the system. Recurring operations and maintenance expenses are excluded. This sort of analysis is usually done when performing a performance versus cost trade-off, primarily because it is simple and expedient. The first-cost analysis usually involves the following cost elements:

Systems installation	Systems production
Nonrecurring engineering	Nonrecurring engineering
Materials capital	Preproduction qualification
Installation costs	Production tooling
Facilities construction	First article production
Monitoring systems	Material
Initial spares	Touch labor
Maintenance equipment	Support
Initial training costs	QC and QA
Tax, freight, and miscellaneous	Test
	Shipping

8.5.2 Time-Based Analysis

This form of analysis studies the cost of developing, installing, and operating the system over time, and is usually designed to value the system, and its financing structure, in terms of a financial return or benefit. A time-based analysis takes into account annual recurring costs as well as the first costs and unit costs listed before. Also, a time-based analysis time phases the investment, expenses, debt, and other elements into the years in which they are realized.

If the system is intended to provide a benefit, such as revenue or cost savings, an analysis over some time period is performed in order to determine the return on investment or net income. The time-based analysis can also be performed to determine appropriate financing or budget strategies.

The returns analysis is further discussed in Section 8.7. There are many valuation criteria and the more common are [7]

(a) the accounting approach;

(b) the operating return;

(c) present worth; and

(d) the internal rate of return.

8.5.3 Life-Cycle Cost (LCC)

The life-cycle cost (LCC) is a time-based analysis used primarily in the military to value the development and deployment of a system for a particular application, such as the use of fiber optics in an airborne control system. It values the cost of a system based on the total cost associated with deployment and operation over the life of the system. LCC, therefore, takes into account all of the elements of a time-based analysis that were listed before. The model, however, may also include the cost savings over time associated with the particular deployment. For example, the higher initial costs of deploying a fiber-optics system on an aircraft versus copper wire may be greatly offset in time by lower fuel and maintenance costs brought about by its lower weight and higher reliability.

LCC analysis is beyond the scope of this book; however, a summary of the methodology is presented in order to demonstrate the general analysis approach. LCC analysis is generally performed as two analyses, each from a different perspective:

1. A "bottom-up" analysis, which determines program cost by analyzing and adding up all the cost elements that go into it, and performing a parametric analysis on those cost elements to predict production and deployment costs over time.
2. A "top-down" analysis, which looks at the major cost elements of the platform or larger program of which the system is a part. By varying certain key parameters, such as MTBF or weight savings, the system is then analyzed by virtue of the effect it has on the LCC cost of the greater program.

Inputs to the bottom-up LCC analysis include the following analysis elements:

(a) A systems-requirements analysis (SRA) and systems-design analysis provide the basic system design inputs.

(b) Reliability/maintainability analysis (RMA) evaluates:

maintenance concept
mean time to fail (MTBF)
mean time to repair (MTTR)

(c) Level-of-repair analysis (LORA) evaluates the amount of maintenance support that is required. Generally, a sensitivity analysis is run to trade off LOR costs against MTBF improvements or throw away unit production cost (UPC). LOR costs include:

inventory administration
inventory and repair material
support equipment
packing and shipping
storage and work space
labor
training
documentation

(d) Logistics support analysis evaluates the supportability of the components within the military logistics system.

(e) Bill of materials and materials planning documents provide the list of components and their pricing.

(**f**) A unit-production-cost (UPC) analysis determines the required unit production cost of the system or subsystem, and develops a production and cost control plan to meet it. The plan is known as "design to unit production cost" (DTUPC).

DTUPC and LCC generally go hand in hand in any government full-scale development and production program. A very sophisticated management and control system is set up during all program phases to track LCC variances. DTUPC implies not only that production cost goals will be met, but that the equipment will be designed for maintenance and support such that LCC goals are met as well.

8.6 FIRST-COST ANALYSIS METHODOLOGY

First-cost analysis is a means of identifying and partitioning cost elements so that a sensitivity analysis can be run as performance, architecture, and channel-capacity variables change.

8.6.1 Analysis Stages

Step 1.   Develop the problem statement and determine what variables the system is to be costed against. In the simplest case, the problem could be to determine the systems cost as designed. Alternatively, a sensitivity analysis could be performed in order to optimize the design against some parameter, capacity or repeater spacing, for example.

For example, the problem could be to determine the cost per DS-1 channel over a specified route distance for various terminal-equipment products with different transmission rates and, therefore, different repeater spacings.

Step 2.   Develop the systems functional block diagram in order to determine the basic systems elements and their interconnections and dependencies. Figure 7.2, Chapter 7, is such a diagram for a fiber trunking system.

Step 3.   Partition the systems block diagram into blocks that represent equipment quantities and independent cost elements as they relate to the problem statement. Figure 8.6 shows such a partitioning for a fiber trunking system into independent cost elements configured so as to study cost versus circuit capacity.

Step 4.   Using the elements from Step 3 and the problem statement from Step 1, develop the analytical relationships between cost elements and the problem statement.

The relationships that follow are an example of how one might describe the cost per DS-1 channel for a 565-Mb/s fiber trunking system, as shown in Figure 8.6. The relationships are derived by starting from the lowest level of multiplexing and defining the cost per channel between that level and the next highest.

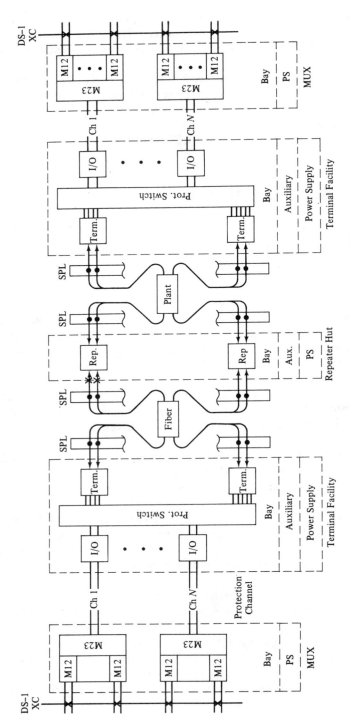

Figure 8.6 Digital fiber-optic trunking system cost model.

Cost per DS-1 (duplex):

$$\$/\text{DS-1} = \frac{\$\text{MUX} + \$\text{FOT}}{N_1} \tag{8-47}$$

$$\$\text{MUX} = 2N_{12}(\$/\text{M12}) + 2N_{23}(\$/M23) + 2N_{\text{MC}}(\$/\text{MC}) \tag{8-48}$$

where $\$/M_{12}$ = DS-1 channel multiplexer card cost (4 DS-1s/card)

$\$/M_{23}$ = M_{23} multiplexer and 28 DS-1s crossconnect common cost (7 M_{12} cards per multiplexer)

$\$/\text{MC}$ = common cost for MUX bays and power supplies

N_1 = number of DS-1s

N_{12} = number of M_{12} cards = INT $(N_1/4 + 3/4)$

N_{23} = number of M_{23} multiplexers and 28 channel cross connect
= INT $(N_{12}/7 + 6/7)$

N_{MC} = number of common bays = INT $[N_{23}/\text{MPB} + (\text{MPB} - 1)/\text{MPB}]$, where MPB = multiplexer shelves per bay

$$\$\text{FOT} = \$\text{Terminal} + \$\text{Repeater} + \$\text{Plant} + \$\text{Overhead} \tag{8-49}$$

where

$$\$\text{Terminal} = 2N_3(\$/\text{CH3}) + 2N_T[\$/\text{T} + 2(\$/\text{CS})] + 2(\$/\text{SE}) + 2(\$/\text{PROT})$$
$$+ 2N_{\text{TC}}(\$/\text{TC}) + 2(\$/\text{TAUX}) + 2(\$/\text{TFAC})$$

$$\$\text{Repeater} = N_R[N_T(\$/\text{R}) + N_{\text{RC}}(\$/\text{RC}) + \$/\text{RAUX} + 4N_T(\$/\text{CS}) + 2(\$/\text{SE})$$
$$+ \$/\text{RFAC}]$$

$$\$\text{Plant} = D[N_F(\$/\text{CFM}) + \$\text{I}/\text{M}] + N_S[\$/\text{SP} + N_F(\$/\text{S})] + \$\text{Test}$$

$$\$\text{Overhead} = \$\text{Eng}/\text{PM} + \$\text{Acceptance} + \$\text{Profit} + \$\text{Misc}$$

where $\$/\text{CH3}$ = DS-3 duplex channel card cost

$\$/\text{T}$ = cost per fiber terminal with $N \times$ DS-3 multiplexer

$\$/\text{CS}$ = cost per splice and connector, labor and materials

$\$/\text{SE}$ = splice-enclosure common cost, labor and materials

$\$/\text{PROT}$ = cost per terminal for protection switching

$\$/\text{TC}$ = cost per terminal bay and power supplies

$\$/\text{TAUX}$ = cost per auxiliary channel equipment

$\$/\text{TFAC}$ = per-end cost for terminal facilities

$\$/\text{R}$ = cost per repeater channel

$\$/\text{RC}$ = repeater bay and common costs

$\$/\text{RAUX}$ = cost per repeater auxiliary electronics

$\$/\text{RFAC}$ = cost per repeater hut

$\$/\text{CFM}$ = cost per cable fiber meter

$\$\text{I}/\text{M}$ = installation cost per meter

$\$/\text{SP}$ = cost per splice placement, labor and materials

$\$/\text{S}$ = cost per splice, labor and materials

$\$/\text{Test}$ = end-to-end fiber plant test and reel tests

N_3 = number of DS-3s = N_{23}

N_T = number of terminals and protection channels

= INT $(N_3/12 + 11/12) + 1$

N_{TC} = number of terminal bays = INT $[N_T/\text{TPB} + (\text{TPB} - 1)/\text{TPB}]$, where TPB = terminals per bay

N_R = number of repeater locations on the route

N_{RC} = number of repeater bays = INT $[N_T/\text{RPB} + (\text{RPB} - 1)/\text{RPB}]$, where RPB = repeaters per bay

D = plant route distance

N_F = number of fibers = $2N_T$ + spares

Step 5. Calculate the number of functional blocks (repeaters, terminals, fibers, etc.) required for each design or performance range to be analyzed. This is achieved through the process described in Chapter 6. The basic information required is as follows:

(a) For each design, list the channel capacity on a number-of-channels-per-fiber-pair basis. This defines the number of terminals and fibers necessary to serve the total requirement.

(b) Based on the number of channels at the protection interface (in this case, the DS-3 level), define the number of protection terminals required. Note that with optics, the entire terminal complement of DS-3 channels is generally switched simultaneously.

(c) Determine the number of repeaters required by performing a link-budget analysis on each design as per Chapter 6.

Step 6. Determine and list the cost for each element in the diagram in relation to the problem statement. Table 8.2 shows the values used for this example. Unit costs are first determined and then redefined in terms of the problem statement. For example, if distance is the test variable, fiber cost would be defined in terms of cost per kilometer. Note that unit costs should be increased by some factor to include tax, shipping, and spares.

TABLE 8.2 COST ANALYSIS EXAMPLE: COST PER DS-1 FOR A 565-MB/S TRUNK

Number of DS-1s =	600
Route Distance =	120,000 meters
Number of Repeaters =	4

M13 Multiplexer equipment ($MUX)

Cost element	Unit cost ($)	Factor tax, ship, spares	Quantity element	Quantity total	Total cost per element ($)
$M12	500	1.25	$N_{12} = 150$	$2N_{12} = 300$	187,500
$M23	11,000	1.25	$N_{23} = 22$	$2N_{23} = 44$	605,000
$MC	5,000	1.09	$N_{MC} = 2$	$2N_{MC} = 4$	21,800
			MPB = 12	$M13 =	814,300

TABLE 8.2 (*continued*)

Fiber terminal equipment ($Terminal)

Cost element	Unit cost ($)	Factor tax, ship, spares	Quantity element	Quantity total	Total cost per element ($)
$/CH3	800	1.25	$N_3 = 22$	$2N_3 = 44$	44,000
$/T	36,000	1.2	$N_T = 3$	$2N_T = 6$	259,200
$/CS	200	1.09	$N_T = 3$	$4N_T = 12$	2,616
$/SE	1,000	1.09		2	2,180
$/PROT	10,000	1.2		2	24,000
$/TC	5,000	1.09	$N_{TC} = 1$	$2N_{TC} = 2$	10,900
			TPB = 6		
$/TAUX	4,000	1.09		2	8,720
$/TFAC	150,000	1		2	300,000
				$Term =	651,616

Fiber repeater equipment ($Repeater)

Cost element	Unit cost ($)	Factor tax, ship, spares	Quantity element	Quantity total	Total cost per element ($)
$/R	20,000	1.25	$N_T = 3$	$N_R N_T = 12$	300,000
$/RC	4,000	1.09	$N_{RC} = 1$	$N_R N_{RC} = 4$	17,440
			RPB = 12		
$/CS	200	1.09	$N_T = 3$	$4N_R N_T = 48$	10,464
$/SE	1,000	1.09	$N_R = 4$	$2N_R = 8$	8,720
$/RAUX	3,000	1.09	$N_R = 4$	$N_R = 4$	13,080
$/RFAC	75,000	1	$N_R = 4$	$N_R = 4$	300,000
				$Rep =	649,704

Fiber plant ($Plant)

Cost element	Unit cost ($)	Factor overage contingency	Quantity element	Quantity total	Total cost per element ($)
$/CFM	0.30	1.2	$N_F = 12$	$DN_F = 1,440,000$	518,400
			No. Spare = 6		
$I/m					
Engr	1.00	1.1		$D = 120,000$	132,000
Prep	0.60	1.1		$D = 120,000$	79,200
Plow	4.00	1.1		$D = 120,000$	528,000
Bore	50.00	1.1		$D_b = 100$	5,500
$/SP	1,000	1.1	$N_S = 24$	$N_S = 24$	26,400
$/S	100.00	1.1	$N_F = 12$	$N_S N_F = 288$	31,680
$TEST/F	200.00	1	$N_F = 12$	$N_R N_F = 48$	9,600
				$Plant =	1,330,780
Overhead Costs ($OH)					600,000
		Total First Cost =			4,046,400
		Cost per DS-1 (600 DS-1s) =			6,744

Step 7. Using the relationships established in Steps 4 and 5, use the costs established in Step 6 and solve. Table 8.2 shows the solution.

8.7 TIME-BASED ANALYSIS

Some of the major elements of a time-based analysis are summarized here to aid in the evaluation of a fiber transmission-system design from the standpoint of return or net value over some time period. Return does not necessarily relate to revenue. It can be related in terms of cost savings or some other benefit as well.

8.7.1 Return On Investment

The financial analysis is generally designed to consider net income, cash flow, debt planning, return on investment, or net present value. The returns analysis, based on Park and Jackson [7] follows:

(a) Accounting approach: In this approach, the net cash flow (NCF) after depreciation and taxes is related to the average outstanding investment (INV):

$$\text{ROI\%} = \frac{\text{NCF}(0\text{--}T)/T}{\text{INV}(0\text{--}T)/2} \times 100 \tag{8-50}$$

where T is the number of years, and $(0 - T)$ represents the cumulative total from year 0 to year T. The shortcoming of this method is that the effect of compound interest is not considered.

(b) Operating return: This method expresses the return as the ratio of the average annual cash return to the original investment:

$$\text{ROI\%} = \frac{\text{PROFIT}(0\text{--}T)/T}{\text{INV}(0\text{--}T)} \times 100 \tag{8-51}$$

This method also does not take into account compound interest.

(c) Present worth or net present value (NPV): This form of analysis evaluates a project's economic viability by equating future net income and future value to present worth. This is done by discounting the future amount (FA) to a present worth (PW) using an interest rate (i) that represents the normal finance rates for the business.

By converting investment and earnings to present worth before summing them, the analysis can include yearly factors for cost-of-money (interest) inflation and component cost trends as well. The present worth (PW) of a future amount (FA) in year T, assuming an interest rate on money i, is

$$\text{PW} = \frac{\text{FA}}{(1 + i)^T} \tag{8-52}$$

An annuity amount based on the present amount (PA) is

$$\text{Annuity} = \frac{i(1 + i)^T}{(1 + i)^T - 1}\text{PA} \tag{8-53}$$

The return based on net present value (NPV), where PW(0–T) is the sum of the present worth values from year 0 to year T, is, therefore,

$$\text{NPV} = \text{PROFIT}_{[\text{PW}(0-T)]} - \text{INV}_{[\text{PW}(0-T)]} \tag{8-54}$$

$$\text{ROI\%} = \frac{\text{NPV}}{\text{INV}_{[\text{PW}(0-T)]}} \times 100 \tag{8-55}$$

(d) Internal rate of return: This is a special case of the present-worth method. It is the rate of return that causes the present worth of the net cash flows for the project to be zero. It depends on the relationship of investment to revenues (less operating costs and taxes). It is often used when large initial capital investments are required, and it is a method of determining the most efficient use of money. The analysis is the same as the present-worth case except that discount rates (interest) are adjusted in a trial-and-error basis until NPV = 0.

Every organization sets up its financial model in a slightly different manner, however, depending on what is to be solved and the particular circumstances of the business; therefore, there are many variations on the returns analysis.

The relationships for earnings (ERN), net cash flow (NCF), and investment (INV) are discussed in the following sections.

8.7.2 Income Statement

In the calculations of net income, cash flow, and debt, all are interrelated since negative cash flow, debt, and interest are balanced in an iterative process to achieve the desired results.

(a) Net income (NI): In a particular year, this is the income after tax, interest, depreciation, and expenses are taken:

$$
\begin{array}{ll}
+\text{PTI} & +\text{pretax income} \\
\underline{-\text{ITAX}} & \underline{-\text{income tax}} \\
\text{NI} & \text{net income}
\end{array} \tag{8-56}
$$

(b) Pretax income (PTI): This is the income value before income tax:

$$
\begin{array}{ll}
+\text{EBDI} & +\text{earnings before depreciation and interest} \\
-\text{DEP} & -\text{depreciation} \\
\underline{-\text{INT}} & \underline{-\text{interest expense}} \\
\text{PTI} & \text{pretax income}
\end{array} \tag{8-57}
$$

(c) Profit (PROFIT): This can be considered the net income before tax and depreciation:

$$
\begin{array}{ll}
+\text{EBDI} & +\text{earnings before depreciation and interest} \\
-\underline{\text{INT}} & -\underline{\text{interest expense}} \qquad\qquad\qquad\qquad (8\text{-}58) \\
\text{PROFIT} & \text{net income before tax and depreciation}
\end{array}
$$

(d) Earnings before depreciation and interest (EBDI): This is revenue less direct expenses:

$$
\begin{array}{ll}
+\text{REV} & +\text{revenue} \\
-\text{CBD} & -\text{credits and bad debt} \\
-\text{PTAX} & -\text{property tax} \\
-\text{OEXP} & -\text{operating expenses} \qquad\qquad\qquad (8\text{-}59) \\
-\underline{\text{LEXP}} & -\underline{\text{lease expense}} \\
\text{EBDI} & \text{Earnings before depreciation and interest}
\end{array}
$$

(e) Revenues (REV): This is the money received from the sale of goods and services taken prior to any taxes or fees. It could also be considered as cost savings in some scenarios.

(f) Credits and bad debt (CBD): These are the expenses due to credits given to customers to compensate for unsatisfactory service and expenses due to unpaid bills and payments to collection agencies.

(g) Operating expenses (OEXP): These are costs generally associated with operations, maintenance, and administration.

(h) Lease expenses (LEXP): Instead of purchasing capital with debt or equity, leasing may be considered where it provides a tax advantage or such. Lease expense is the monthly or annual payment to the leasing institution. The total lease expense for any year is the sum of all lease payments falling within that year.

(i) Income tax (ITAX): This is a percentage of pretax income taxed by government agencies depending on the tax rate (TR) for the particular business.

$$
\text{ITAX} = \frac{\text{PTI} \times \text{TR}\%}{100} \tag{8-60}
$$

(j) Real estate and property tax (PTAX): This is a tax levied on the value of property or real estate, and, in some cases, outside plant and terminal equipment.

(k) Depreciation (DEP): This is generally taken as a linear depreciation over a period of years, depending on the useful life of the capital, the type of business, and other factors. The depreciation amount taken each year, on a dollar amount of capital invested in one year, is equal to the capital investment (CAP) divided by the depreciation period (TD). The depreciation period starts in the investment year. The total depreciation, DEP(T), for any year T is, therefore,

$$
\text{DEP(T)} = \frac{\text{CAP(1)} + \text{CAP(2)} + \cdots + [\text{CAP(T)}/2]}{\text{TD}} \tag{8-61}
$$

8.7.3 Cash-Flow Analysis

Cash flow provides the visibility of available cash (a measure of profit) from the business. Where cash flow goes from negative to positive can be defined as the break-even point. Cash flow is calculated on an annual basis and influences debt and equity infusion, which are adjusted in order to achieve positive cash flow in the starting years.

(a) Cash flow (CF): In any year, cash flow is equal to

$$
\begin{array}{ll}
+\text{EBDI} & +\text{earnings before depreciation and interest} \\
-\text{INT} & -\text{interest expense} \\
+\text{DI} & +\text{debt infusion} \\
-\text{DP} & -\text{debt repayment} \\
-\text{CAP} & -\text{capital expenditure} \\
-\text{ITAX} & -\text{income taxes} \\
+\text{INV} & +\text{investment} \\
\hline
\text{CF} & \text{cash flow}
\end{array}
\tag{8-62}
$$

(b) Net cash flow (NCF): Over some time period, net cash flow is used in the ROI relationships:

$$
\begin{array}{ll}
+\text{EBDI} & +\text{earnings before depreciation and interest} \\
-\text{INT} & -\text{interest expense} \\
-\text{ITAX} & -\text{income tax} \\
-\text{INV} & -\text{investment} \\
\hline
\text{NCF} & \text{Net Cash Flow}
\end{array}
\tag{8-63}
$$

(c) Debt infusion (DI): This is simply cash acquired, each year that it is needed, through loans. The amount of debt required to be taken in any one year can be found by taking the cash-flow analysis to that point and determining the amount of negative cash flow that must be offset by debt. If cash flow is positive, then the amount of debt retained depends on the debt-to-equity ratio desired and other factors.

(d) Debt repayment (DP): This is the repayment of principal on loans. Interest is covered separately in the interest expense (INT).

(e) Investment or equity infusion (INV): This usually occurs at startup when the owners want to finance part of the business through direct investment or sale of stock.

(f) Capital expenditure (CAP): This is the amount of money invested or spent in any particular year for fixed assets, often including design, development, and installation.

8.7.4 Debt Analysis

Based on the amount of total debt (D) used to finance the business, this analysis computes the interest expense (INT) to be used in the income statement, and schedules the debt repayment (DP).

(a) Total debt (D): This is the total outstanding debt computed on an annual basis. The total debt in year T equals the total unpaid debt from previous year debt $D(T-1)$ and current year debt infusion $DI(T)$ less current year repayments $DP(T)$.

$$\begin{array}{ll} +D(T-1) & +\text{unpaid debt} \\ +DI(T) & +\text{added debt} \\ \underline{-DP(T)} & \underline{-\text{debt payment}} \\ D(T) & \text{debt in year } T \end{array} \tag{8-64}$$

(b) Debt infusion (DI): This is the added debt (new loans) incurred each year.

(c) Debt repayment (DP): This is the portion of payment each year on the loans to cover the principal. The annual repayment on a loan, $DP(T)$, can be computed as the loan amount (DI) divided by the period of the loan (T_i).

$$DP(t) = \frac{DI}{T_i} \tag{8-65}$$

(d) Average debt (D_{avg}): As a good approximation, annual interest expenses can be based on the average debt per year. Average debt for year T equals year T total debt and previous year total debt, all divided by 2.

$$D_{avg}(T) = \frac{D(T) - D(T-1)}{2} \tag{8-66}$$

(e) Interest (INT): The interest expense in year T, INT (T), is the annual interest rate, i, multiplied by the average debt for that year.

$$INT\ (T) = D_{avg}(T) \times i \tag{8-67}$$

where $i = \dfrac{\%\ \text{interest rate}}{100}$

REFERENCES

1. *Reliability Prediction Procedure for Electronic Equipment*, Bell Communications Research Technical Advisory 10425, Bell Communications Research, Inc., Piscataway, NJ.

2. MIL-HDBK-217C.

3. CCITT Recommendation G.821.

4. J. Chipman and M. Scholten, "When Errors Occur in Lightwave Systems," *Lightwave* (April 1986): 18–22.

5. *Generic Metropolitan Interoffice Digital Lightwave System Requirements and Objectives,* Bell System Technical Reference PUB 43806, American Telephone and Telegraph, Basking Ridge, NJ, 1982.

6. American National Standards Institute, *Optical Fiber Digital Transmission Systems Considerations for Users and Suppliers,* ANSI/EIA TSB-19-1986, New York, 1986.

7. W. R. Parks and D. E. Jackson, *Cost Engineering Analysis,* Wiley, New York, 1984.

Index

P

D